AF740136

TDLS 2009

Frank K. Tittel • Markus W. Sigrist
Editors

TDLS 2009

Selected Papers of the 7th International Conference on Tunable Diode Laser Spectroscopy, (TDLS 2009), Zermatt, Switzerland, 13-17 July 2009

Springer

Editors
Prof. Dr. Frank K. Tittel
Rice University
Electrical and Computer Engineering MS-366
6100 Main Street
Houston, TX 77005-1892
fkt@rice.edu

Prof. Dr. Markus W. Sigrist
ETH Zürich
Institute for Quantum Electronics
Schafmattstrasse 16, HPF D19
CH-8093 Zürich, Switzerland
sigrist@iqe.phys.ethz.ch

Reprinted from Applied Physics B, Vol. 100, issue 2

ISBN 978-3-642-02291-3 e-ISBN 978-3-642-02292-0
DOI 10.1007/978-3-642-02292-0
Springer Heidelberg Dordrecht London New York

Library of Congress Control Number: 2011925552

© Springer-Verlag Berlin Heidelberg 2011
This work is subject to copyright. All rights are reserved, whether the whole or part of the material is concerned, specifically the rights of translation, reprinting, reuse of illustrations, recitation, broadcasting, reproduction on microfilm or in any other way, and storage in data banks. Duplication of this publication or parts thereof is permitted only under the provisions of the German Copyright Law of September 9, 1965, in its current version, and permission for use must always be obtained from Springer. Violations are liable to prosecution under the German Copyright Law.
The use of general descriptive names, registered names, trademarks, etc. in this publication does not imply, even in the absence of a specific statement, that such names are exempt from the relevant protective laws and regulations and therefore free for general use.

Cover design: eStudio Calamar S.L.

Printed on acid-free paper

Springer is part of Springer Science+Business Media (www.springer.com)

Preface

The Seventh International Conference on Tunable Diode Laser Spectroscopy, TDLS-2009, was held in Zermatt, Switzerland during the period 13-17 July 2009. The conference was officially organized under the auspices of the A. M. Prokhorov General Physics Institute, the Russian Academy of Sciences, Connecticut College, New London, CT, USA, and the ETH Zurich, Institute of Quantum Electronics, Zurich, Switzerland.

The purpose of these meetings is to report and discuss the progress that has been made in the technical development of diode lasers and their applications in areas of fundamental research as well as real world applications of the technology. This is accomplished in a format that encourages strong interactions between researchers and device developers and semi-conductor laser manufacturers. The relevant technologies include the original lead salt diode lasers, near infrared diode lasers and the latest developments with mid-infrared quantum and interband cascade lasers as well as with nonlinear optical schemes.

The TDLS-2009 International Advisory Committee consisted of Mark Allen (USA), A. R. W. McKellar (CANADA), Markus W. Sigrist (SWITZERLAND) and Frank K. Tittel (USA). The committee recommended a list of invited speakers which formed the scientific core of the program for the conference; the 2009 conference had a total of five invited speakers. Based upon the information contained in the abstracts submitted for contributed papers, the Conference organizers A.W. Mantz (New London, USA), A.I. Nadzhdinskii (Moscow, Russia) and M.W. Sigrist (Zurich, Switzerland) and the International Advisory Committee selected thirteen papers for 15 minute oral presentations. A total of 96 posters were presented in six poster sessions intended to foster informal and more detailed discussions among authors and other participants. A total of 114 participants from Belarus, Belgium, Canada, France, Gambia, Germany, Ireland, Japan, Norway, Pakistan, Russia, Switzerland, Syria, UK, and USA attended TDLS 2009.

This is the second time that selected papers for the TDLS conference were published in a special issue of Applied Physis B. Guest editors for TDLS 2009 were Professor Markus W. Sigrist (ETH Zurich) and Professor Frank K. Tittel (Rice University, Houston). Referees for the manuscripts appearing in this issue were selected by the International Advisory Committee from participants in the conference as well as other experts.

This book contains 25 papers carefully selected from the contributions to TDLS 2009. The first part contains papers discussing the technology and cutting-edge development of sources and detection schemes, followed by papers on basic physical phenomena. Finally several papers present specific applications in sensing.

Zurich/Houston, February 2011

Markus W. Sigrist and Frank K. Tittel

Contents

Optical feedback cavity-enhanced absorption spectroscopy (OF-CEAS) in a ring cavity

D.J. Hamilton · M.G.D. Nix · S.G. Baran · G. Hancock · A.J. Orr-Ewing

Received: 17 August 2009 / Revised version: 21 October 2009 / Published online: 12 November 2009
© Springer-Verlag 2009

Abstract Optical feedback cavity-enhanced absorption spectroscopy (OF-CEAS) has been demonstrated by coupling a distributed feedback diode laser to a ring cavity. Frequency-selected light decaying from the ring cavity is retro-reflected, inducing a counter-propagating intra-cavity beam, and providing optical feedback to the laser. At specific laser-to-cavity distances, all cavity mode frequencies return to the diode laser with the same phase, allowing spectra to be accumulated across the range of frequencies of the current-tuned laser. OF-CEAS has been used to measure very weak oxygen isotopologue ($^{16}O^{18}O$ and $^{16}O^{17}O$) absorptions in ambient air at wavelengths near 762 nm using the electric-dipole forbidden O_2 *A*-band. A bandwidth reduced minimum detectable absorption coefficient of 2.2×10^{-9} cm^{-1} Hz$^{-1/2}$ is demonstrated.

PACS 42.55.Px · 42.60.Da · 42.62.Fi

1 Introduction

Cavity ring-down spectroscopy (CRDS) and the related cavity-enhanced absorption spectroscopy (CEAS) are ultrasensitive absorption spectroscopy techniques finding wide ranging applications as analytical tools. In the first demonstration of CRDS [1], a pulsed dye laser was used to excite an optical cavity and the subsequent rate of decay of light from the cavity analysed to obtain wavelength-dependent absorption losses. The optical path length enhancement of $(1 - R)^{-1}$ offered by this technique, where R is the cavity mirror intensity reflectivity, allows very weak absorptions to be measured. Since its initial application, numerous improvements and variations on the original technique have been made [2–8]. The coupling of an optical cavity to a continuous wave laser (CW-CRDS [9, 10]) permits a greater build up of light intensity inside the cavity than for pulsed laser CRDS. A fast optical switch then allows a ring-down to be observed. Increased signal intensity improves the spectrometer sensitivity while the reduction in laser bandwidth possible with CW lasers improves the spectral resolution of this technique. Further, active locking of the laser frequency to a resonant frequency of the cavity increases the injection efficiency, detection limit and data acquisition rate [11], but at the cost of greater experimental complexity. An alternative technique to CW-CRDS is CEAS [12, 13] in which absorptions are detected by instead measuring the total transmitted light intensity through the cavity at resonant mode frequencies. Thus the enhanced path length and high sensitivity of CRDS is preserved while the signal processing is simplified since neither fast transient data acquisition nor fitting of exponential decays are required. Comparative studies of CEAS and CRDS have shown that the former method is, in general, inherently less sensitive [14, 15], primarily due to the nature of the measurement because the cavity output intensity is subject to laser intensity fluctuations.

The development of tuneable, compact and relatively inexpensive continuous wave semiconductor lasers, driven by the telecoms industry, has fueled the design of CW CRD

D.J. Hamilton · M.G.D. Nix · A.J. Orr-Ewing (✉)
School of Chemistry, University of Bristol, Cantock's Close, Bristol BS8 1TS, UK
e-mail: a.orr-ewing@bris.ac.uk
Fax: +44-117-9250612

S.G. Baran · G. Hancock
Department of Chemistry, Physical and Theoretical Chemistry Laboratory, University of Oxford, South Parks Road, Oxford OX1 3QZ, UK

Present address:
M.G.D. Nix
School of Chemistry, University of Leeds, Leeds LS2 9JT, UK

spectrometers with the potential to perform *in field* measurements of key atmospheric species [16, 17]. Recently, the value of optical feedback (OF) locking of a diode laser to an optical cavity has been recognised. OF CRDS [18] is increasingly used because of benefits over standard CW-CRDS such as greater signal intensity and kilohertz acquisition rates [19]. In principle it offers a less expensive set-up because fast optical switches and high levels of optical isolation are not required. Greater sensitivity should also result from increased intra-cavity light intensity.

The effect of optical feedback on narrowing the bandwidth of semiconductor lasers is well understood [20, 21]. If a distributed feedback diode laser (DFB DL) is coupled to a high-finesse cavity and a fraction of the very narrow line width (~10 kHz) intra-cavity radiation is allowed to return to the diode, it has the effect of seeding the laser, narrowing the laser bandwidth to below the line width of the cavity resonance and locking the laser to the frequency of the cavity mode. With the current applied to the laser continuously changing, however, after a time the locking range is exceeded and the laser can no longer operate at the cavity mode frequency. The laser frequency emitted must jump to either the frequency of a neighbouring mode or the free running condition, where the laser frequency scans continuously with applied current. OF-CRDS measurements are performed without a fast optical switch (such as an acousto-optic modulator, AOM) by chirping the laser with a square wave superimposed on the laser current. Throughout this ON pulse the laser scans, but the scan rate slows as equilibrium is reached. The laser locks to successive cavity modes for increasing periods of time, and towards the end of the current pulse, a steady state of cavity intensity is reached. The end of the ON pulse might be designed to extinguish the laser emission completely or to maintain emission but cause an abrupt change in the laser frequency, so that it detunes from the cavity mode. In either case, a cavity ring-down event can then be observed. Using this technique, detailed studies of absorption by trace gases [18] and extinction by aerosol particles have been undertaken [19, 22].

OF-CEAS also exploits the frequency narrowing and locking behaviour of semiconductor lasers to excite fully successive cavity modes, but instead of the square-wave ON pulse described above, involves a slow ramp of the laser frequency to monitor transmission at successive longitudinal modes. The resultant mode-by-mode CEAS method was first demonstrated by Morville et al. [23] using a V-shaped optical cavity. This technique is well suited to in situ studies of atmospheric constituents, being capable of measuring single or multiple absorption profiles during a single scan of the laser. Consequently, it has a growing range of applications, including in situ measurements of atmospheric methane [24], geothermal gases [25], water [26] and carbon dioxide isotopologues [27], and laboratory measurements of NO_2[28] and CO_2 isotopologues with applications in breath analysis [29].

Currently the most widely used optical feedback design for CRD/CEAS measurements is the V-cavity [18] created by folding a conventional linear cavity about its centre with the addition of a third mirror. This is a variation of the original tilted confocal cavity used to stabilise the frequency of a semiconductor laser [30]. Creating a cavity with three mirrors means the geometry is such that the angle of incidence of the input laser beam on the central folding mirror is not normal, so most of the radiation incident on the cavity input mirror is reflected away from the laser beam path and does not return to the laser. Only a small subset of the frequencies within the laser bandwidth is resonant with cavity modes and can couple into the cavity. The frequency selected light ringing down from inside the cavity can, however, feed back to the laser, narrowing the output bandwidth and increasing the efficiency of injection of light into the cavity. In the current work, possible advantages are explored of the use of a triangular ring cavity rather than a V-shaped folded linear cavity.

Our ring cavity is illustrated in Fig. 1 and also uses three mirrors to ensure, by geometry, that the reflection of the incident laser beam directly from the cavity input mirror does not return to the laser. Once light of the resonant frequency enters the cavity, it propagates as a travelling wave around the cavity. Ring cavity resonant frequencies (ω_n) correspond to a phase change of 2π after a single round-trip of length L_R. Angular frequencies ($\omega = 2\pi c/\lambda$ for a wavelength λ) resonant within the cavity are given by (1), where

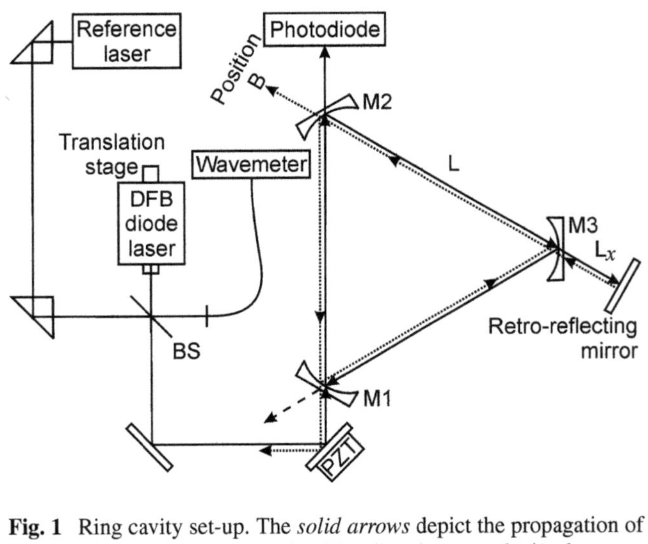

Fig. 1 Ring cavity set-up. The *solid arrows* depict the propagation of the main laser beam in the cavity. The *dotted arrows* depict the propagation of the retro-reflected beam. Abbreviations used: BS = beam splitter; M1, M2, M3 = first, second, third cavity mirror; L = path length between reflections; L_x = path length between third cavity mirror (M3) and retro-reflecting mirror; PZT = piezoelectric transducer. Note that the wavemeter is fibre coupled. Position B is used as a photodiode location to monitor the standing wave generated in the cavity by the retro-reflecting external mirror

n is any integer and c is the speed of light in the intra-cavity medium.

$$\omega_n = \frac{2n\pi c}{L_R} \quad (1)$$

To allow optical feedback to occur, a mirror is placed after either the second (M2) or third (M3) cavity mirror to retro-reflect escaping light back into the cavity so that it can propagate along the TEM_{00} cavity mode in the opposite direction. When this counter-propagating light exits from the cavity through the first input mirror (M1) it follows the cavity input path back to the laser diode where it can lock the laser's emitted frequency to that of the excited cavity mode. Without the retro-reflecting mirror in place, an additional source of optical feedback is seen, causing an erratic enhancement of random cavity modes. This feedback is caused by backscatter from the reflective surfaces of any of the three mirrors, returning light to the laser by exciting the counter-propagating TEM_{00} cavity mode. With careful cleaning of the cavity mirrors this backscatter is, however, insignificant relative to the light introduced into the counter-propagating TEM_{00} cavity mode by the retro-reflecting mirror.

Potential advantages of a ring cavity over a folded linear cavity may now be identified. First, the ring cavity does not display odd and even modes associated with standing waves present in a folded linear cavity [23]. Second, the retro-reflecting mirror allows the level of optical feedback experienced by the laser to be increased from near zero, without introducing attenuating optical elements between the laser and the cavity; such elements have the undesirable effect of reducing the power incident on the cavity input mirror (M1) and transmitted to the detector.

The absorption lines measured during this study to test the performance of OF-CEAS in a ring cavity lie in the atmospheric oxygen A-Band, corresponding to the $b^1\Sigma_g^+ \leftarrow X^3\Sigma_g^-(0,0)$ transition. This transition is very weak, being electric dipole (because of the $g \leftarrow \times \rightarrow g$ selection rule), spin (singlet $\leftarrow \times \rightarrow$ triplet) and symmetry ($\Sigma^+ \leftarrow \times \rightarrow \Sigma^-$) forbidden and is driven by a magnetic dipole interaction. The A-band was studied in detail by Dicke and Babcock [31], and later by Babcock and Herzberg [32]; these investigations determined molecular constants for $^{16}O^{16}O$ as well as for its two main rare isotopologues, $^{16}O^{18}O$ and $^{16}O^{17}O$. The most recent, high-accuracy studies were those undertaken by Robichaud et al. [33–36] using CRDS. $^{16}O^{16}O$ is a symmetric molecule containing two identical spin-0 nuclei, and thus in the ground state only odd N'' rotational angular momentum quantum states can be populated [37]. For $^{16}O^{16}O$ only half of the expected rotational lines therefore appear in the absorption spectrum. Lines arising from the less abundant isotopologues appear as weak features between the $^{16}O^{16}O$ absorption features: here, the $^{R}R(2)$ and $^{R}Q(1)$ lines of the $^{16}O^{18}O$ and $^{16}O^{17}O$ isotopologues (respective relative abundances of 0.00399 and 0.00074) are observed on the wings of much stronger $^{16}O^{16}O$ (relative abundance 0.99526) rotational lines [38] in ambient air samples. The oxygen A-band is of considerable importance in atmospheric science: the region from 755–770 nm contains very little absorption from other species, and the A-band is therefore used to retrieve vertical pressure and temperature profiles and to measure cloud top heights and pressures, as useful aids to weather prediction.

2 Experimental details

The ring cavity, shown in Fig. 1, consists of three high-reflectivity dielectric mirrors (Layertec) with radii of curvature of 1 m and a highly reflective region centred at 790 nm for 0° angle of incidence (AOI). In the configuration of an equilateral triangle, each cavity arm length is ~18.3 cm, and light is incident on each mirror at 30° AOI, shifting the highly reflective region by ~20 to ~770 nm. In our case of $R > 0.9999$ typical ring down times lie between 6 and 7 μs. To prevent the passage of dust through the intra-cavity beam the optical cavity and the retro-reflecting mirror are located inside a sealed chamber, containing ambient air at atmospheric pressure. The laser beam enters the chamber through wedged fused-silica windows.

A distributed feedback (DFB) diode laser (Sarnoff) emitting >10 mW of tunable 761 nm radiation is housed in a thermo-electrically cooled mount (Thorlabs), itself located on a translation stage (Thorlabs, 25 mm travel, micrometer precision). Temperature and current can be used to tune the emission frequency and are controlled by a Sacher Lasertechnik diode laser driver. When recording spectra, an asymmetric current ramp (20 Hz) is applied to the laser and the cavity output recorded for 1 s using a photodiode (New Focus, DC, 125 MHz) and a data acquisition card with a 14-bit vertical resolution (Gage CS1450). The resultant data file thus contains the result of 20 ramps of the laser.

The diode laser emission is steered by two silver mirrors and focussed into the centre of the first arm of the cavity by a single anti-reflection coated aspheric lens (numerical aperture 0.30), screwed directly into the laser mount. One of the silver mirrors is mounted on a PZT to adjust the laser to cavity distance with sub-wavelength precision. A thin glass plate between the laser and the first steering mirror serves to deflect a small portion of the beam into a wavemeter (Angstrom) and to introduce a visible reference beam, used for cavity alignment. All components are mounted on a vibration isolated optical table.

The emission from the diode laser is strongly polarised, and the DL mount is oriented so that vertically polarised light is emitted; this light is s-polarised relative to the optical cavity. Due to the non-normal angle of incidence upon each cavity mirror this s-polarised light experiences a higher finesse than any p-polarised light circulating in the cavity. The s and p modes resonate in the ring cavity at different frequencies but with the same free spectral range [39]. At angles of incidence less than the Brewster's angle, p-polarised light (polarised in the ring plane) does not experience a π-phase shift upon reflection, whereas s-polarised light does (here, we adopt the conventions for phase and reference frame of [40] and neglect the effects of the multiple interface structure of dielectric mirrors). Thus, optical cavities with an odd number of mirrors will exhibit frequency splitting of the modes because of an accumulated phase shift difference of π, corresponding to a frequency difference of half a free spectral range. p-polarised light then resonates at frequencies shifted by half a free spectral range from modes of s-polarised light, as shown by Saraf et al. [39]. A polarising prism can be placed before the photodiode to select only the higher finesse s-polarised modes for analysis.

3 Theory

3.1 Molecular absorption from relative mode intensities

During the ramp of the applied laser current, if an excited cavity resonance lies at the frequency at which an intracavity gas absorbs, the beam intensity inside the cavity and so the detected intensity is reduced. From the measured cavity mode intensity and a measured or fitted background transmitted intensity, the absorption at discrete and well-defined cavity mode frequencies can be determined. From an analysis of the superposition of the amplitudes of the fields transmitted by the ring cavity, we obtain an expression for the cavity transfer function, the ratio of transmitted to incident light intensities (I and I_{in}, respectively), at the maxima of cavity modes for the ring-cavity configuration shown in Fig. 1:

$$\frac{I}{I_{\text{in}}} = \frac{T^2 e^{-\alpha L}}{(1 - R^{3/2} e^{-3\alpha L/2})^2} \tag{2}$$

Here, α is the intra-cavity absorption coefficient at the cavity mode frequency, and the three cavity mirrors are equidistantly separated by a distance L (so the cavity round-trip distance $L_R = 3L$).

In the experimental measurements described here, weak absorption spectra are obtained superimposed on an underlying absorption corresponding to the wings of much stronger spectral lines. As is described later, the background absorption losses are determined by measuring the transmitted baseline intensity (I_b) at selected mode frequencies and interpolating using a baseline fitting procedure. At a cavity mode frequency ω_n lying within the lineshape of the weak absorption feature of interest, the molecular absorption coefficient $\alpha_m(\omega_n)$ above the baseline level $\alpha_b(\omega_n)$ (such that the total absorption coefficient $\alpha = \alpha_m + \alpha_b$) is then obtained from a ratio of the interpolated baseline intensity to the transmitted value (I_b/I):

$$\alpha_m(\omega_n) = -\frac{1}{c\tau_b(\omega_n)} - \frac{2}{3L} \ln\left\{1 - \sqrt{\frac{I_b(\omega_n)}{I(\omega_n)}}\left(1 - e^{-3L/2c\tau_b(\omega_n)}\right)\right\} \tag{3}$$

In (3), $\tau_b(\omega_n)$ is the ring-down time corresponding to the baseline absorption losses $\alpha_b(\omega_n)$ at the frequency of the measurement. By taking the ratio of intensities in (3), the need to monitor the incident light intensity (2) is circumvented. The term $3L/2c\tau_b$ is much smaller than unity, allowing the simplification of (3) to the form:

$$\alpha_m(\omega_n) = \frac{1}{c\tau_b(\omega_n)}\left\{\sqrt{\frac{I_b(\omega_n)}{I(\omega_n)}} - 1\right\} \tag{4}$$

The spectral resolution of the mode-by-mode CEAS technique is limited by the free spectral range of the optical cavity.

3.2 Optical feedback phase and the laser-to-cavity distance

For reasons discussed in detail by Laurent et al. [20] and Morville et al. [23], the performance of an OF-CEAS spectrometer is optimized if the relative phase ϕ of the feedback light reaching the laser after returning from the cavity lies within a small range of values about $\phi = 0$. With an arbitrary laser to cavity distance, it is unlikely that light of different modal frequencies, corresponding to different values of n in (1), will have the same phase after propagation from the cavity back to the laser. The subtly different wavelength for each cavity mode gives a different number of field oscillations between two arbitrary points. It is therefore necessary to find a distance from the cavity to the laser where all feedback light has identical phase. The analysis must consider the source of optical feedback, i.e. the retro-reflecting mirror in our ring-cavity design, and the frequency of the feedback light itself. By following the analysis of the phase of optical feedback from a tilted confocal cavity [20] and a V-cavity [23] and applying the method to our ring cavity, we obtain that the path from the laser, through the cavity, and to the retro-reflecting mirror must equal a half integer number of cavity round-trip lengths, as shown in (5).

$$L_{L\to C} + NL + L_x = k\frac{L_R}{2} \tag{5}$$

Here $L_{L\to C}$ is the distance between the laser and the input cavity mirror (M1), L_x is the distance between the retro-reflecting mirror and the cavity and NL is the path length through the cavity between the laser input mirror and the exit mirror behind which is located the retro-reflecting mirror. N can take the value 1 or 2 depending on whether the retro-reflecting mirror is positioned after the first or the second cavity mirror. For example, Fig. 1 shows the $N = 2$ configuration. L_R is the optical cavity round-trip distance and k is any integer. Under this condition all cavity mode frequencies undergo a phase change of 2π during their return trip to the laser and can lock the laser with maximum efficiency.

The condition expressed by (5) allows both L_x and $L_{L\rightarrow C}$ to be optimised. Weakly divergent light exiting from the cavity towards the retro-reflecting mirror is best returned after a short distance to maximise injection into the counter-propagating TEM_{00} mode. A planar silver mirror is sufficient for the retro-reflection. The laser to cavity distance can then take values separated by half the cavity round-trip distance. These are chosen to optimise the mode matching of the laser beam into the TEM_{00} cavity mode by matching the beam diameter on the cavity input mirror and the position of the intra-cavity beam waist. A ring cavity consisting of three identical spherically curved mirrors has a beam waist positioned at the centre of each cavity arm, but because of the non-normal incidence on the mirrors, each beam waist has an elliptical cross section [41].

The laser to cavity distance can be positioned correctly to within 5 μm by examining the effect of the variation of the phase of optical feedback across ~175 cavity modes, corresponding to a frequency range of ~3.2 cm^{-1} ($\Delta\nu_{FSR} = 0.01824\ cm^{-1}$). All modes in the scan then have peaks of excitation intensity within the cavity mode-locked range and mode-by-mode CEAS can be performed. Successful coupling of the optical feedback requires all mode frequencies returning from the cavity to interfere constructively with the emitted light, thus a node in the beam field must be present at the laser. Thermal drift and vibrations alter the laser to cavity distance, moving the node from the laser diode. Continuous readjustment of the laser to cavity distance is therefore necessary, and a feedback loop similar to those described in [42, 48] was implemented which differentiates each cavity mode, obtains an error signal proportional to the phase error by integrating the result and moves the laser beam steering mirror mounted on a PZT to adjust the laser to cavity distance. This feedback operation was performed using a LabVIEW program instead of electronic hardware and successfully corrects thermal drift of the laser to cavity distance, thereby allowing averaging of the cavity output signal over many successive scans of the laser frequency.

3.3 Modelled and measured counter-propagating mode intensities

The generation of the counter-propagating beam allows optical feedback to occur, locking the laser to excited cavity mode frequencies. However, the level of feedback experienced by the laser at each locked frequency is not uniform.

As the laser scans in wavelength across a cavity mode, light of the resonant frequency enters the optical cavity and propagates around the ring, decaying from each mirror. Of the light reflected by the external retro-reflecting mirror back towards the cavity, a small fraction ($\leq 1 - R$) incident on the cavity re-entry mirror (M3) can re-enter the cavity and couple back into the cavity in the opposite direction. The counter-propagating beam forms a weak standing wave by interference with a small proportion of the initial input beam circulating in the cavity. The standing wave field has a fixed node at its source, the retro-reflecting mirror, and extends into and around the ring cavity. Two points exist inside the cavity, at $L_R/2$ and L_R from the retro-reflecting mirror, where the standing waves for all mode frequencies have a node. The standing wave gives rise to a weak travelling component that decays from the cavity through mirror M1 and feeds back to the laser.

The rate of optical feedback is determined by the amplitude of the standing wave and is not equal for all cavity modes. When the laser is sited according to (5) every mode has identical phase after returning to the laser diode, but this in-phase condition is not met at the inner face of each cavity mirror. Consecutive cavity modes have an additional phase change of 2π, adding two nodes to the standing wave component of the intra-cavity light field. The different field amplitude at each cavity mirror means that different reflectivities are experienced by each cavity mode, with angular frequency ω_n. The magnitude of the standing wave on each cavity mirror can be modelled, an estimated reflectivity applied for each mirror and a cavity mode intensity calculated from the average reflectivity. This process gives a pattern with a rough periodicity in the intensity of the counter-propagating beam. If a detector is sited at position B, see Fig. 1, the intensity of the counter-propagating beam can be measured and similarities are observed in the simulated pattern, as shown in Fig. 2.

It should be noted that the mode-to-mode intensity variation is of the standing wave in the cavity and of the resultant travelling component escaping from the cavity and providing feedback. The travelling wave used for detection of molecular absorptions by CRDS/CEAS experiences *the same* mode-to-mode reflectivity independent of frequency. The variation of the rate of optical feedback on these modes used for CEAS is, however, seen as a slight mode-to-mode intensity variation during CEAS measurements. This effect could therefore limit the minimum detectable absorption coefficient achievable through CEAS with a ring shaped cavity.

The relative intensity of the counter-propagating beam in the cavity, induced by the retro-reflecting mirror, relative to the forward-propagating beam can be as much as ~0.03, as measured by comparison of both cavity outputs. If allowed

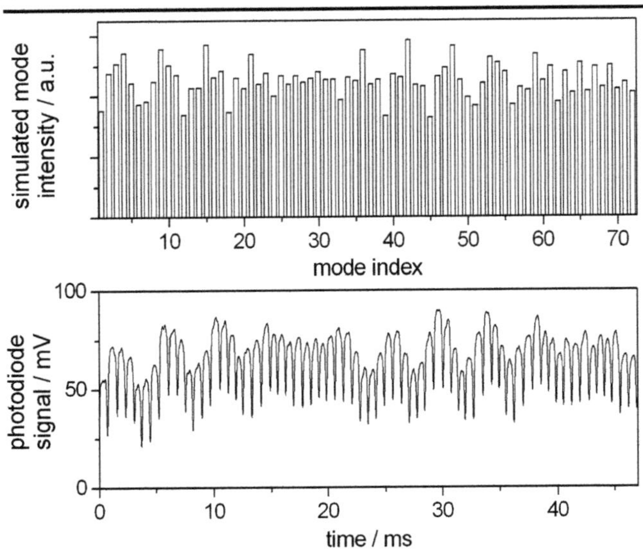

Fig. 2 *Upper panel*: simulated counter-propagating cavity mode intensities incorporating the different possible field amplitudes of the intra-cavity standing wave on each cavity mirror. The cavity arm lengths, L, used are accurate to 10^{-3} m, when compared to the experimental apparatus, but 10^{-9} m precision is required for an exact model. *Lower panel*: measured counter-propagating mode structure at position B. A periodicity similar to that calculated in Fig. 2 is seen despite use of approximate cavity arm lengths in the simulation

to return to the laser, this level of feedback causes mode hopping, and sequential cavity modes are not reliably excited during a mode-by-mode OF-CEAS scan, most notably during the high-current regime towards the end of the rising current ramp. It is therefore necessary to reduce the intensity of the counter-propagating beam so that the rate of optical feedback is sufficient for the laser to lock to each cavity mode, for just less than the time taken by the laser to scan through a single free spectral range. This can be achieved by adjusting a variable neutral density filter located between the optical cavity output mirror (M3) and the retro-reflecting mirror. An alternative method is to misalign slightly in the vertical direction the reflection from the retro-reflecting mirror. Due to the significant horizontal ellipticity of the laser output, horizontal misalignment of the retro-reflecting mirror has the undesirable tendency to excite (and thus cause the laser to lock to) higher order modes in the ring plane, such as TEM_{10} and TEM_{20} modes. Neither of these techniques requires any form of optical attenuator between the laser and the cavity, such as a polarising prism, which would reduce the power incident on the optical cavity.

The ring cavity, together with other optical feedback designs, maintains an ease of switching from CEAS to CRDS measurements. No hardware modifications are necessary; instead the triangular wave superimposed on the applied laser current is switched to a square wave and the OF phase locking software maintaining the laser to cavity distance is configured to differentiate only the first few cavity modes excited. The final mode, over which the laser locks for an extended period of time, shows little phase information on its top; instead the mode intensity is dominated by amplitude fluctuations of the laser emission, so this mode is ignored in the phase locking program. An advantage of using the ring cavity to perform OF-CEAS/CRDS is the absence of odd and even modes, found when using a V-cavity design [23]. The terms odd and even describe the alternating reflectivity experienced by consecutive cavity modes due to the phase of the standing wave inside the cavity at the folding cavity mirror. The different reflectivities cause alternate cavity modes to experience alternating intensity build-ups and ring-down times. This effect can be corrected during the data analysis stage by treating sets of data points arising from alternate modes separately, and combining the two sets of results once transformed to absolute absorption units. Alternatively the laser to cavity distance can be adjusted so that only odd or even modes are in phase upon returning to the laser and can feed back. While this latter methods simplifies the data collection, it increases the mode spacing to twice the free spectral range of the cavity, halving the spectral resolution. This reduced resolution may be offset by creating a longer optical cavity because $\Delta\nu_{\mathrm{FSR}} \propto L^{-1}$.

For completeness we note that other techniques exist that have demonstrated OF-CRD/CEAS measurements. These include the use of a Brewster-angle optic within the cavity [42] and a technique exploiting the residual birefringence of cavity mirror coatings [43]. OF-CRDS with a linear cavity has also been attempted in our laboratory using a pinhole between the diode laser and the optical cavity to spatially filter the frequency selected intra-cavity beam decaying from the cavity from the direct reflection from the cavity input mirror [44]. After extensive modelling of relative beam paths between the laser and the cavity and application of possible optical designs, no sufficiently robust solution to this latter strategy was found. Using a ring shaped cavity Zare and workers have also employed OF cavity-enhanced spectroscopy methods to measure Rayleigh scattering cross sections; D.S. Kuramoto, R.N. Zare, Private communication (2009).

4 Detection of weak absorptions

With a high finesse optical cavity, R values typically greater than 0.9999 (corresponding to a minimum finesse of 31400, $\tau_0 \approx 6.5$ μs) and optical feedback, very weak absorptions can be measured for gaseous samples. Here we illustrate the performance of mode-by-mode OF-CEAS in the ring cavity using three absorption lines in the oxygen A-band in air under ambient conditions. A simulation of all oxygen absorptions present in the frequency range of the diode laser (~761–762 nm) is shown in Fig. 3. The enlargement of the region centred at 13130 cm^{-1} lies between two well

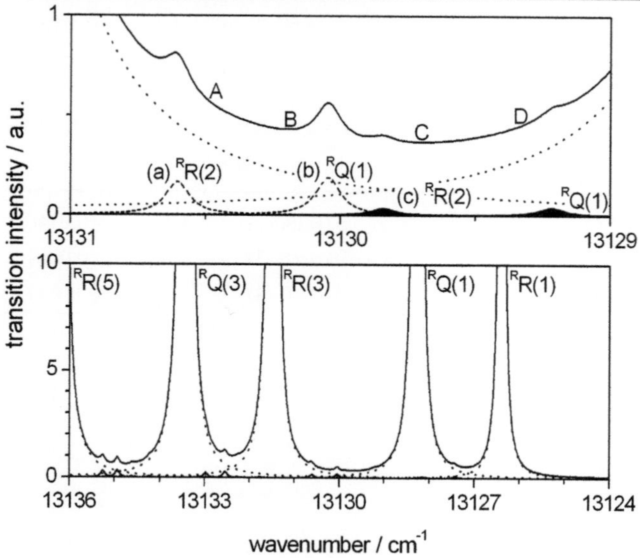

Fig. 3 HITRAN 2008 simulation of the absorptions due to molecular oxygen ($P = 1$ atm, $T = 22°C$) [38]. Rotational line assignments are given in the form $^{\Delta N}\Delta J(N'')$. *Lower panel*: from 13136 to 13124 cm^{-1}. Note the very weak absorption lines of $^{16}O^{17}O$ and $^{16}O^{18}O$. The *solid line* is the composite of absorption by the three isotopologues considered here. *Upper panel*: enlargement centred on the region recorded using OF-CEAS. Lines marked (a), (b) and (c) were measured, see Table 1. The *upper solid curve* is the composite spectrum comprising $^{16}O^{16}O$ (*dotted*), $^{16}O^{18}O$ (*dashed*) and $^{16}O^{17}O$ (*filled*). Frequencies marked A to D are used to obtain the background ring-down time, τ_b, as detailed in the text

Table 1 Transition frequencies of four isotopologue absorptions between 13131 and 13129 cm^{-1}, the isotopic abundance and rotational line assignment [38]. Absorption lines labelled (a), (b) and (c) were measured in air under ambient conditions

Isotopologue	Abundance	Rotational line assignment, $^{\Delta N}\Delta J(N'')$	Transition frequency/cm^{-1}
$^{16}O^{18}O$	0.399%	^{R}R(2)	13130.605 (a)
		^{R}Q(1)	13130.045 (b)
$^{16}O^{17}O$	0.074%	^{R}R(2)	13129.839 (c)
		^{R}Q(1)	13129.217

separated $^{16}O^{16}O$ lines and contains four isotopologue absorptions, with details provided in Table 1. The relatively strong $^{16}O^{16}O$ absorptions lie sufficiently close together that their wings overlap and significant molecular absorption is present at all frequencies. The simulation shown in Fig. 3 does not consider the underlying continuum in this band due to collision-induced absorption (CIA) [45].

A ramp of the applied current over 65 modes (~1.2 cm^{-1}) at 20 Hz allows the laser to lock to each mode for ~0.7 ms, sufficient time to allow complete excitation. An asymmetric ramp for the applied laser current improves the effective work cycle by using a shorter interval to return the current to its initial value. Excitation of the cavity modes occurs during the slower rising ramp of the pulse, which occupies a longer period of time than the down ramp. Figure 4 shows the mean cavity output signal after one second of averaging at a laser scan rate of 20 Hz. The $^{16}O^{18}O$ isotopologue absorptions (a) and (b) (see Table 1) at ambient temperature, pressure and concentration in air are seen as dips in the transmitted intensity through the cavity.

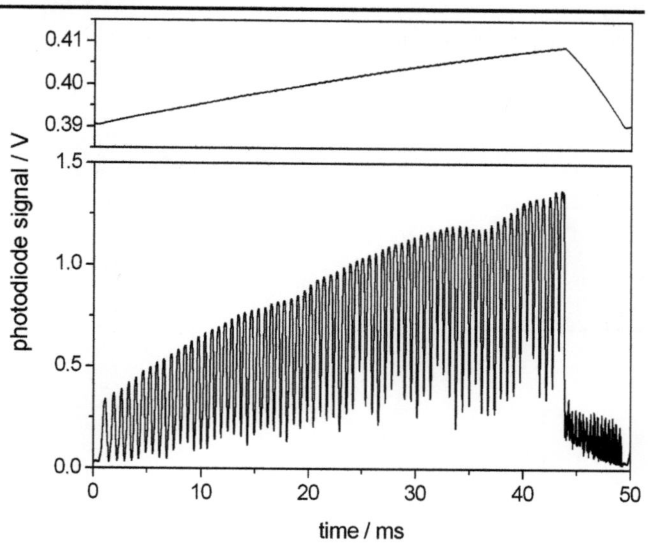

Fig. 4 *Lower panel*: mode-by-mode OF-CEAS cavity output signal showing two weak absorptions by $^{16}O^{18}O$ (lines (a) and (b) in Table 1), at 18 and 36 ms. The current ramp applied during the scan causes an increase in the emitted laser power (shown in the *upper panel*) and a corresponding increase in cavity mode intensity, seen as a general rising trend in the *lower panel*

The cavity output signal is recorded for 1 s and the resultant data file contains the result of 20 ramps of the laser. Individual spectra are analysed using a peak detection algorithm (LabVIEW) based on a quadratic least squares fit and the maximum intensity of each mode obtained. The mean intensity, $I(\omega_n)$, and standard deviation, SD[$I(\omega_n)$], of each mode at its discrete resonant frequency is determined. An analytically fitted polynomial baseline provides the background cavity mode intensities, $I_b(\omega_n)$. This baseline is chosen to exclude rare isotopologue absorptions although it includes any CIA and $^{16}O^{16}O$ absorptions as required by (3) or (4). The background ring-down time, τ_b, used for absolute calibration is also inclusive of cavity losses caused by CIA and the wings of $^{16}O^{16}O$ absorption lines, but exclusive of isotopologue absorptions, which is achieved by taking ring-down measurements at mode positions marked A, B, C and D in Fig. 3. A polynomial fit through these four points was used to determine the background ring-down time of all modes excited by the laser scan. This method thus includes cavity losses due to mirror transmission and all underlying absorption. The resultant absorption spectrum containing lines (b) and (c) is shown in Fig. 5 as a function of the mode index (a running number which is incremented by the free spectral range of the cavity). A spectrum of the $^{16}O^{18}O$ ^{R}R(2) line (a) is presented in Fig. 6.

Table 2 Comparison of measured integrated line intensities (σ_{INT}) and collisional-broadened half widths (γ_{air}) of three oxygen isotopologue lines with the values given in the HITRAN database.[38] Standard errors are given by the uncertainty in the fit and displayed in parentheses in units of the least significant digit

		OF-CEAS	HITRAN	OF-CEAS	HITRAN
		$\sigma_{INT}/10^{-27}$ cm^2 molec.$^{-1}$ cm^{-1}		γ_{air}/cm^{-1} atm^{-1}	
(a)	$^{16}O^{18}O$ $^RR(2)$	4.79(14)	5.73	0.0438(17)	0.0546
(b)	$^{16}O^{18}O$ $^RQ(1)$	5.68(25)	6.73	0.0469(23)	0.0564
(c)	$^{16}O^{17}O$ $^RR(2)$	1.02(17)	1.18	0.0411(74)	0.0546

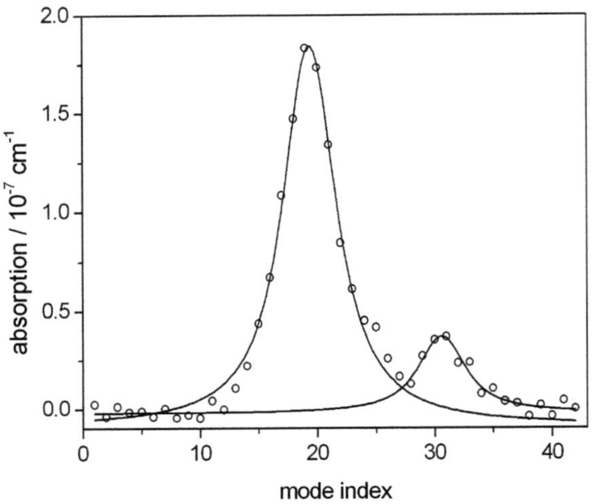

Fig. 5 Absorption spectrum of two weak oxygen lines in the $b^1\Sigma_g^+ \leftarrow X^3\Sigma_g^-(0,0)$ transition measured under ambient conditions. The stronger is the $^RQ(1)$ line of $^{16}O^{18}O$ at 13130.045 cm^{-1}. The weaker absorption is the $^RR(2)$ line of $^{16}O^{17}O$, at 13129.839 cm^{-1}. *Circles* indicate absorptions determined from cavity modes; the *solid lines* are Voigt profile fits

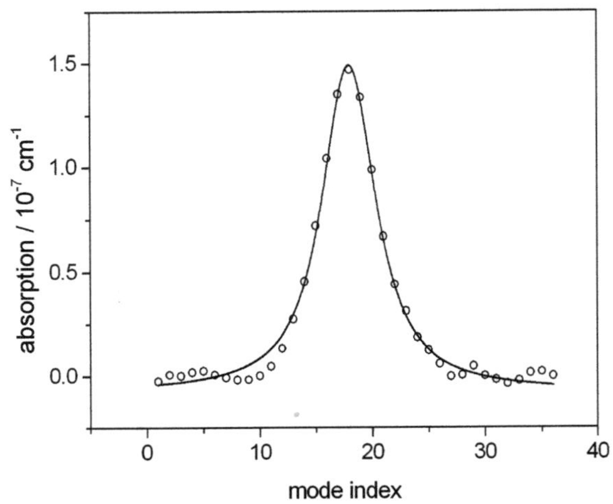

Fig. 6 Absorption spectrum of the $^RR(2)$ line of $^{16}O^{18}O$ at 13130.605 cm^{-1} in the $b^1\Sigma_g^+ \leftarrow X^3\Sigma_g^-(0,0)$ transition measured under ambient conditions. The experimental measurements (*circles*) obtained in 1 s are fitted to a Voigt profile (*solid line*)

A Voigt profile is fitted to each measured absorption line by floating all parameters except the Doppler (Gaussian) contribution to the width, which is fixed at the expected value for the temperature of the gas. The total area of the absorption multiplied by the free spectral range of the cavity, i.e. the mode separation, gives an integrated absorption coefficient, α_{INT} (cm^{-2}). This can be analysed using (6) to obtain the integrated line intensity, σ_{INT} (cm^2 molec.$^{-1}$ cm^{-1}):

$$\alpha_{INT} = \sigma_{INT}[O_2] \tag{6}$$

$[O_2]$ is the concentration of oxygen in the laboratory and is dependent on the ambient air pressure and temperature. An air pressure of 1000 mbar at 22°C corresponds to a concentration of oxygen in the air of 5.14×10^{18} molec. cm^{-3}. Pressure fluctuations of $\pm 2\%$ are possible as a result of changing weather conditions, corresponding to a probable molecular number density of between 5.03×10^{18} molec. cm^{-3} and 5.24×10^{18} molec. cm^{-3}. The spectrum was taken during a period of good settled weather in which the Bristol air pressure was ~1020 mbar and so the latter value was used for analysis. Table 2 shows the results for integrated line intensities and pressure broadened Lorentzian half widths obtained from the Voigt fits to the data in Figs. 5 and 6.

The collisionally broadened contribution to the line width is ~three times greater than the Doppler contribution at ambient air pressure. As the entries in Table 2 demonstrate, the measured half widths of lines (a), (b) and (c) derived from the Lorentzian components of the Voigt fits are ~15–25% lower than those given in the HITRAN database. The systematic uncertainties in our values will be fairly high, however, because of the signal-to-noise ratios for the very weak absorption lines, and the limited number of frequency points defining the lineshape. The combination of these factors restricted the lineshape analysis to fitting a Voigt function, when a more complicated lineshape function might be more appropriate. Differences also exist between the measured integrated line intensity and the literature values, with the measured values being 14–16% lower than those quoted in the HITRAN database. A number of factors may contribute to these differences: in particular, the removal of the complex underlying $^{16}O^{16}O$ absorption [45–47] using a polynomial fit to the baseline to extract the isotopologue absorption may introduce an error in derived line intensities. The cavity mode powers were also not normalised with respect to the incident laser power; instead, any variation with diode laser current was accommodated in the baseline polynomial fit. The mode intensities and their corresponding ring-down times are not currently recorded simultaneously as in [48], and instead a time delay exists between the measurements. Larger discrepancies of up to ~190% between measured and HITRAN data were observed by Motto-Ros et al. [48] while studying rare oxygen isotopologue lines in the O_2 B-band ($b \leftarrow X$, $1 \leftarrow 0$), using OF-CEAS with a linear cavity and Brewster-angle coupling method. In their case, the differences were attributed to difficulties in performing ab initio calculations of parameters for weak transitions for the entries in the HITRAN 2004 database. These calculations were made in 1998 [49] and have not been improved upon for the 2008 edition. The oxygen A-band literature values contained in HITRAN 2008 are the result of experimental studies, using isotopically enriched samples, and are therefore expected to be much more reliable [36]. These studies, mentioned in the Introduction, fitted the absorption profiles using Galatry line shapes to incorporate the effects of Dicke narrowing and line mixing.

We conclude that the small discrepancies between our line width and line intensity data and those in the current version of the HITRAN database are likely to arise from errors in our measurements resulting from the use of unenriched isotopologue samples, and the difficult calibration of background cavity losses (dominated by overlapping CIA features and the wings of the much stronger $^{16}O_2$ absorption lines). Nevertheless, the main purpose of the current work, to test the performance of a new design of OF-CEAS spectrometer, is successfully demonstrated and the instrument is shown to have sufficient sensitivity to observe $^{16}O^{18}O$ and $^{16}O^{17}O$ in ambient air in 1 s duration measurements via absorption on a weak, electric-dipole forbidden transition.

A common estimation of the sensitivity of any absorption technique is obtained from the minimum detectable absorption coefficient, α_{min}. For a single pass Beer-Lambert type absorption the minimum detectable absorption coefficient is largely dependent on the smallest detectable change in the beam intensity after passing through the sample. The sensitivity of CEAS is enhanced by the multi-pass effect of the high finesse cavity (by a factor of $(1-R)^{-1}$, with $R = 0.999919$ as determined from $\tau_0 = 6.5$ μs for our apparatus), and (7) [6] provides a method to calculate α_{min} values:

$$\alpha_{min} = \frac{1}{c\tau_0}\frac{\Delta I_{min}}{I_0} \tag{7}$$

I_0 is the mean output intensity, ΔI_{min} is taken as $\sqrt{2}\mathrm{SD}[I(\omega_n)]$ and the minimum detectable absorption coefficient presented is the mean over all mode frequencies in the spectrum. Each spectrum is the mean result of several scans of the laser. 15 averages acquired in 0.5 s (laser scan rate: 30 Hz) yield a minimum detectable absorption coefficient of 1.2×10^{-8} cm^{-1}, corresponding to a bandwidth reduced sensitivity of $\alpha_{min} = 2.2 \times 10^{-9}$ cm^{-1} Hz$^{-1/2}$. Consistent with this estimate, the residuals (experimental − fitted absorption) of the Voigt fit in Fig. 6 display a standard deviation of 3.9×10^{-9} cm^{-1} for a 1 s spectrum accumulation time. If a spectral line were to be measured against a background of zero absorption instead of the wings of $^{16}O^{16}O$ lines and the CIA, α_{min} would be reduced to 2.0×10^{-9} cm^{-1} Hz$^{-1/2}$.

5 Conclusion

We have described the development and implementation of an OF-CEAS scheme using a ring-shaped cavity. A counter-propagating wave induced in the optical cavity by a retro-reflecting mirror provides optical feedback and forms a standing wave with a small proportion of the forward-propagating beam. The average field amplitude of the standing wave upon each mirror has been modelled and the expected mode structure compares well to the measured counter-propagating cavity output. The levels of feedback experienced at each cavity resonant frequency are not identical, and without careful mirror cleaning and fine control over the amount of re-injected counter-propagating light, relative mode amplitudes can be distorted by varying levels of feedback. The travelling wave that dominates the internal field in the ring cavity experiences the same reflectivity for each cavity resonant frequency and CRD measurements are therefore unaffected by the different feedback levels.

The required laser to cavity distance that allows all cavity mode frequencies to return to the laser with the same phase of optical feedback has been given. Fine adjustments of the laser to cavity distance necessary to correct for thermal drift and acoustic noise are controlled by analysing the shapes of the cavity modes in computer software and relaying a feedback signal to a piezo-electric optical mount.

Three weak absorptions in the oxygen A-band, caused by the rare isotopologues $^{16}O^{18}O$ and $^{16}O^{17}O$, at wavenumbers close to 13130 cm^{-1} have been measured in ambient air. The measured integrated line intensity and collisional-broadened half width of each absorption line were determined and compared to values reported in the HITRAN 2008 database. A bandwidth reduced minimum detectable absorption coefficient of 2.2×10^{-9} cm^{-1} Hz$^{-1/2}$ was achieved while measuring rare isotopologue absorptions superimposed upon a non-zero baseline of overlapping $^{16}O^{16}O$ absorption features and continuum CIA. With a non-absorbing background and mirrors of higher reflectivity the sensitivity of this technique could be further improved.

Acknowledgements Funding for this work, and for a studentship for DJH is gratefully acknowledged from EPSRC grant EP/E018297. SGB thanks the Analytical Trust Fund of the Royal Society of Chemistry for a studentship. AJOE thanks the Royal Society and Wolfson Foundation for a Research Merit Award.

References

1. A. O'Keefe, D.A. Deacon, Rev. Sci. Instrum. **59**, 2544 (1988)
2. J.J. Scherer, J.B. Paul, A. O'Keefe, R.J. Saykally, Chem. Rev. **97**, 25 (1997)
3. M.D. Wheeler, S.M. Newman, A.J. Orr-Ewing, M.N.R. Ashfold, J. Chem. Soc., Faraday Trans. **94**, 337 (1998)
4. G. Berden, R. Peeters, G. Meijer, Int. Rev. Phys. Chem. **19**, 565 (2000)
5. S.S. Brown, Chem. Rev. **103**, 5219 (2003)
6. M. Mazurenka, A.J. Orr-Ewing, R. Peverall, G.A.D. Ritchie, Annu. Rep. Prog. Chem., Sect. C **101**, 100 (2005)
7. C. Vallance, New J. Chem. **29**, 867 (2005)
8. G. Friedrichs, Z. Phys. Chem. **222**, 31 (2008)
9. D. Romanini, A.A. Kachanov, N. Sadeghi, F. Stoeckel, Chem. Phys. Lett. **264**, 316 (1997)
10. B.A. Paldus, J.S. Harris, J. Martin, J. Xie, R.N. Zare, J. Appl. Phys. **82**, 3199 (1997)
11. B.A. Paldus, C.C. Harb, T.G. Spence, B. Wilke, J. Xie, J.S. Harris, R.N. Zare, J. Appl. Phys. **83**, 3991 (1998)
12. R. Engeln, G. Beerden, R. Peeters, G. Meijer, Rev. Sci. Instrum. **69**, 3763 (1998)
13. A. O'Keefe, J.J. Scherer, J.B. Paul, Chem. Phys. Lett. **307**, 343 (1999)
14. M. Mürtz, B. Frech, W. Urban, Appl. Phys. B **68**, 243 (1999)
15. R. Peeters, G. Beerden, A. Ólafsson, L.J.J. Laafhoven, G. Meijer, Chem. Phys. Lett. **337**, 231 (2001)
16. W.R. Simpson, Rev. Sci. Instrum. **74**, 3442 (2003)
17. M. Pradhan, M.S.I. Aziz, R. Grilli, A.J. Orr-Ewing, Environ. Sci. Technol. **42**, 7354 (2008)
18. J. Morville, D. Romanini, A.A. Kachanov, M. Chenevier, Appl. Phys. B **78**, 465 (2004)
19. T.J.A. Butler, J.L. Miller, A.J. Orr-Ewing, J. Chem. Phys. **126**, 174302 (2007)
20. Ph. Laurent, A. Clairon, Ch. Breant, IEEE J. Quantum Electron. **25**, 1131 (1989)
21. C.E. Wieman, L. Hollberg, Rev. Sci. Instrum. **62**, 1 (1991)
22. T.J.A. Butler, D. Mellon, J. Kim, J. Litman, A.J. Orr-Ewing, J. Phys. Chem. A **113**, 3963 (2009)
23. J. Morville, S. Kassi, M. Chenevier, D. Romanini, Appl. Phys. B **80**, 1027 (2005)
24. D. Romanini, M. Chenevier, S. Kassi, M. Schmidt, C. Valant, M. Ramonet, J. Lopez, H.-J. Jost, Appl. Phys. B **83**, 659 (2006)
25. S. Kassi, M. Chenevier, L. Gianfrani, A. Salhi, Y. Rouillard, A. Ouvrard, D. Romanini, Opt. Express **14**, 11442 (2006)
26. E.R.T. Kerstel, R.Q. Iannone, M. Chenevier, S. Kassi, H.-J. Host, D. Romanini, Appl. Phys. B **85**, 397 (2006)
27. R. Wehr, S. Kassi, D. Romanini, L. Gianfrani, Appl. Phys. B **92**, 459 (2008)
28. I. Courtillot, J. Morville, V. Motto-Ros, D. Romanini, Appl. Phys. B **85**, 407 (2006)
29. S.G. Baran, G. Hancock, R. Peverall, G.A.D. Ritchie, N.J. van Leeuwen, Analyst **134**, 243 (2009)
30. B. Dahmani, L. Hollberg, R. Drullinger, Opt. Lett. **12**, 876 (1987)
31. G.H. Dicke, H.D. Babcock, Proc. Natl. Acad. Sci. USA **13**, 670 (1927)
32. H.D. Babcock, L. Herzberg, Astrophys. J. **108**, 167 (1948)
33. D.J. Robichaud, J.T. Hodges, P. Maslowski, L.Y. Yeung, M. Okumura, C.E. Miller, L.R. Brown, J. Mol. Spectrosc. **251**, 27 (2008)
34. D.J. Robichaud, J.T. Hodges, L.R. Brown, D. Lisak, P. Maslowski, L.Y. Yeung, M. Okumura, C.E. Miller, J. Mol. Spectrosc. **248**, 1 (2008)
35. D.J. Robichaud, J.T. Hodges, D. Lisak, C.E. Miller, M. Okumura, J. Quantum. Spectrosc. Radiat. Transfer **109**, 435 (2008)
36. D.J. Robichaud, L.Y. Yeung, D.A. Long, M. Okumura, D.K. Havey, J.T. Hodges, C.E. Miller, L.R. Brown, J. Phys. Chem. A (2009). doi:10.1021/jp901127h
37. W. Heisenberg, Z. Phys. **41**, 239 (1927)
38. L.S. Rothman, I.E. Gordon, A. Barbe, D.C. Benner, P.F. Bernath, M. Birk, V. Boudon, L.R. Brown, A. Campargue, J.-P. Champion, K. Chance, L.H. Coudert, V. Dana, V.M. Devi, S. Fally, J.-M. Flaud, R.R. Gamache, A. Goldman, D. Jacquemart, I. Kleiner, N. Lacome, W.J. Lafferty, J.-Y. Mandin, S.T. Massie, S.N. Mikhailenko, C.E. Miller, N. Moazzen-Ahmadi, O.V. Naumenko, A.V. Nikitin, J. Orphal, V.I. Perevalov, A. Perrin, A. Predoi-Cross, C.P. Rinsland, M. Rotger, M. Simeckova, M.A.H. Smith, K. Sung, S.A. Tashkun, J. Tennyson, R.A. Toth, A.C. Vandaele, J. Vander Auwera, J. Quantum. Spectrosc. Radiat. Transfer **110**, 533 (2009)
39. S. Saraf, R.L. Byer, P.J. King, Appl. Opt. **46**, 3850 (2007)
40. F.A. Jenkins, H.E. White, *Fundamentals of Optics* (McGraw-Hill, Singapore, 1981)
41. T. Skettrup, T. Meelby, K. Færch, S.L. Frederiksen, C. Pedersen, Appl. Opt. **39**, 4306 (2000)
42. V. Motto-Ros, J. Morville, P. Rairoux, Appl. Phys. B **87**, 531 (2007)
43. J. Morville, D. Romanini, Appl. Phys. B **74**, 495 (2002)
44. D.J. Hamilton, MSci Thesis. University of Bristol (2008)
45. G.D. Greenblatt, J.J. Orlando, J.B. Burkholder, A.R. Ravishankara, J. Geophys. Res. **95**, 18577 (1990)
46. K.M. Smith, D.A. Newnham, J. Geophys. Res. **105**, 7383 (2000)
47. H. Tran, J.-M. Hartmann, J. Geophys. Res. **113**, D18104 (2008)
48. V. Motto-Ros, M. Durand, J. Morville, Appl. Phys. B **91**, 203 (2008)
49. R.R. Gamache, A. Goldman, L.S. Rothman, J. Quantum. Spectrosc. Radiat. Transfer **59**, 495 (1998)

External-cavity quantum cascade lasers with fast wavelength scanning

T. Tsai · G. Wysocki

Received: 5 October 2009 / Revised version: 8 December 2009 / Published online: 20 January 2010
© Springer-Verlag 2010

Abstract Fast wavelength scanning of an external-cavity quantum cascade laser (EC-QCL) has been developed using a modified Littrow-type cavity configuration. Scan rates up to 5 kHz and tuning ranges up to 7 cm^{-1} have been realized using folded cavity arrangement with a fast piezoactuated intracavity mirror to vary the effective diffraction grating angle. High-resolution molecular spectroscopy, limited primarily by the effective laser linewidth, has been demonstrated in pulsed and CW mode using direct absorption spectroscopy of ammonia and ethylene in the 10-μm wavelength region. A high-resolution broadband spectral scan employing mode-hop-free tuning of a set of discrete laser modes within the fast frequency scan has been developed. The method has been demonstrated by performing high-resolution broadband spectrum of ethylene under atmospheric and reduced pressures.

1 Introduction

Quantum cascade lasers (QCLs) are extremely useful sources for spectroscopic applications in the mid-infrared (mid-IR) spectral region, in which a majority of molecules possess their strongest fundamental rotational–vibrational absorption bands [1–3]. QCLs are small and robust and can provide hundreds of milliwatts of mid-IR optical power. These characteristics make them ideal sources for spectroscopic sensing applications in the field such as air pollution monitoring or breath-analysis for medical diagnostics [1, 2, 4]. Compared to other spectroscopic instruments, such as Fourier-transform infrared spectrometers (FTIR) or diffractive monochromators, laser-based systems offer superior spectral resolution and significantly higher source spectral brightness in a compact size (i.e., FTIR spectral resolution is related to the maximum path difference obtainable by its moving mirror, and therefore, a higher resolution requires a larger instrument). QCL emission linewidths are on the order of <0.001 cm^{-1} and are easily achievable [1, 5, 6]. Thus, their applications in high-resolution molecular spectroscopy and ultrasensitive chemical detection are of particular interest to scientists as well as to industries.

In addition to high spectral resolution, many applications require broadband coverage of spectral frequencies to perform reliable identification of chemical species. Such broadband coverage is especially important for multimolecular analysis that requires simultaneous measurement of several narrow spectral lines distributed over broad spectral range; or for identifying complex molecules with broadband unresolved spectral features. In general, interferometers and monochromators can easily achieve broadband spectral coverage and might be preferred in applications that do not require high resolution or high intensity light sources. Unlike interferometers and monochromators, whose wide spectral coverage is provided by their broadband incoherent light sources (infrared broadband emitters, e.g., tungsten filament bulb), tunability of laser-based spectrometers is often limited by the narrow spectral profile of the laser gain. However, since QCLs generate photons through intersubband transitions within a quantum well, their spectral gain profiles can be tailored through precise quantum engineering of the active regions [7]. One of the first QCL active region designs that provided broader laser gain profiles is the so-called bound-to-continuum design demonstrated by Faist et al. [8]. Such broad-gain material can be used to produce an array of multiple single-mode distributed feedback (DFB) QCLs las-

T. Tsai · G. Wysocki (✉)
Department of Electrical Engineering, Princeton University, Princeton, NJ 08544, USA
e-mail: gwysocki@princeton.edu

ing within the entire range of available wavelengths [9]; or, if used in an external cavity (EC), a single laser can generate any frequency within the gain curve [10, 11]. Even broader gain profiles can be obtained by combining multiple active regions into a single QCL chip [12]. With such heterogeneous cascade devices, tuning ranges with a single QCL chip can reach over 430 cm^{-1}. In many cases, such tunability is sufficient to perform spectroscopy in applications focused on a particular spectral region. For example, a single device demonstrated recently by Hugi et al. [13] can cover the entire second atmospheric transparency window used for atmospheric chemistry studies.

If a DFB structure is used to obtain single frequency operation, the total tuning range is primarily limited by the thermal tuning of the effective refractive index within the QCL waveguide [7, 14]. Thermal tuning of the DFB QCL is performed either by varying the QCL substrate temperature (10–20 cm^{-1} tuning range) or by changing the QCL bias current (2–3 cm^{-1} tuning range). Substrate temperature tuning is very slow, limited by the response time of the temperature control system and the thermal design of the QCL submount and package. Bias current tuning can be much faster and has been used to perform fast spectroscopy. A short pulse on the order of >100 ns is applied to the QCL, and due to thermal change of the refractive index, wavelength tuning within one pulse can be exploited to perform spectroscopy at kHz to MHz spectral scan repetition rates. However, the intrapulse tuning range is limited to ~2–3 cm^{-1} and requires fast photodetectors and electronics to acquire the data [15, 16]. On the other hand, by using an external-cavity quantum cascade laser (EC-QCL), one can vary the diffraction grating angle to tune the laser wavelength over the entire available tuning range (>150 cm^{-1}) [17]. Our EC-QCL system developed earlier tunes through a piezoelectric actuation (rotation and translation) of the grating, and due to the large mass of the actuated optomechanical elements, this EC-QCL tunes with scan rates lower than 50 Hz. Some sensing applications require substantially higher scan rates, e.g., for trace-gas sensing in turbulent systems, such as smokestacks and furnaces. A QCL must tune over a broad spectral range at kHz scan rate so that the turbulent system can be approximated as a steady state [18–20]. Additionally, a majority of spectroscopic sensing instruments are affected primarily by low-frequency ($1/f$) noise, which can be suppressed if higher scan rates are used. In this work, we focused on the development of a new EC-QCL system that can obtain up to two orders of magnitude higher wavelength scan rates compared to our previous EC systems and is capable of performing high-resolution spectroscopy with a resolution determined by the effective laser linewidth.

2 Fast tuning EC-QCL system design

As mentioned earlier, fast (kHz) scan rates are necessary to perform spectroscopy of turbulent media. Additionally, since we are using optomechanical means to tune, higher scan rates also require that the frequency of the mechanical resonances of the scanning system should be shifted to the highest possible values to avoid excitation of any unwanted mechanical vibrations. Such a fast optomechanical scanning can only be performed when small masses are actuated. Therefore, we developed a modified Littrow configuration employing a folded cavity arrangement with a compact fast piezoactuated intracavity tilt mirror used for fast wavelength tuning. A schematic of the new fast tuning EC-QCL arrangement is shown in Fig. 1.

The gold-coated (on a glass substrate) intracavity mirror (M) with diameter of 10 mm and thickness of 2 mm varied the incident angle (α) at the diffraction grating; and thus, by changing the tilt angle, the wavelength of the laser feedback could be scanned. To achieve fast tuning speeds, the intracavity mirror was mounted onto a fast piezoelectric tilt platform (Piezosystem Jena PSH 4/1) that could tilt at speeds up to 5 kHz. Significant electrical power of 120 W would be required to achieve the full tilt range at such high frequencies. Due to power limitations of our driving electronics, there is a trade-off between tuning speed and tuning range. The full 4-mrad tilt range could only be performed with scan rates up to 1 kHz. Therefore, for the proof-of-concept demonstration, we chose to work with 1-kHz scan rates. The fast tuning could be performed anywhere within the EC-QCL total spectral tuning range by a coarse adjustment of the diffraction grating that was mounted on a motorized rotation stage. The total tuning range of the EC-QCL was primarily determined by the gain chip and the quality of the antireflection

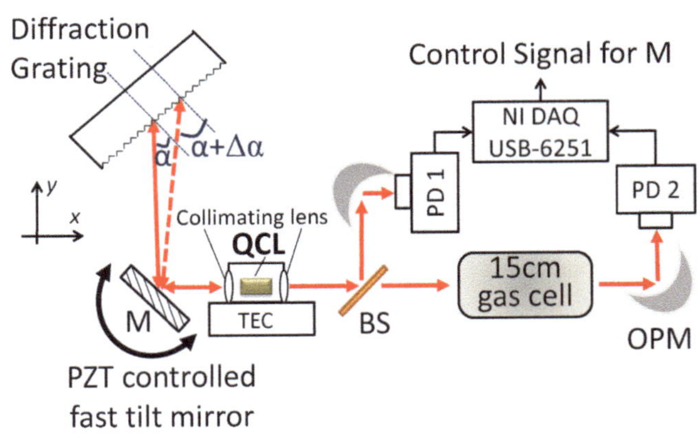

Fig. 1 Fast tuning external-cavity quantum cascade laser (EC-QCL) and direct absorption spectroscopy experimental setup. By varying the tilt mirror, M, the grating angle, α, is varied in the external cavity. In the spectroscopic setup, BS is a beamsplitter used to direct a reference beam into photodetector 1 (PD1). Both photodetectors are thermoelectrically cooled HgCdZnTe based detectors. OPM is off-axis parabolic mirror

(AR) coatings deposited on the chip facets. For this work, two different QCL chips operating at ~10 μm were used: one chip operated only in a pulsed mode and a second QCL was used for continuous wave (CW) experiments. The QCL chips were mounted, AR and HR (high reflection) coated, and assembled in an optomechanical package with collimating lenses by DayLight Solutions Inc., which kindly provided the QCLs for our fast tuning experiments. The first laser chip used for pulsed mode experiments exhibited a total tuning range of 45 cm^{-1}; and the second CW QCL exhibited a coarse tuning range of 35 cm^{-1}. Those tuning ranges, though quite narrow, were more than sufficient to study the fast tuning capability which required ~7 cm^{-1} of the total tuning range.

To demonstrate the capabilities of the fast tuning EC-QCL as a spectroscopic source, direct absorption spectroscopies of ammonia (NH_3) and ethylene (C_2H_4) were performed with the pulsed laser chip and CW chip respectively.

3 Spectroscopic measurements with fast EC-QCL

3.1 Molecular spectroscopy in pulsed mode

The intracavity tilt mirror changes the effective grating incident angle (α), and thus, the wavelength (λ) of the retroreflected wave is tuned based on the standard grating equation for the first diffraction order in Littrow configuration where d is the distance between grooves:

$$\lambda = 2d\sin(\alpha). \tag{1}$$

Given the total tilt range of 4 mrad and the diffraction grating equation, the tuning range for a central frequency of 993 cm^{-1} was calculated to be 6.89 cm^{-1} and agreed with the experimentally observed ~7 cm^{-1}. However, this tuning range was not continuous, because besides the grating selected wavelength, there are also two other frequency selective elements to consider: the Fabry–Perot (FP) cavity formed by the QCL chip facets; and the external FP cavity which is formed between the grating and the QCL output facet. Variation of the tilt angle changes the effective EC length, but it does not change the resonant wavelength in the cavity considerably. This can be interpreted as a discrete change in L at each groove, which corresponds to the increment equal to the groove depth. Therefore, the local change in L is always equal to $\lambda/2$. Since it is the retroreflected light that is coupled back into the laser, the result is an optical path length difference of λ, and thus, the resonant wavelength within the cavity does not tune by varying the effective EC length with the tilt mirror. In other words, translation of the grating in the direction transverse to the beam axis would not provide any tuning of the resonant wavelength in the Littrow cavity. Therefore a continuous tuning of the lasing frequency can be performed only when the chip FP resonance and EC FP resonance are synchronously tuned together with the grating filter. By performing the tilt angle scanning without synchronous variation of the EC length and the bias current (i.e., thermally tuning the optical QCL chip length), one will observe mode hops primarily associated with the QCL chip FP modes, which are more spectrally sparse then EC FP modes. The number of useful spectral points is then reduced to data collected at intervals separated by QCL free spectral range (FSR) rather than a continuous set of points across the entire tuning range as desired for high-resolution spectroscopy.

In pulsed mode, the thermal chirp associated with each QCL current pulse causes variations in the laser mode structure, which broadens the effective laser linewidth to 0.2 cm^{-1} or even larger values [17, 21]. Thus, pulsed EC-QCLs are usually used for spectroscopic measurements of broad absorption features. Such applications can additionally benefit from the fast tuning capability provided by the EC-QCL configuration developed in this work. To suppress the sparse QCL FP mode structure mentioned above, we applied an additional modulation of the laser core temperature by adding a subthreshold current waveform to the laser pulse train. The subthreshold current waveform and the pulse train were not synchronized to provide effective randomization of the tuning process, which led to a significant broadening of the laser linewidth. We estimated the effective linewidth by performing direct absorption spectroscopy of 0.1% of ammonia mixed with a buffer gas (N_2) in a 15-cm gas cell at a reduced pressure of 100 Torr. Deconvolution of the laser linewidth from the Doppler broadened NH_3 line at 991.7 cm^{-1} yielded a laser linewidth of ~0.35 cm^{-1} (Gaussian laser line profile was assumed). This linewidth was not sufficient to perform high-resolution spectroscopy of small molecules at reduced pressures, but given the simplicity of the laser operation and the resolution comparable with basic FTIR instruments, the fast pulsed EC-QCL source could be successfully used for spectroscopy of pressure broadened lines of small molecules or broadband spectroscopic features in applications that require rapid scanning (i.e., in turbulent media, such as furnaces or exhaust lines).

To demonstrate its spectroscopic capabilities, we performed a direct absorption spectroscopy experiment using NH_3 mixture with N_2 at atmospheric pressure (760 Torr) in a 15-cm gas cell at room temperature. We targeted three NH_3 lines at ~992 cm^{-1}, which at atmospheric pressure, were broadened and formed a single ~2 cm^{-1} wide absorption feature. In this experiment, only photodetector 2 was used (Fig. 1), and each spectra was the average of 100 scans, which resulted in a total acquisition time of 100 ms. Baseline was collected by first recording the intensity at photodetector 2 without any gas in the cell. In postprocessing, the

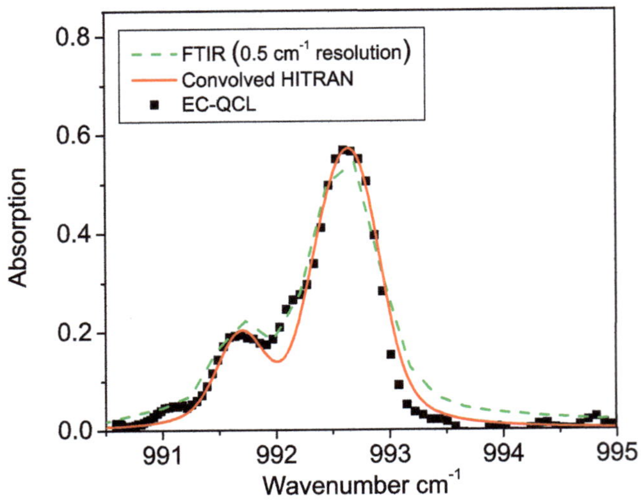

Fig. 2 Absorption spectrum of NH_3 mixture in N_2 at 760 Torr performed with the fast pulsed EC-QCL at 1-kHz scan rate averaged over 100 scans. The data fitted by a forward model based on line-by-line HITRAN simulation convolved with a Gaussian instrumental line shape yielded 0.32% NH_3 concentration and 0.38-cm^{-1} laser linewidth. The measurements show good agreement with an FTIR measurement

absorption was calculated with the baseline factored out. To take full advantage of the kHz scan capability, the baseline correction must be performed on scan-to-scan basis (e.g., by fitting with a polynomial) and averaging should be performed on baseline-corrected single scan spectra [22]. The EC-QCL frequency calibration was performed by the FTIR by measuring the total (integrated) tuning range and the central frequency of the laser emission. With this information and knowledge of the sine wave parameters applied to the tilt mirror, the frequency of each point could be determined.

The measured absorption spectrum in Fig. 2 was recalculated to absorbance and fitted by a model based on line-by-line calculation of the absorbance spectrum using line parameters from the HITRAN 2008 database [23] convolved with Gaussian instrumental line shape. Although this model can be considered accurate only in the linear region of absorption process, it is sufficient to provide approximation of laser capabilities. The concentration of the sample and width of the instrumental function were used as fitting parameters in the model. The nonlinear fitting algorithm yielded concentration of 0.32% of NH_3 and the laser linewidth of 0.38 cm^{-1}, which is in good agreement with our measurement of the gas sample at reduced pressure. To further verify the EC-QCL absorption spectrum, an FTIR measurement was performed on the same sample. With the FTIR set to comparable spectral resolution (in this case 0.5 cm^{-1}) both the FTIR and the EC-QCL absorption spectra show very good agreement as depicted in Fig. 2.

3.2 High resolution spectroscopy using fast cw EC-QCL

Another QCL chip capable of CW operation up to submount temperatures of 20°C was used to perform high-resolution spectroscopy with the fast EC-QCL arrangement. Unlike pulsed mode, CW operation of EC-QCLs does not suffer from the linewidth broadening due to the intrapulse frequency chirp. Therefore, substantially narrower laser linewidths can be achieved with CW QCLs [6, 16]. Since the narrow laser linewidth should be preserved for high-resolution spectroscopy, we did not use the subthreshold current modulation approach to suppress the mode structure as was done with pulsed EC-QCL system. Thus, the CW EC-QCL tends to lase at FP modes separated by the QCL chip FSR. To better understand the mode structure and the wavelength tuning process, the tilt mirror was slowly stepped through its entire voltage range, and the wavelength and intensity of the output radiation were recorded with a wavelength meter (Bristol Instruments 621) and photodetector, respectively (Fig. 3). As shown in Fig. 3, by changing only the diffraction grating angle, the laser wavelength varied noncontinuously, and the mode hops could be clearly identified in its frequency and intensity characteristics. The output intensity within one mode increased and then decreased to a point that the laser hopped to another FP mode that exhibited lower loss. The average mode hop was about 0.5 cm^{-1}, which suggested strong coupling to the QCL chip FP modes with FSR of ~0.52 cm^{-1} (for 3-mm chip length and 3.2 effective refractive index of the core).

Despite the mode-hops, the overall trend of the frequency tuning should follow the standard diffraction equation (see (1)). With a center frequency of 950.32 cm^{-1} and a grating of 150 grooves/mm, (1) yielded a grating tuning rate of ~−0.0475 cm^{-1}/V (red dashed line in Fig. 3a), which agreed well with the experimentally found rate of −0.051 cm^{-1}/V.

Since the data presented in Fig. 3 was collected as the tilt mirror was slowly stepped through its 150-V range, we were able to extend the tilt range to 4.825 mrad, which is more than the manufacturer specification of 4 mrad. Due to the geometry of the folded cavity arrangement, the variation of the effective grating angle is doubled, and so the 0.825-mrad difference actually translates to over 1 mrad. This difference causes the tuning range discrepancy between the slow measurements taken in Fig. 3 compared to the slightly lower tuning range taken at 1 kHz scan rate. As seen in Fig. 3a, a slow scan would measure 16 high resolution points with ~0.5 cm^{-1} spacing. At 1 kHz scan rate, the tilt angle was slightly reduced and met the manufacturer specifications of 4 mrad. The EC-QCL could then be tuned over 11 distinct FP modes, which gave a tuning range of 5.7 cm^{-1}. From the grating equation, the tuning range about the center frequency of 950.32 cm^{-1} was calculated to be 5.92 cm^{-1}, which agreed with the experimental range of 5.7 cm^{-1}.

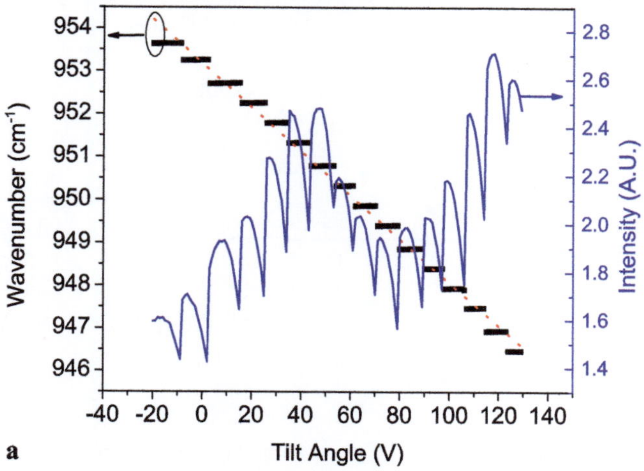

a

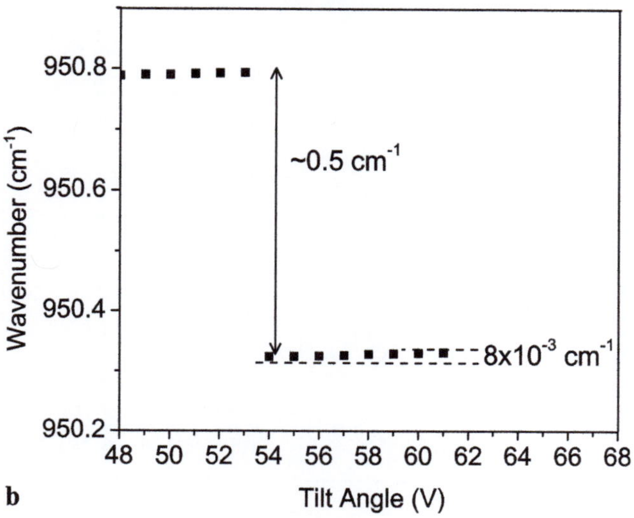

b

Fig. 3 (**a**) Laser frequency, theoretical grating tuning rate, and output intensity recorded during a single scan of the tilt mirror within the total available tilt range. Data was collected with the QCL current held constant at 820 mA and temperature at 288.3 K. (**b**) Magnification of the frequency tuning curve shows a mode hop of $\sim$0.5 cm^{-1} and frequency variation of 8×10^{-3} cm^{-1} for each lasing mode

This laser frequency characterization as seen in Fig. 3b also demonstrated that each fast scan would be a set of high resolution spectral points with 0.008 cm^{-1} precision at each frequency step. This small variation could be the result of a slight misalignment in which the tilt mirror pivot point was not positioned on-axis with the laser beam, which would result in residual tuning of the EC length and thus lasing frequency determined by the EC FP resonance. Using simple geometrical relations we estimated that the observed tuning would correspond to a shift of $\sim$2.5 mm of the mirror tilt axis with respect to laser beam. To confirm this observation, the same experiment was repeated with precisely realigned EC (with precision better than 0.5 mm). The measurement showed that the residual tuning was no longer measurable with the wavemeter. It is worth noting that even with relaxed alignment precision, the 0.008 cm^{-1} tuning within each mode does not influence significantly the precision that is required for molecular spectroscopy in gas phase.

To demonstrate the applicability of this cw EC-QCL as spectroscopic light source, we performed direct absorption spectroscopy measurements of ethylene. The laser was scanned with 1 kHz scan rate and within each fast scan 11 high resolution spectral data points (each with resolution better than $\sim$0.008 cm^{-1}) separated by $\sim$0.5 cm^{-1} were acquired. First, the spectrum was collected with no gas in the cell to calibrate the response of photodetector 1 with respect to photodetector 2 in the sample channel (Fig. 1). The scans were then repeated with the ethylene mixture, and the baseline was continuously acquired by photodetector 1. Additionally, the absorption was simultaneously calculated during the data collection from the baseline provided by photodetector 1 and the sample data from photodetector 2. For each mode, only the absorption data point that corresponded to the highest light intensity was used to generate the spectrum in postprocessing. Data acquired by the photodetector 1 was used to determine which data point corresponded to the highest light intensity, and the same point was used in the absorption data. The optical frequency of each mode was determined by using data in Fig. 3a. The absorption data corresponding to the 11 modes generated a sparse set of spectral samples of the absorption spectrum. An example spectrum of 0.1% C_2H_4 mixed with a buffer gas (N_2) in a 15-cm gas cell at room temperature is shown in Fig. 4. The spectrum was calculated using an average of 100 scans, which gave a total acquisition time of 100 ms. Figure 4 shows a good correlation between the collected data with the FTIR measurement and HITRAN simulation where the average of the residuals between the collected data and the simulation was equal to 0.005.

Some applications can benefit from such rapid laser frequency scans even if only a small number of high-resolution data points are available [18, 24]. A fast scan over well-defined spectral points combined with multispecies principle component spectral analysis can provide a powerful tool for measurements in environments with rapidly changing transmission properties.

3.3 Simultaneous mode-hop-free tuning of fast EC-QCL modes

In the past we demonstrated mode-hop-free tuning of conventional EC-QCLs, and this technology proved itself in spectroscopic sensing applications both in the laboratory and in the field [25, 26]. In this work we have developed the fast tuning EC-QCL source that within one rapid (<1 ms) scan operates sequentially on multiple high-resolution modes separated by $\sim$0.5 cm^{-1}. In order to perform high-resolution broadband spectroscopy with high-density spectral points within the spectrum, we have in-

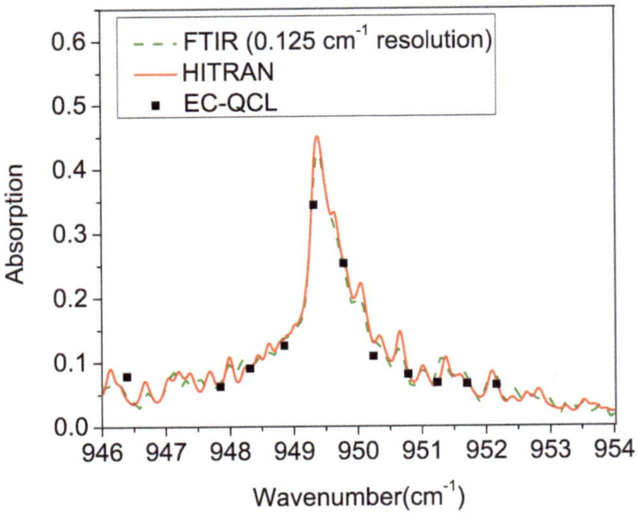

Fig. 4 Direct absorption spectroscopy measurements of 0.1% C_2H_4 in N_2 at 760 Torr acquired using fast CW EC-QCL at 1 kHz scan rate. The measurements are compared with an FTIR spectrum and simulation based on HITRAN database

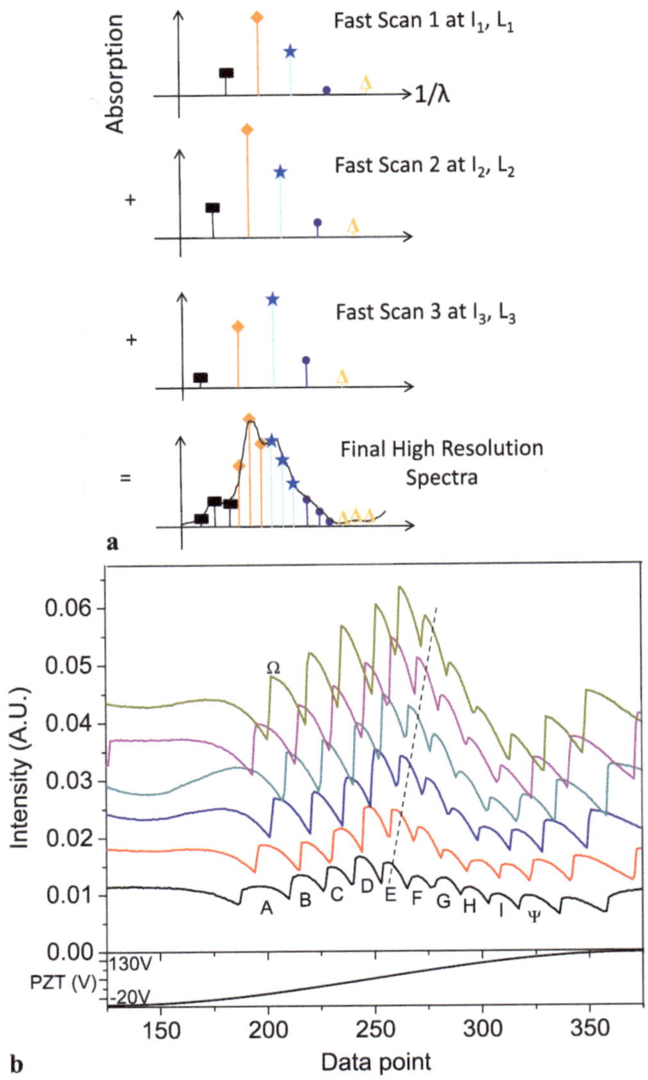

Fig. 5 (**a**) Schematic picture representing the spectral stitching method where each color/symbol represents a different mode. (**b**) Laser intensity collected for different fast 1-kHz scans (shown by *different colored lines*) taken at a different QCL bias current tuned synchronously with the external-cavity length to demonstrate simultaneous mode-hop-free tuning. The *dotted line* serves as a guide to show the actual mode tracking, and the *inset below* shows how the PZT voltage is varied sinusoidally through the scan

vestigated simultaneous mode-hop-free tuning of all high-resolution modes by synchronously varying the QCL bias current and the external-cavity length. To vary the external-cavity length, a second piezoactuator with maximum travel distance of 80 μm was used to translate the fast tilt mirror in the x direction (Fig. 1). This external length variation can be performed relatively quickly because for this application it is sufficient to provide tuning over a very short range corresponding to the change by one chip FSR ($\sim$0.5 cm^{-1}). For our current system configuration with EC length of $\sim$35 mm, this corresponded to a total EC length change of $\sim$21 μm. By synchronous variation of the bias current and the external-cavity length at a timescale much slower than the single fast scans, each mode available within the fast scan exhibits mode-hop-free tuning on the scan-to-scan basis. This allowed us to combine the consecutive spectra into one broadband scan as shown in Fig. 5a. Since the bias current and external-cavity length were varied linearly, the modes shifted in an approximately linear fashion. This is shown in Fig. 5b as a dotted line which follows the same mode in the scans collected at different bias currents and EC lengths. The PZT tilt mirror (grating angle) was varied sinusoidally within 150-V range. For further analysis, only points corresponding to the swing between the minimum and the maximum of the sinusoid have been considered. By using a transmission spectrum through the germanium etalon, we determined that each mode tuned over $\sim$0.58 cm^{-1}, which was sufficient to cover one FSR of the QCL chip FP resonator. In this work, the experiment was performed by collecting the raw data at kHz scan rate and the final mode-hop-free spectra was constructed in post-processing.

Similarly to the sparse CW spectroscopy presented in the previous section, we only considered the data taken at the peak intensity of each mode. Modes were identified through an algorithm that used the derivative of the data with respect to the applied tilt mirror voltage to search for steep slopes associated with mode-hops. The peak intensity was found by fitting third-order polynomial to the intensity time series as collected by photodetector 1 (Fig. 1). The local maximum of that fitted polynomial provided the point at which to take the absorption information collected by photodetector 2. If the maximum in the fitted polynomial occurred between collected data points, the absorption information was

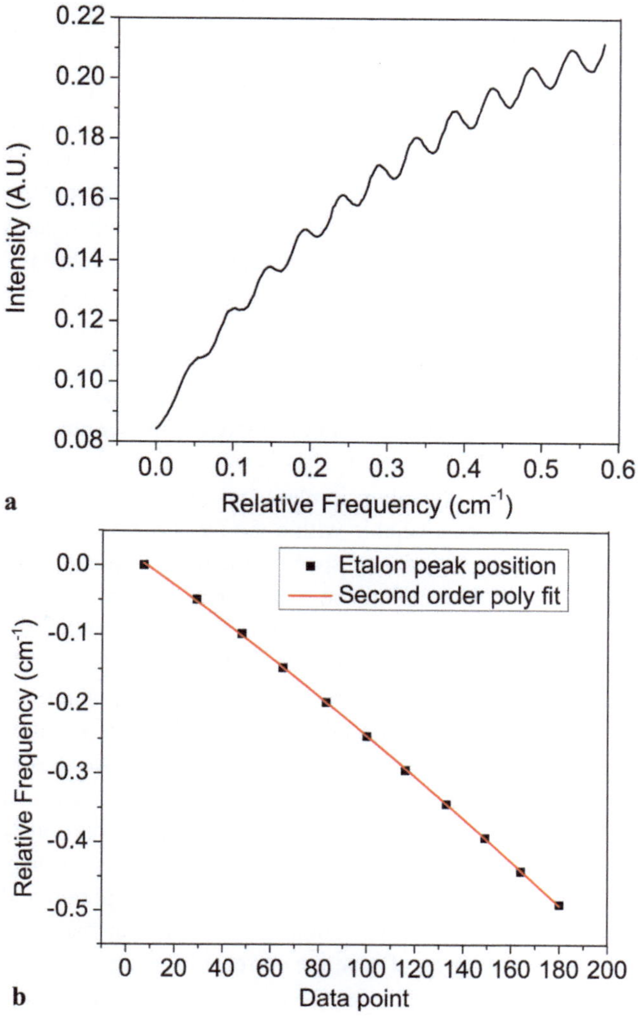

Fig. 6 (**a**) Transmission of EC-QCL through germanium etalon with FSR of 0.0491 cm^{-1}. This scan was taken at a constant tilt angle while the bias current and cavity length were synchronously tuned. (**b**) Second-order polynomial laser frequency calibration curve

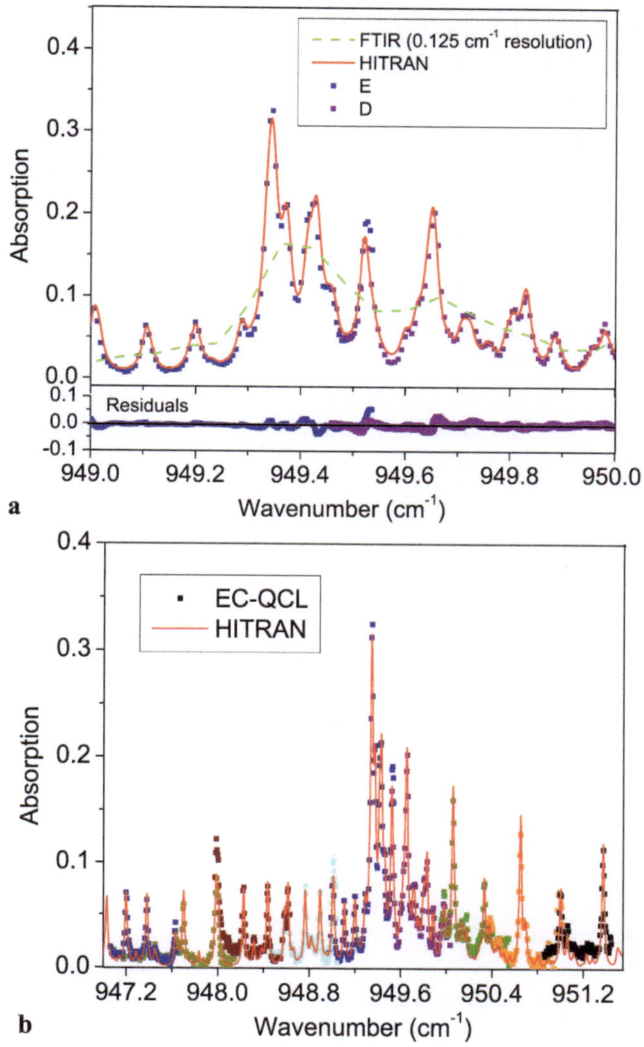

Fig. 7 Spectra of 0.159% C_2H_4 in N_2 in a 15-cm gas cell at 100 Torr acquired with simultaneous mode-hop-free tuning of all fast EC-QCL modes compared with HITRAN simulation for: (**a**) two neighboring modes within the fast EC-QCL scan (E and D as noted in Fig. 5b and the FTIR measurement of the same sample; and (**b**) all nine modes within the fast EC-QCL scan

interpolated linearly. In the case that the local maximum was not present, the spectral position of the maximum intensity value was used. The baseline was determined in the same way as in the previous experiment. As noted in Fig. 5b with the letters A–I, only 9 out of the 11 available modes were simultaneously tracked, because the last mode, Ψ, and the first mode, Ω, were not present over the entire mode-hop-free range. Therefore, to perform reliable analysis of the tuning process we have analyzed only 9 modes, A–I, with full 0.58 cm^{-1} mode-hop-free tuning range. We used a germanium etalon to determine the frequency calibration for the center mode E. The transmission spectrum through the etalon collected with this mode is shown in Fig. 6.

Figure 6 clearly demonstrates the mode-hop-free character of the mode tracking. The QCL current was varied by 120 mA and the EC length was varied by ~24 μm using the second piezoactuator. The calibration curve shown in Fig. 6b is slightly nonlinear, which is related to the nonlinearity of the piezoactuator used for EC length tuning. For precise and absolute frequency calibration within the entire broadband spectral scan, we have utilized positions of multiple absorption lines in the collected data based on HITRAN simulation. Given the nonlinear character of the frequency scan, a second-order polynomial was used for frequency calibration. The postprocessing algorithm was applied simultaneously to each identified laser mode, and the resulting broadband spectrum was automatically generated as shown in Fig. 7.

Figure 7 is a result of combining 100 consecutive spectra taken at different bias currents and synchronously changed cavity lengths. By synchronously varying the bias current and cavity length, we performed mode-hop-free tuning on all nine modes, A–I, simultaneously and stitched over the

QCL chip FSR with 100 points (as depicted in Fig. 5a). All consecutive spectra were an average of 100 single fast scans recorded with 1 kHz scan rate at constant bias current and cavity length. The bias current and cavity lengths were varied every 110 ms (additional 10 ms was used to assure reliable data acquisition and storage). As a result the total data acquisition time for the spectrum shown in Fig. 7 was 11 seconds. This experiment was performed at reduced pressure of 100 Torr where the multiple ethylene lines were distinct in contrast to the situation in Fig. 4 where the lines were congested due to pressure broadening at 760 Torr. To better illustrate the EC-QCL resolving individual absorption lines, Fig. 7a shows magnification of the complete combined spectra shown in Fig. 7b for two neighboring laser modes at $\sim$949.3 cm^{-1}. From Fig. 7a, the superior resolution of the fast tunable EC-QCL is readily apparent compared to the table top FTIR at its highest resolution of 0.125 cm^{-1} (shown as the dashed line). Figure 7b shows the combined spectra acquired simultaneously with all nine modes available within the fast 1 kHz scan (each color of the dots represents a different mode). The correlation between the collected data and HITRAN simulation results in coefficient of determination $R^2 = 0.935$, which demonstrates an excellent agreement between the measured data and the HITRAN 2008 simulation.

4 Conclusion

Spectroscopy in turbulent systems, such as trace gas sensing in smokestacks and furnaces, require QCL sources that can tune over a wide range of frequencies at kHz scan rates. In this paper we presented an optical breadboard prototype EC-QCL based on a modified Littrow configuration with folded cavity. A fast intracavity tilt mirror is used for frequency tuning at kHz rates which meets the requirements for fast spectroscopic applications.

We developed and demonstrated three different operation methods of the fast EC-QCL, which can perform high resolution molecular spectroscopy. A pulsed EC-QCL with subthreshold modulation, which suppressed a parasitic mode structure, was used to collect continuous spectra with frequency resolution of 0.38 cm^{-1} ($\sim$12 GHz). This resolution is only sufficient for fast spectroscopy of pressure broadened molecular absorption lines or broadband absorbers in gas, liquid, or solid phase. The spectral scanning over 7 cm^{-1} with 1 kHz scan rates is significantly faster than the data acquisition rates achievable with compact FTIR spectrometers while still providing a comparable spectral resolution.

Significantly higher spectral resolution was obtained with CW fast EC-QCL. A fast 1-kHz scan provided a high resolution but sparse data set with spectral data points separated by $\sim$0.5 cm^{-1}. While 11 high-resolution data points may not be enough to resolve low-pressure spectra of simple molecules, trace gas sensing applications that require fast wavelength scanning are typically under elevated pressure conditions where the narrow absorptions lines are significantly broadened. It has also been shown before that small number of high spectral resolution and optimally selected spectral sampling points can be reliably used to perform multispecies detection [18, 24]. As an example, the ethylene spectrum at atmospheric pressure, at which dense absorptions lines form simpler and wider absorption features, could be easily resolved with the sparse data set. Moreover, the high mechanical tuning speed (kHz) can significantly decrease the $1/f$ noise, which has been difficult to suppress with our previous conventional EC-QCLs [11].

Finally, by performing simultaneous mode-hop-free tuning of the modes available within the fast EC-QCL scan, a high-resolution broadband spectrum can be acquired. The time required to complete the scan increases proportionally to the number of points collected for each mode, but the advantage of having each individual scan performed at kHz rate is not affected. At much slower scan rates at which the quasistatic approximation fails, the baseline fluctuates during the scan and is difficult to determine [22]. In such cases, erroneously identified spectral features result in artifacts and false molecular quantification. The fast scan rate capability allows us to analyze a rapidly changing system as a steady state within each fast scan, which enables a simplified and more reliable baseline correction on a scan-to-scan basis and allows for averaging of baseline corrected spectra.

This fast EC-QCL can achieve a broadband tuning range at kHz rates to perform high-resolution spectroscopy in turbulent systems, which opens up a whole new set of spectroscopic applications previously difficult to accomplish with DFB QCLs and conventional EC-QCLs.

Acknowledgements The authors acknowledge support from NSF Engineering Research Center MIRTHE, a subaward from a DoE STTR grant from Aerodyne Research Inc., and would like to thank Dr. Tim Day from Daylight Solutions Inc. for providing laser chip and lens assembly for the project.

References

1. F.K. Tittel, D. Richter, A. Fried, Mid-Infrared Laser Applications in Spectroscopy, in *Solid-State Mid-infrared Laser Sources*, ed. by I.T. Sorokina, K.L. Vodopyanov. Springer Topics in Applied Physics (Springer, Berlin, 2003)
2. F.K. Tittel, Y. Bakhirkin, A.A. Kosterev, G. Wysocki, Rev. Laser Eng. **34**, 275 (2006)
3. R.F. Curl, F.K. Tittel, Annu. Rep. Prog. Chem. Sect. C, Phys. Chem. **98**, 217 (2002)
4. A.A. Kosterev, F.K. Tittel, IEEE J. Quantum Electron. **38**, 582 (2002)
5. F. Capasso, R. Paiella, R. Martini, R. Colombelli, C. Gmachl, T.L. Myers, M.S. Taubman, R.M. Williams, C.G. Bethea, K. Unterrainer, H.Y. Hwang, D.L. Sivco, A.Y. Cho, A.M. Sergent,

H.C. Liu, E.A. Whittaker, IEEE J. Quantum Electron. **38**, 511 (2002)
6. R. Maulini, D.A. Yarekha, J.-M. Bulliard, M. Giovannini, J. Faist, E. Gini, Opt. Lett. **30**, 2584 (2005)
7. C. Gmachl, F. Capasso, D.L. Sivco, A.Y. Cho, Rep. Prog. Phys. **64**, 1533 (2001)
8. J. Faist, M. Beck, T. Aellen, E. Gini, Appl. Phys. Lett. **78**, 147 (2001)
9. B.G. Lee, M.A. Belkin, R. Audet, J. MacArthur, L. Diehl, C. Pflugl, F. Capasso, D.C. Oakley, D. Chapman, A. Napoleone, D. Bour, S. Corzine, G. Hofler, J. Faist, Appl. Phys. Lett. **91**, 231101 (2007)
10. R. Maulini, M. Beck, J. Faist, E. Gini, Appl. Phys. Lett. **84**, 1659 (2004)
11. G. Wysocki, R.F. Curl, F.K. Tittel, R. Maulini, J.M. Bulliard, J. Faist, Appl. Phys. B **81**, 769 (2005)
12. R. Maulini, A. Mohan, M. Giovannini, J. Faist, E. Gini, Appl. Phys. Lett. **88**, 201113 (2006)
13. A. Hugi, R. Terazzi, Y. Bonetti, A. Wittmann, M. Fischer, M. Beck, J. Faist, E. Gini, Appl. Phys. Lett. **95**, 061103 (2009)
14. R. Kohler, C. Gmachl, A. Tredicucci, F. Capasso, D.L. Sivco, S.N.G. Chu, A.Y. Cho, Appl. Phys. Lett. **76**, 1092 (2000)
15. E. Normand, M. McCulloch, G. Duxbury, N. Langford, Opt. Lett. **28**, 16 (2003)
16. T. Beyer, M. Braun, A. Lambrecht, J. Appl. Phys. **93**, 3158 (2003)
17. G. Wysocki, R. Lewicki, R.F. Curl, F.K. Tittel, L. Diehl, M. Troccoli, G. Hofler, D. Bour, R. Maulini, M. Giovannini, J. Faist, Appl. Phys. B **92**, 305 (2008)
18. G. Wysocki, A. Kosterev, F. Tittel, Appl. Phys. B **80**, 617 (2005)
19. H. Teichert, T. Fernholz, V. Ebert, Appl. Opt. **42**, 2043 (2003)
20. G.B. Rieker, H. Li, X. Liu, J.B. Jeffries, R.K. Hanson, M.G. Allen, S.D. Wehe, P.A. Mulhall, H.S. Kindle, Meas. Sci. Technol. **18**, 1195 (2007)
21. M.C. Phillips, T.L. Myers, M.D. Wojcik, B.D. Cannon, Opt. Lett. **32**, 1177 (2007)
22. E. Schlosser, T. Fernholz, H. Teichert, V. Ebert, Spectrochim. Acta, Part A **58**, 2347 (2002)
23. L.S. Rothman, I.E. Gordon, A. Barbe, D.C. Benner, P.F. Bernath, M. Birk, V. Boudon, L.R. Brown, A. Campargue, J.P. Champion, K. Chance, L.H. Coudert, V. Dana, V.M. Devi, S. Fally, J.M. Flaud, R.R. Gamache, A. Goldman, D. Jacquemart, I. Kleiner, N. Lacome, W.J. Lafferty, J.Y. Mandin, S.T. Massie, S.N. Mikhailenko, C.E. Miller, N. Moazzen-Ahmadi, O.V. Naumenko, A.V. Nikitin, J. Orphal, V.I. Perevalov, A. Perrin, A. Predoi-Cross, C.P. Rinsland, M. Rotger, M. Simecková, M.A.H. Smith, K. Sung, S.A. Tashkun, J. Tennyson, R.A. Toth, A.C. Vandaele, J. Vander Auwera, J. Quant. Spectrosc. Radiat. Transfer **110**, 533 (2009)
24. P. Weibring, C. Abrahamsson, M. Sjöholm, J.N. Smith, H. Edner, S. Svanberg, Appl. Phys. B, Lasers Opt. **79**, 525 (2004)
25. D. Weidmann, G. Wysocki, Opt. Express **17**, 248 (2009)
26. R. Lewicki, J.H. Doty, R.F. Curl, F.K. Tittel, G. Wysocki, Proc. Natl. Acad. Sci. **106**, 12587 (2009)

Time-resolved spectral characteristics of external-cavity quantum cascade lasers and their application to stand-off detection of explosives

B. Hinkov · F. Fuchs · Q.K. Yang · J.M. Kaster · W. Bronner · R. Aidam · K. Köhler · J. Wagner

Received: 15 September 2009 / Revised version: 1 December 2009 / Published online: 14 January 2010
© Springer-Verlag 2010

Abstract We present the time-resolved spectral analysis of the settling process in a pulsed external-cavity quantum-cascade laser (EC-QC laser) emitting in the wavelength range around 7.4 µm. Due to mode competition between the two cavities formed by the laser-chip on one hand and the external-cavity on the other hand, a time-dependent spectral emission on a ns time-scale can be observed. Depending on the spectral characteristic of the external cavity with respect to the chip gain curve time-delays of the external cavity emission of 15–35 ns within the 100 ns drive pulse can be observed.

Regarding the application, we discuss the results in the context of using the EC-QC laser system together with a high-performance IR-camera for stand-off detection of surface-contaminations with explosives such as TNT. The detected densities can be as small as 10 µg/cm^2 corresponding to traces left by a fingerprint.

1 Introduction

For a lot of applications tunable room-temperature laser sources in the mid-infrared spectral range are needed. In 1971, Kazarinov et al. [1] proposed a laser based on intersubband transitions between quantized states of the conduction band of a semiconductor. The first demonstration of a quantum-cascade (QC) laser by Faist et al. [2] followed in 1994. Due to the progress in the laser-design, -growth, -processing and -mounting, tremendous improvements regarding laser output power, wall-plug efficiency and spectral purity could already be achieved. Meanwhile, QC lasers operating at room-temperature with quite high efficiencies have become reality [3].

Infrared counter-measures [4–6] is just one example for the application of QC lasers. A wide tunability of the laser source is necessary for other applications like the detection of complex molecules showing broad absorption bands [7–10] in the infrared, or the stand-off detection of explosives [11–15]. Therefore, much effort is put into research of new concepts for tunable QC laser systems.

In this paper, we present the time-resolved spectral analysis of an external-cavity system in Littrow-configuration. We will show that in pulsed mode the laser starts lasing in the spectral range of the internal cavity and couples after 15–35 ns to the external cavity. The time-constant of this process strongly depends on the spectral position of the EC-QC laser emission. In addition, we will present a rate-equation model which describes the experimental findings quantitatively.

Finally, some results of our imaging stand-off system for the detection of surface contaminations with explosives will be shown. A discussion of our experimental results on the time-resolved behavior with respect to the drive conditions of the stand-off detection system will also be included.

The paper is organized as follows: First, we will give information on the QC laser chip in Sect. 2, followed by presenting our external-cavity (EC) setup in Littrow-configuration in Sect. 3. The next chapter (Sect. 4) will deal with our experimental results including the time-resolved measurements and the behavior under changing emission wavelengths of the external cavity. In Sect. 5 a rate-equation model will be presented. Finally, the consequences for the

B. Hinkov (✉) · F. Fuchs · Q.K. Yang · J.M. Kaster · W. Bronner · R. Aidam · K. Köhler · J. Wagner
Fraunhofer Institute for Applied Solid State Physics (IAF), Tullastraße 72, 79108 Freiburg, Germany
e-mail: borislav.hinkov@iaf.fraunhofer.de

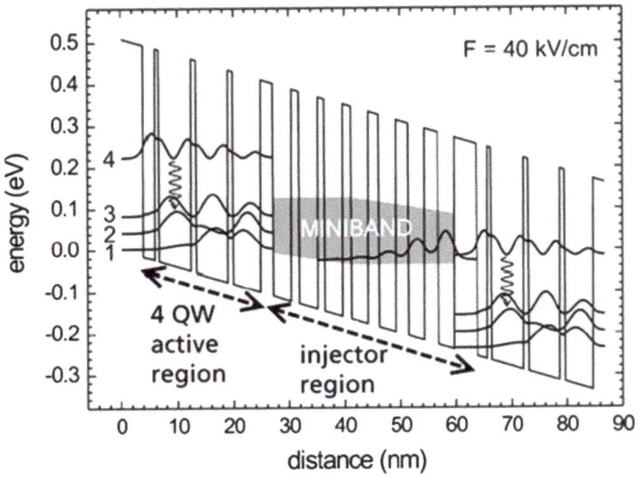

Fig. 1 Typical conduction band profile of a QC laser with an active region comprising a 4-quantum-well design

drive conditions of our stand-off detection system will be discussed (Sect. 6) and first results of measurements of surface contaminations of TNT will be shown.

2 Quantum-cascade lasers

QC lasers consist of a multiple quantum-well structure in the active region comprising hundreds of nm-thin layers. This complex design leads to quantized states in the conduction band. Between these states the laser transitions take place, leading to emission wavelengths in the mid-infrared [16–21] to THz [22–24] spectral range between 3 μm and >400 μm. QC lasers are unipolar in contrast to the classical bipolar diode-lasers. By band-structure engineering the emission wavelength of a QC laser can easily be changed by varying the quantum well thickness.

Figure 1 shows the typical conduction band profile of a QC laser with a 4 quantum-well active-region design.

The QC lasers used in this work comprise $Ga_{0.47}In_{0.53}As$ quantum wells and $Al_{0.48}In_{0.52}As$ barriers with a central emission wavelength of 7.4 μm. They are grown lattice-matched on an InP substrate by molecular beam epitaxy. This structure is embedded between two GaInAs waveguide-layers and is capped by the upper cladding and a highly Si-doped InP contact-layer both grown by metal organic vapor phase epitaxy. Double-trench waveguides with 8 μm wide ridges are processed and a several μm thick electroplated Au-layer is deposited on top for better heat extraction. More details on the QC laser design can be found in [19, 25]. Laser growth and processing is described in detail in [20, 26].

The QC lasers are cleaved into 1500 μm long ridges and are mounted epitaxial side-up onto Au-coated standard c-mounts. Figure 2 shows the SEM picture of the cleaved front facet of a typical QC laser ridge.

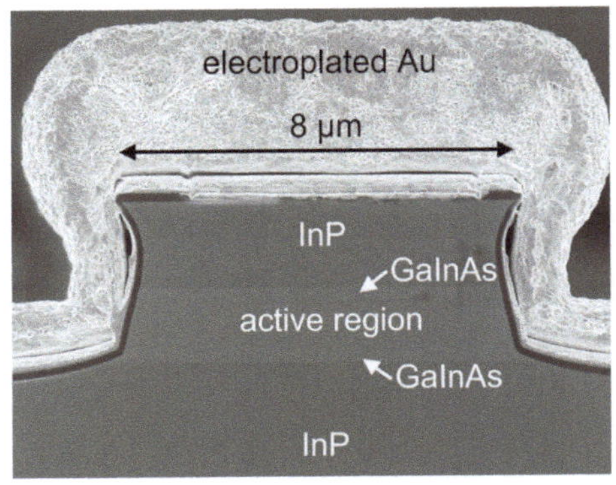

Fig. 2 SEM picture of the front facet of a QC laser with a 8 μm wide ridge. Relevant layers are indicated

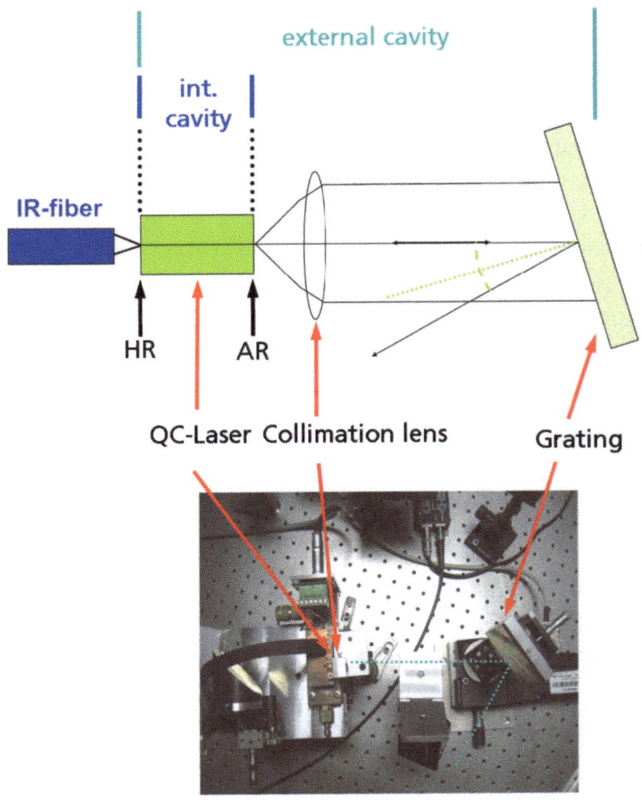

Fig. 3 EC setup in Littrow-configuration

3 Experimental setup

Placing a QC laser chip in an external-cavity setup opens the possibility for broad wavelength tuning. Figure 3 shows a typical EC setup in Littrow-configuration.

The high-reflection (HR) coated laser back-facet (>80%) forms the first EC-mirror in this configuration. The laser emission is coupled out through the anti-reflection (AR) coated laser front-facet collimated by an aspherical lens with a numerical aperture of $NA = 0.9$ and directed towards a blazed diffraction-grating optimized for an emission-wavelength of 8 μm. The diffraction-grating acts as wave-

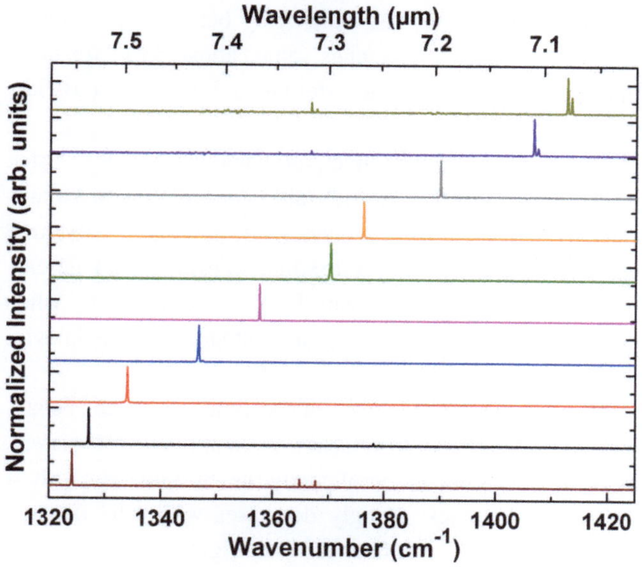

Fig. 4 Tuning-range of >80 cm^{-1} of a typical 15 × 1500 μm^2 QC laser driven 25% above laser threshold with 100 ns pulses at a repetition-rate of 10 kHz at 290 K

length-selective cavity mirror with an effective reflectivity of 80%. By tilting the angle of the grating the wavelength of the laser emission can be tuned. The total external cavity length is 22.15 cm which is about two orders of magnitude larger than that of the bare QC laser chip itself (0.15 cm). For better coupling to the external cavity a broadband AR coating is deposited on the laser front-facet with a residual reflectivity <1%.

The HR coated laser back-facet acts as an outcoupling mirror. The emission is either collimated by a f/4 parabolic mirror or coupled directly without any focusing optics into a IR-fiber with a core-diameter of 450 μm by positioning the end of the fiber at a distance of ~100 μm with respect to the laser facet. The fiber couples the radiation into a Bruker Vertex80v FTIR-spectrometer with a maximum resolution of 0.075 cm^{-1}.

Figure 4 shows the tuning-range of >80 cm^{-1} of a 15 × 1500 μm^2 broad gain QC laser under typical drive conditions: 100 ns current-pulses with a repetition-rate of 10 kHz at 290 K driven 25% above laser threshold.

Alternative designs with broad gain curves (bound-to-continuum design [27]) can extend the tuning-range significantly. Applying such a design, we already extended the tuning range to >175 cm^{-1} [28]. Combining several broad gain designs with different central emission wavelengths into a (heterogeneous) design leads to tuning ranges beyond 400 cm^{-1} as has been shown recently [21].

For measuring time-resolved spectra gated detection can be combined with the high-resolution FTIR-spectrometer. This "boxcar technique" allows the definition of a time window positioned at an arbitrary position during the 100 ns laser pulse. The length of the time window can be as small

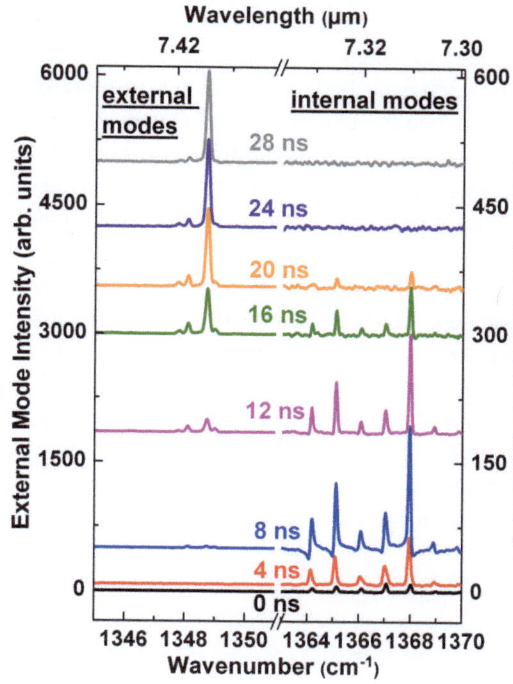

Fig. 5 Time-resolved spectra of the external (*left*, 1348.5 cm^{-1}) and internal (*right*, 1364–1368 cm^{-1}) cavity modes

as 3 ns. The signal integrated within this time window is processed by the spectrometer. With this technique the time-dependent evaluation during a long laser-pulse or the settling of the laser emission at the beginning of the drive pulse can be investigated. More details on the "boxcar-technique" can be found in [26].

In the following measurements, we positioned the time-window at different positions during the laser pulse with respect to the onset of the spectrally integrated optical signal.

4 Results: time-resolved spectral behavior

In this section, we will briefly sum up the most important results of our measurements. The focus will be on a theoretical rate-equation model for quantitative analysis. More details on the experimental results can be found in [29].

Figure 5 shows the typical time-resolved behavior of the coupled cavity system. The internal and external cavity modes are spectrally separated by tuning the external cavity away from the gain maximum of the QC laser chip.

At the beginning of the current pulse, lasing on the Fabry–Pérot modes of the internal cavity can be observed, whereas no emission occurs in the wavelength-range of the external cavity. The internal mode intensity increases and reaches the maximum after 8 ns. This time is defined to be the internal cavity *rise time*. After a gradual decay beginning at 8 ns the internal modes vanish completely after 24 ns.

In contrast, the external cavity modes start lasing after about 10 ns, beginning with the decay of the internal modes.

The intensity of the external modes further increases and reaches its steady state after 22 ns.

This characteristic behavior is shown in more detail in Fig. 6. We plotted the integrated mode intensity (area under each spectrum curve) versus the time-position of the time-window. For better comparison of both obtained curves, the experimental data points of the internal cavity have been multiplied by a factor of 5.

Figures 7 and 8 show the wavelength-dependence of the settling process. For better illustration, the intensities of the curves are normalized. As expected the wavelength position defined by the external cavity has no influence on the onset of the internal cavity emission. On the other hand, detuning the external cavity has a large effect on the onset of the external cavity emission. Both can be seen in Fig. 7. For a better observation of the abrupt onset, logarithmic intensity scales are used. The time-delay between the internal and external emission reaches values above 35 ns for maximum detuning, where the external cavity barely starts lasing.

Due to the lower effective gain when tuning the external cavity away from the QC laser-chip gain-maximum a status can evolve leading to coexistence of the internal and external cavity emission. This unstable situation results in an increasing undershoot of the intensity for the external cavity when compared to the intensity evolution at small external cavity detuning.

Besides these differences in the time-dependent behavior, an influence on the emission-intensity is also observed. When detuning the external cavity the emission strength of the external modes strongly decreases while the *internal* cavity modes are not influenced (see Fig. 9).

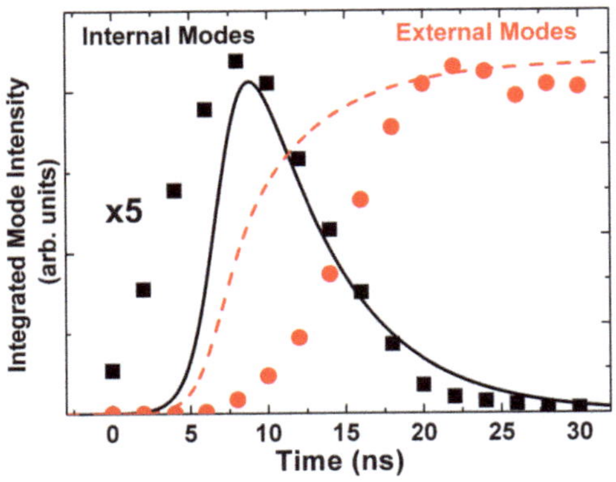

Fig. 6 Integrated mode intensity plot versus time delay of the gate. The opposite behavior of internal (*black squares*, ~1366 cm^{-1}) and external (*red dots*, 1348.5 cm^{-1}) cavity modes is shown. The *solid* and *dashed line* are the results of the rate-equation model

5 Rate-equation model of the external-cavity system

In order to understand the underlying physical processes in a coupled cavity system, we performed a rate-equation analysis (see (1)–(3)). Equation (1) describes the *carrier*-dynamics, whereas (2)–(3) stand for the *photon*-dynamics in the corresponding ((2) = internal, (3) = external) cavity.

$$\frac{dN}{dt} = \underbrace{J_e/e}_{\text{injected carriers}} - \underbrace{N/\tau_e}_{\text{(non-)radiative electron-loss}} \underbrace{- \upsilon \cdot (g_{\text{FP}} S_{\text{FP}} + g_{\text{EC}} S_{\text{EC}})}_{\text{opt. stimulated transitions}}, \quad (1)$$

Fig. 7 Time-resolved spectral behavior of the internal (*lower part*) and external (*upper part*) cavity modes on logarithmic intensity scales. The wavenumbers and the time position of the onset of the external cavity modes are indicated

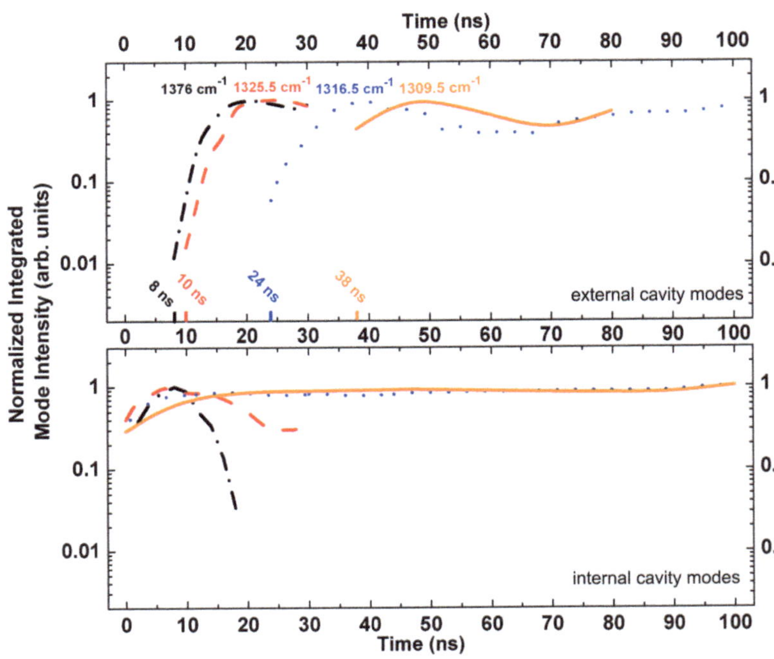

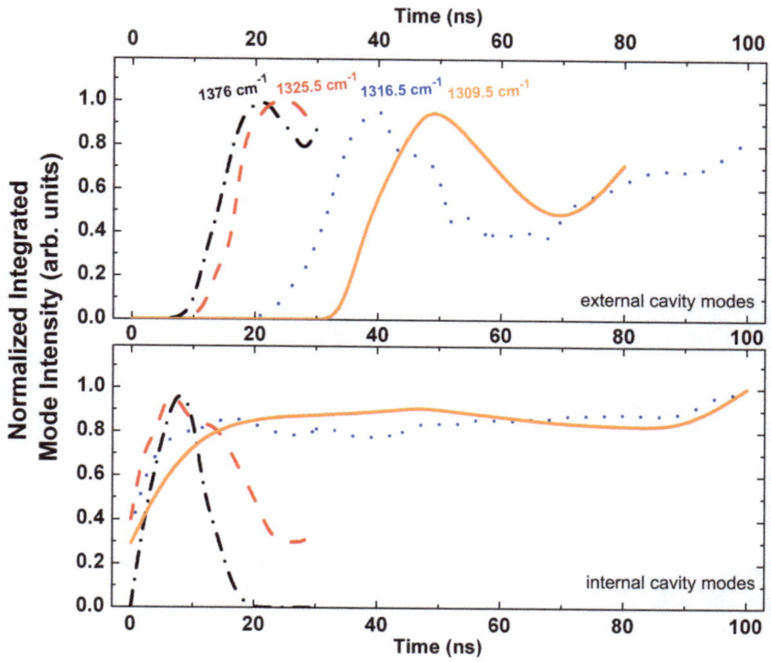

Fig. 8 Time-resolved spectral behavior of the internal (*lower part*) and external (*upper part*) cavity on a linear intensity scale. The wavenumbers of the external cavity modes are indicated

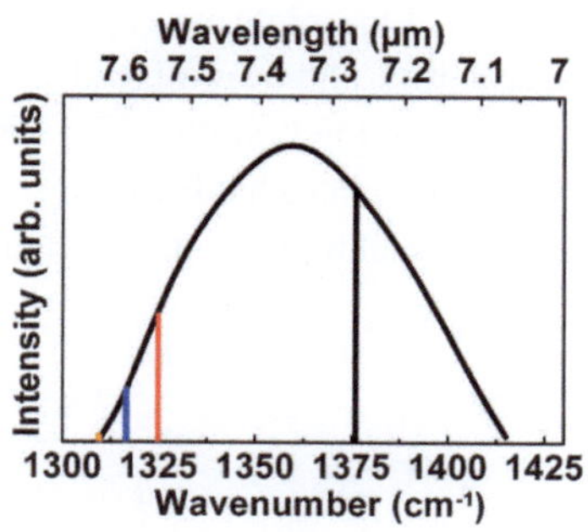

Fig. 9 Measured gain-curve for the investigated QC lasers

$$\frac{dS_{\mathrm{FP}}}{dt} = \underbrace{\Gamma v g_{\mathrm{FP}} S_{\mathrm{FP}}}_{\text{opt. gain (FP)}} - \underbrace{v S_{\mathrm{FP}}(\alpha_i + \alpha_{m,\mathrm{FP}})}_{\text{total opt. losses (FP)}} + \underbrace{\beta_{sp} N/\tau_{sp,\mathrm{FP}}}_{\text{spontaneous emission (FP)}}, \tag{2}$$

$$\frac{dS_{\mathrm{EC}}}{dt} = \underbrace{\Gamma v g_{\mathrm{EC}} S_{\mathrm{EC}}}_{\text{opt. gain (EC)}} - \underbrace{v S_{\mathrm{EC}}(\alpha_i + \alpha_{m,\mathrm{EC}})}_{\text{total opt. losses (FP)}} + \underbrace{\beta_{sp} N/\tau_{sp,\mathrm{EC}}}_{\text{spontaneous emission (EC)}}, \tag{3}$$

where N ($= 1.5 \times 10^{11}$ cm^{-2}) stands for the sheet carrier-density, J_e for the injection current-density, e for the electron charge and τ_e ($= 3$ ps) for the electron transition time of one period of the QC laser structure. $v_g = c/n$ is the group-velocity of light, $g_{\mathrm{FP/EC}}$ is the gain-coefficient and $S_{\mathrm{FP/EC}}$ ($= 0.2 \times 10^{11}$ cm^{-2}/ $= 0.8 \times 10^{11}$ cm^{-2}) the photon-density. Γ stands for the optical confinement of the electromagnetic wave with the laser active region, α_i and $\alpha_{m,\mathrm{FP/EC}}$ stand for the cavity- and mirror-losses, respectively, and β_{sp} ($\sim$0.001) stands for the spontaneous emission contributing to the laser mode with the spontaneous emission lifetime $\tau_{sp,\mathrm{FP/EC}}$ ($\sim 1.4 \times 10^{-7}$ s). The indices "FP" and "EC" correspond to the internal and external cavity.

The results of the rate-equation model can be seen in Fig. 6 as a solid and dashed line for the internal and external cavity, respectively. The line-shape is reproduced in very good agreement with the experimental data. However, the time-delay of around 8–10 ns in the experimental data is not well reproduced within this model. The simulation yields only a time-separation of 1–2 ns.

The uncertainty introduced by the 3 ns time window of the boxcar technique has almost no effect on the theoretically observed time-separation when introduced to the model. The time difference is mainly attributed to the strongly different round-trip times in both cavities. The bare QC laser chip (internal cavity) has a cavity length of 1.5 mm in contrast to the 22.15 cm of the external cavity, leading to different resulting cavity round-trip times being derived with (4) and (5):

$$\tau_{\mathrm{int}} = \frac{2 \cdot l_{\mathrm{int}}}{c_0/n}, \tag{4}$$

$$\tau_{\mathrm{ext}} = \frac{2 \cdot l_{\mathrm{int}}}{c_0/n} + \frac{2 \cdot l_{\mathrm{ext}}}{c_0}, \tag{5}$$

with $l_{\mathrm{int}} = 1.5$ mm, $l_{\mathrm{ext}} = 22$ cm being the bare internal and external cavity lengths, respectively, and the refractive index $n = 3.26$; the following round-trip times are obtained: $\tau_{\mathrm{int}} =$ 32.6 ps, $\tau_{\mathrm{ext}} = 1.5$ ns.

In [30], the introduction of a factor ρ_{cav} defined as the ratio between the different round-trip times has been proposed:

$$\rho_{cav} = \frac{\tau_{int} + \tau_{ext}}{\tau_{int}}. \qquad (6)$$

First simulations have shown that including this factor in a more refined model leads to better agreement with the experimental results. The time separation of the theoretical simulation then is extended to 8–10 ns which is in very good agreement with the experimental observation. The refined model is beyond the scope of this work and will be published separately [31].

6 Imaging stand-off system for the detection of explosives

As shown recently [15], we also work on the application of QC lasers for the detection of surface-contaminations with explosives. In this section, we will present results of our imaging stand-off detection system demonstrating that we can identify concentrations corresponding to a fingerprint (several μg/cm^2). We will also show that the results of the time-resolved investigations presented in Sects. 4–5 play an important role for choosing the right drive-conditions of the QC laser in this system.

6.1 Stand-off detection system

The stand-off detection system mainly consists of three parts: the EC-QC laser system, a unit for reduction of laser speckle and a high-performance IR-camera. After illuminating the surface of interest with our QC laser, we analyze the backscattered light with a IR-camera. As shown in Fig. 10, two images are taken: one at an absorption band of TNT as surface contamination ("sensing mode") and a second one at a wavelength without absorption of TNT ("reference mode"). We obtain a ratio image by dividing both images after subtracting the thermal background ($ratio = \frac{sense}{reference}$).

For imaging an optimized high-performance Mercury–Cadmium–Telluride- (MCT-) based IR-camera was used, being able to detect radiation in the wavelength-range of 4.2–10.5 μm. The camera frame rate was 400 Hz with an integration time of 100 μs (duty-cycle, 4%). The average output-power was maximized by driving the QC laser with 100 ns pulses at a duty-cycle of 50% (repetition-rate, 5 MHz) during the 100 μs integration time of the camera. As shown in the time-resolved analysis (Sect. 4), the EC-QC laser system needs 15–20 ns to couple to the external cavity. Thus 100 ns drive pulses provide sufficiently high spectral purity of the emitted radiation.

A detailed description of the EC-QC laser setup can be found in Sect. 3.

Since we are using an active imaging system comprising a coherent light source (QC laser), the formation of laser speckle is an important issue. When illuminating a rough surface with a coherent light source interference-patterns can be observed in the backscattered light. The variation in intensity due to this effect can be very strong, obscuring the change in intensity due to an absorbing thin film on the investigated surface. To avoid the formation of laser speckle, the spatial or temporal coherence of the QC laser source has to be reduced. Different approaches are under discussion [32–35].

We fabricated a diamond beam scrambler consisting of two polycrystalline diamond plates grown at IAF by chemical vapor deposition. One plate was stationary while the second one was rotating and was placed at a distance of ~400 μm with respect to the first one. The RMS roughness (~3 μm) of the diamond plates leads to a variation in the refractive index, resulting in a phase-mixing and reduction of

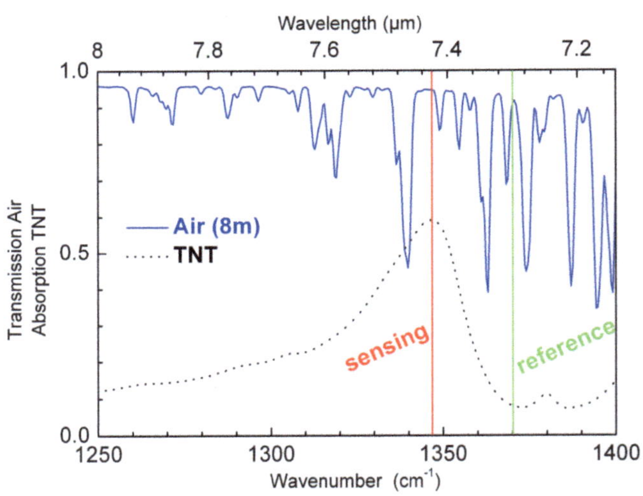

Fig. 10 Transmission-spectrum of 8 m air and absorption band of TNT with indicated sensing (*red*) and reference (*green*) mode

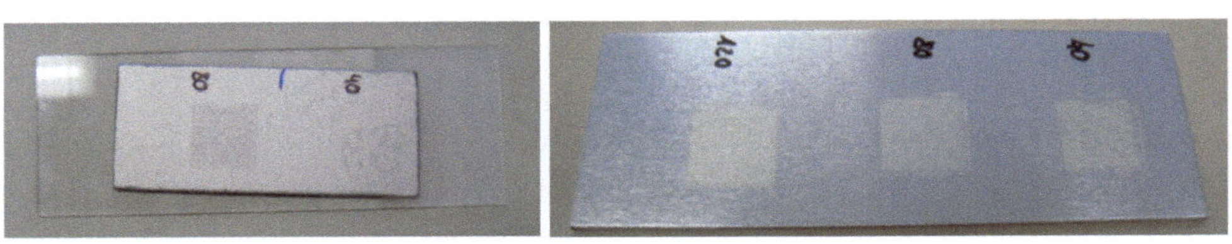

Fig. 11 *Left*, standard car paint sample with a thin TNT-layer (80 and 40 μg/cm^2). *Right*, aluminum-substrate also partially covered with a thin TNT-film with three different concentrations (120, 80 and 40 μg/cm^2)

the coherence. The EC-QC laser emission is coupled to the speckle-reduction unit by positioning one end of the multi-mode IR-fiber at the QC laser back-facet (see Sect. 3). The other end of the fiber is positioned at a distance of 500 µm with respect to the stationary diamond diffusor.

More details on the despeckling-unit and the stand-off detection system can be found in [15].

6.2 Stand-off detection of TNT on aluminum-substrate and standard car paint

We investigated samples on two types of substrate-materials. Aluminum was used to represent the group of mirror-like, high-reflective materials. In addition, we used standard car paint to investigate strongly absorbing substrate-materials. The test-samples were fabricated at Fraunhofer Institute for Chemical Technology (ICT) using a special printing technique allowing to deposit thin films with a defined TNT-concentration in the µg/cm^2-range [36]. Figure 11 shows typical samples.

In Fig. 12, typical results for the aluminum-plate substrate are shown. The QC laser spot is pointed between the regions with 40 µg/cm^2 and 80 µg/cm^2 of TNT which can easily be identified as dark squares in the resulting ratio-image. The intensity decreases in the sensing mode, and therefore results in a value below 1 for the ratio-image.

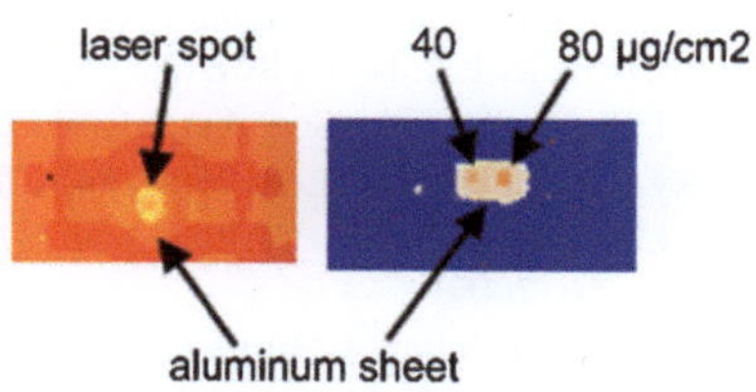

Fig. 12 *Left*, IR-camera image of the aluminum-sample with laser illumination. *Right*, ratio-image after division. The TNT-contaminations with two different concentrations (40 µg/cm^2 and 80 µg/cm^2) can easily be identified as *dark squares*

In contrast to the measurements on aluminum-substrate the results for the car paint measurements show a different behavior as can be seen in Fig. 13.

Instead of a dark, a *light* square is identified in the region contaminated with TNT. This effect can be explained by the strong absorption of the substrate material. Nevertheless, an identification of the TNT-deposition can easily be achieved even when dealing with concentrations as low as 10 µg/cm^2, corresponding to a concentration which is typically left by a fingerprint.

A more detailed analysis of the substrate-reflectance dependent contrast-reversal effect, comprising measurements of the reflectance of the samples and a theoretical analysis, can be found in [15].

7 Summary

We presented the time-resolved analysis of the settling-process of an EC-QC laser system. A time-delay between the internal and external cavity mode emission of 10–30 ns is observed, which strongly depends on the spectral position of the external cavity. Upon external cavity feedback near the gain maximum, it takes around 15–20 ns for the system to couple to the external cavity. A rate-equation model was proposed to analyze the experimental results.

As an example for the application of a tunable QC laser, we showed first results of our imaging stand-off system on the detection of surface-contaminations with TNT by IR-backscattering-spectroscopy.

Acknowledgements The authors would like to thank P. Ganser and H.-P. Menner for QC laser growth, K. Schäuble, K. Schwarz, U. Weinberg and R. Moritz for expert technical assistance and W. Schweikert and F. Schnürer from Fraunhofer ICT for sample preparation. Helpful discussions with R. Maulini and U. Zeitner are gratefully acknowledged. The work has been supported by the German Ministry for Education and Research (BMBF) through the project IRLDEX.

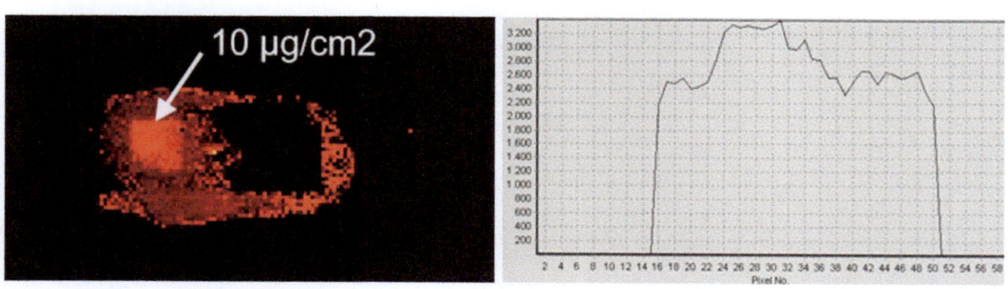

Fig. 13 *Left*, ratio-image after division. The 10 µg/cm^2 TNT-concentration is identified as a *light square* in contrast to the *dark square* when analyzing an aluminum substrate (see Fig. 12). *Right*, vertical cross-section through the ratio-image (see *left figure*) with a TNT contamination of 10 µg/cm^2, showing the TNT as region with a higher value (contrast) when compared with the surrounding car paint substrate

References

1. R.F. Kazarinov, R.A. Suris, Sov. Phys. Semicond. **5**(4), 707 (1971)
2. J. Faist, F. Capasso, D.L. Sivco, C. Sirtori, A.L. Hutchinson, A.Y. Cho, Science **264**, 553 (1994)
3. Y. Bai, S. Slivken, S. Darvish, M. Razeghi, Appl. Phys. Lett. **93**, 021103 (2008)
4. J. Wagner, N. Schulz, B. Rösener, M. Rattunde, Q. Yang, F. Fuchs, C. Manz, W. Bronner, C. Mann, K. Köhler, M. Raab, E. Romasev, H.D. Tholl, in *Proc. SPIE—Technologies for Optical Countermeasures V*, vol. 7115 (2008)
5. S. Hugger, F. Fuchs, R. Aidam, W. Bronner, R. Lösch, Q. Yang, N. Schulz, J. Wagner, E. Romasew, M. Raab, H.D. Tholl, in *Proc. SPIE—Security and Defence*, vol. 7325(17) (2009)
6. J. Wagner, S. Hugger, B. Rösener, F. Fuchs, M. Rattunde, Q. Yang, W. Bronner, R. Aidam, K. Köhler, M. Raab, E. Romasew, H.D. Tholl, in *Proc. SPIE—Europe Security and Defence*, vol. 7483(14) (2009)
7. G. Wysocki, R. Lewicki, R.F. Curl, F.K. Tittel, L. Diehl, F. Capasso, M. Troccoli, G. Hofler, D. Bour, S. Corzine, R. Maulini, M. Giovannini, J. Faist, Appl. Phys. B **92**(3), 305 (2008)
8. A. Kosterev, G. Wysocki, Y. Bakhirkin, S. So, R. Lewicki, M. Fraser, F. Tittel, R.F. Curl, Appl. Phys. B, Lasers Opt. **90**, 165 (2008)
9. I. Dunayevskiy, A. Tsekoun, M. Prasanna, R. Go, C.K. Patel, Appl. Opt. **46**(25), 6397 (2007)
10. T. Day, M. Weida, D. Arnone, S. Crivello, D. Caffey, V. Cook, R. Pritchett, in *Proc. of SPIE*, vol. 7222 (2009)
11. C.W. van Neste, L.R. Senesac, T. Thundat, Appl. Phys. Lett. **92**(23), 234102 (2008)
12. R. Furstenberg, C.A. Kendziora, J. Stepnowski, S.V. Stepnowski, M. Rake, M.R. Papantonakis, V. Nguyen, G.K. Hubler, R.A. McGill, Appl. Phys. Lett. **93**, 224103 (2008)
13. M. Gaft, L. Nagli, in *Proc. of SPIE—Electro-Optical Remote Sensing, Detection and Photonic Technologies and their Applications*, vol. 6739 (2007), p. 73903
14. S. Wallin, A. Petterson, H. Östmark, A. Hobro, Anal. Bioanal. Chem. **395**(2), 259 (2009)
15. B. Hinkov, F. Fuchs, J.M. Kaster, Q. Yang, W. Bronner, R. Aidam, K. Köhler, Broad band tunable quantum cascade lasers for stand-off detection of explosives. Proc. SPIE, vol. 7484(5) (2009)
16. S.Y. Zhang, D.G. Revin, J.W. Cockburn, K. Kennedy, A.B. Krysa, M. Hopkinson, $\lambda \sim 3.1$ μm room temperature InGaAs/AlAsSb/InP quantum cascade lasers. Appl. Phys. Lett. **94**(3), 031106 (2009)
17. Q.K. Yang, C. Manz, W. Bronner, N. Lehmann, F. Fuchs, K. Köhler, J. Wagner, Appl. Phys. Lett. **90**, 121134 (2007)
18. Q. Yang, R. Lösch, W. Bronner, S. Hugger, F. Fuchs, R. Aidam, J. Wagner, Appl. Phys. Lett. **93**, 251110 (2008)
19. D. Hofstetter, M. Beck, T. Aellen, J. Faist, Appl. Phys. Lett. **78**(4), 396 (2001)
20. B. Hinkov, F. Fuchs, W. Bronner, K. Köhler, J. Wagner, Current- and temperature-induced beam steering in 7.8 μm emitting quantum-cascade lasers. J. Quantum Electron. **44**(11) (2008)
21. A. Hugi, R. Terazzi, Y. Bonetti, A. Wittmann, M. Fischer, M. Beck, J. Faist, E. Gini, Appl. Phys. Lett. **95**(6), 061103 (2009)
22. R. Köhler, A. Tredicucci, F. Beltram, H.E. Beere, E.H. Linfield, A.G. Davies, D.A. Richie, R.C. Lotti, F. Rossi, Nature **417**, 156 (2002)
23. A. Wade, G. Fedorov, D. Smirnov, S. Kumar, B.S. Williams, Q. Hu, J.L. Reno, Magnetic-filed-assisted terahertz quantum cascade laser operating up to 225 K. Nat. Photonics **3** (2009)
24. G. Scalari, M.I. Amanti, M. Fischer, R. Terazzi, C. Walther, M. Beck, J. Faist, Appl. Phys. Lett. **94**(4), 041114 (2009)
25. M. Beck, D. Hofstetter, T. Aellen, J. Faist, U. Oesterle, M. Ilegems, E. Gini, H. Melchior, Science **295**(5553), 301 (2002)
26. F. Fuchs, B. Kirn, C. Mann, Q. Yang, W. Bronner, B. Raynor, K. Köhler, J. Wagner, in *Proc. of SPIE, Optical Methods in the Life Sciences*, vol. 6386 (2006)
27. J. Faist, M. Beck, T. Aellen, Appl. Phys. Lett. **78**(2), 147 (2001)
28. F. Fuchs, B. Hinkov, S. Hugger, J.M. Kaster, R. Aidam, W. Bronner, K. Köhler, Q. Yang, K. Degreif, F. Schnürer, W. Schweikert, in *Proc. SPIE—Photonics West*, vol. 7608(8) (2010)
29. B. Hinkov, Q. Yang, F. Fuchs, W. Bronner, K. Köhler, J. Wagner, Appl. Phys. Lett. **94**(22) (2009)
30. R. Maulini, M. Beck, J. Faist, E. Gini, Appl. Phys. Lett. **84**(10), 1659 (2004)
31. Q. Yang, B. Hinkov, F. Fuchs, W. Bronner, K. Köhler, J. Wagner, Rate equation of an external cavity quantum cascade laser (2009, in preparation)
32. R.F. van Ligten, *Speckle removal and improvement in optical slicing in holography*, vols. 1, 2. Nouv. Rev. Opt. Appl. (1970)
33. S. Lowenthal, D. Joyeux, J. Opt. Soc. Am. **61**, 847 (1971)
34. M. Bashkansky, J. Reintjes, Opt. Lett. **25**(8), 545 (2000)
35. V. Yurlov, A. Lapchuk, S. Yun, J. Song, I. Yeo, H. Yang, S. An, Appl. Opt. **48**(1), 80 (2009)
36. G. Bunte, M. Heil, D. Röseling, J. Hürttlen, H. Pontius, H. Krause, Propellants Explos. Pyrotech. **34**, 245 (2009)

5-μm vertical external-cavity surface-emitting laser (VECSEL) for spectroscopic applications

M. Rahim · A. Khiar · F. Felder · M. Fill · H. Zogg ·
M.W. Sigrist

Received: 24 September 2009 / Revised version: 16 February 2010 / Published online: 11 June 2010
© Springer-Verlag 2010

Abstract Mid-IR tunable VECSELs (Vertical External-Cavity Surface-Emitting Lasers) emitting at 4–7 μm wavelengths and suitable for spectroscopic sensing applications are described. They are realized with lead-chalcogenide (IV–VI) narrow band gap materials.

The active part, a single 0.6–2-μm thick PbTe or PbSe gain layer, is grown onto an epitaxial Bragg mirror consisting of two or three $Pb_{1-y}Eu_yTe/BaF_2$ quarter-wavelength layer pairs. All layers are deposited by MBE in a single run employing a BaF_2 or Si substrate, no further processing is needed. The cavity is completed with an external curved top mirror, which is again realized with an epitaxial Bragg structure. Pumping is performed optically with a 1.5-μm laser.

Maximum output power for pulsed operation is currently up to $>1\ W_p$ at $-173°C$ and >10 mW at 10°C. In continuous wave (CW) operation, 18 mW at 100 K are reached. Still higher operating temperatures and/or powers are expected with better heat-removal structures and better designs employing QW (Quantum-Wells).

Advantages of mid-IR VECSELs compared to edge-emitting lasers are their very good beam quality (circular beam with $<1°$ cone diameter), simple structure, and their easy tunability without mode-hopping. Wavelengths ranging from <3 μm up to >15 μm are accessible with $Pb_{1-y}X_yZ$ (X = Sr, Eu, Sn, Z = Se, Te) and/or including QW.

M. Rahim · A. Khiar · F. Felder · M. Fill · H. Zogg (✉)
Thin Film Physics Group, ETH Zurich, 8005 Zurich, Switzerland
e-mail: zogg@phys.ethz.ch
url: http://www.tfp.ethz.ch

M.W. Sigrist
Institute for Quantum Electronics, ETH Zurich, Schafmattstrasse 16, 8093 Zurich, Switzerland

1 Introduction

Vertical external-cavity surface-emitting lasers (VECSEL) are currently of high interest and exhibit attractive properties like narrow beam divergence and wavelength tunability [1–3]. They are often optically pumped, and their high power and power scalability are unmatched by other techniques. Most VECSEL described up to now emit in the 800–2400 nm IR-region and are realized with GaAs or GaSb based lattice-matched structures. GaSb based interband cascade lasers [4] as well as quantum cascade lasers are suited for still longer wavelengths [5], however, they are edge emitters with corresponding very astigmatic high divergence angles.

We recently reported optically-pumped VECSELs emitting at wavelengths up to as long as 5.5 μm [6–9]. The devices were realized with narrow bandgap IV–VI (lead-chalcogenide) materials. The structures consisted of a flat bottom 2–3 pair Bragg mirror with $Pb_{1-y}Eu_yTe$ quarter-wavelength ($\lambda/4$) layers for the high-index, and BaF_2 ($n = 1.45$) or EuTe ($n = 2.4$) for the low-index material, a curved top mirror, and a PbTe or PbSe based active layer. Note that an optical design has always to be done for a certain temperature range of operation due to the considerable shift of emission wavelength with temperature T (the wavelengths, unlike III–V materials, decrease with increasing T).

Here, we describe three improved structures. Two of them lase up to or even above RT, and one is designed with QW. Growth on a Si-substrate (which is even lattice- and thermal expansion-mismatched) yields to a record emission power of $>1\ W_p$.

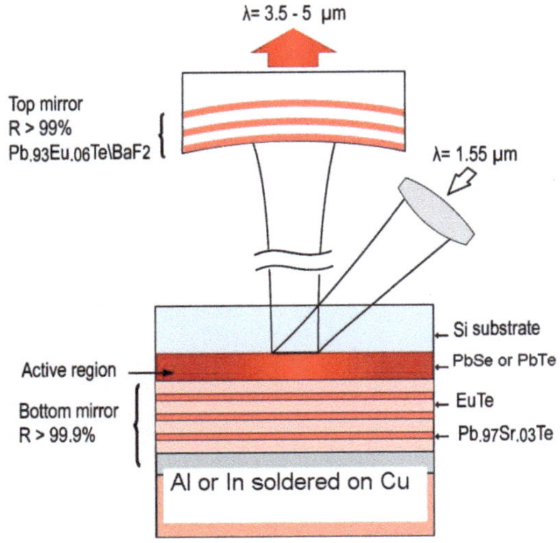

Fig. 1 Schematic representation of a IV–VI (lead chalcogenide) mid-IR VECSEL with resonant design

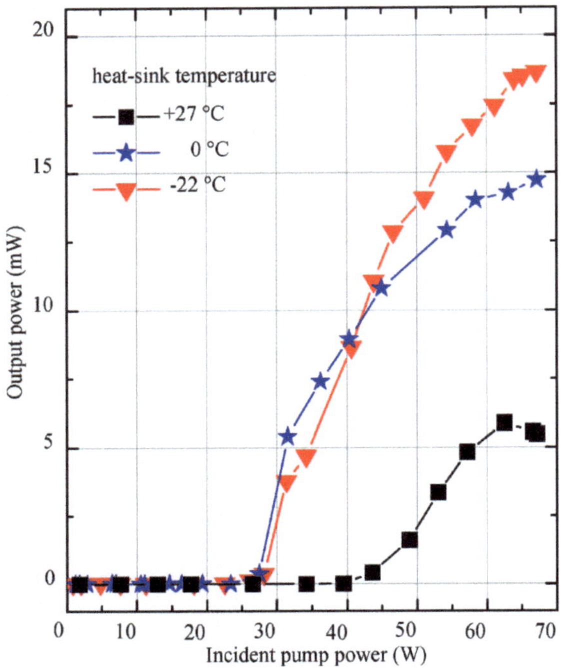

Fig. 2 Light in/light out characteristics at three different temperatures for a *RT* PbSe mid-IR VECSEL when pumped with a 1.55 μm wavelength laser. Note that the absorbed pump power is only about 60% of the incident power

2 Experimental

Figure 1 shows a typical cross section of a IV–VI mid-IR VECSEL. The active layer and one Bragg mirror are grown by solid-state molecular beam epitaxy (MBE) on a BaF_2 or Si substrate. The transparent substrate is located inside the cavity, and a heat spreader (evaporated Al, or, more efficient, a Cu-heat sink onto which the sample is In-soldered) is employed. A curved top mirror ($r = 25$ mm) completes the cav-

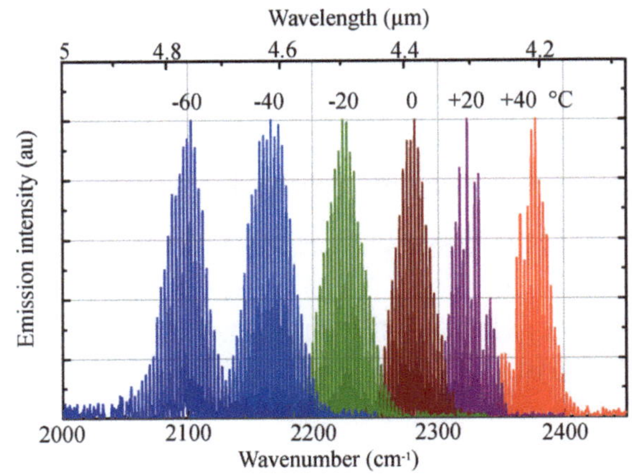

Fig. 3 Laser spectra at different temperatures for a *RT* PbSe mid-IR VECSEL. Each spectrum is individually normalized

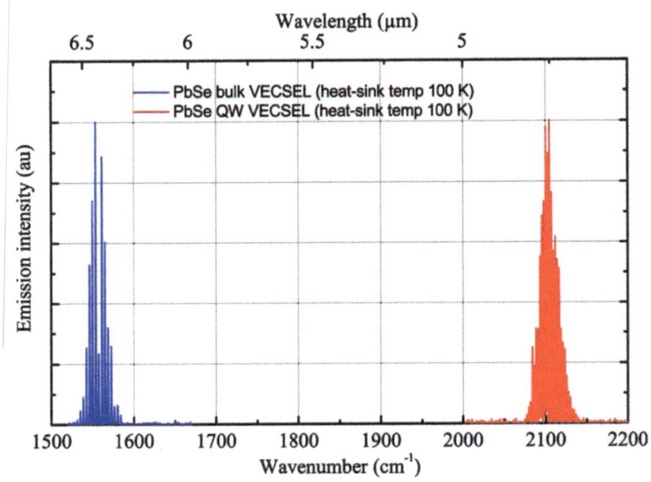

Fig. 4 Normalized laser spectra at 100 K for a QW PbSe mid-IR VECSEL (*right*) in comparison to a "bulk" PbSe active layer (*left*)

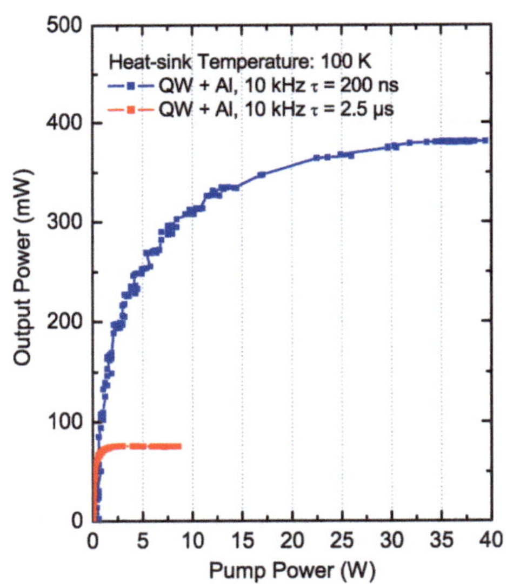

Fig. 5 Light in/light out characteristics for a QW PbSe mid-IR VECSEL when illuminated with a 1.55 μm wavelength laser. Note that the absorbed pump power is only about 60% of the incident power

ity. Pumping is done optically with a 1.55-µm or 2-µm pump laser with a beam focused to ~200 µm diameter. While high-power 1.55-µm lasers are easily available, this is not yet the case for GaSb based 2-µm lasers. However, the quantum deficit is lower, leading to lower pump power at 2 µm pump wavelength. Spectra are recorded with a Bruker FTIR spectrometer.

2.1 Above *RT* PbSe on BaF_2-substrate VECSEL

A resonant design with a one-wavelength thick active PbSe layer was employed. Contrary to our earlier designs, the design thickness was chosen for intended room-temperature operation (the design in Ref. [6] on a BaF_2 substrate was for cryogenic operation). Figure 2 shows light in/light out characteristics and spectra around *RT* when illuminated with 100 ns pulses with 10 kHz repetition frequency [8]. Emission is multimode with a spacing between the lines corresponding to the optical thickness of the substrate (Fig. 3). Note the broad temperature tuning range, 4.8–4.2 µm for heat sink temperatures of −60°C to +40°C. This is due to the huge temperature coefficient of the band gap of narrow gap lead chalcogenides.

2.2 QW PbSe on BaF_2-substrate VECSEL

Quantum wells (QW) have lower threshold with respect to bulk layers and blue-shift the emission wavelengths, the amount of the shift depends on the width of the wells. Figures 4 and 5 show an example. Five PbSe QW, each ~8 nm thick, are embedded in a $Pb_{.95}Eu_{.05}Se$ host of (in total) one-wavelength optical thickness near the maximum value of the

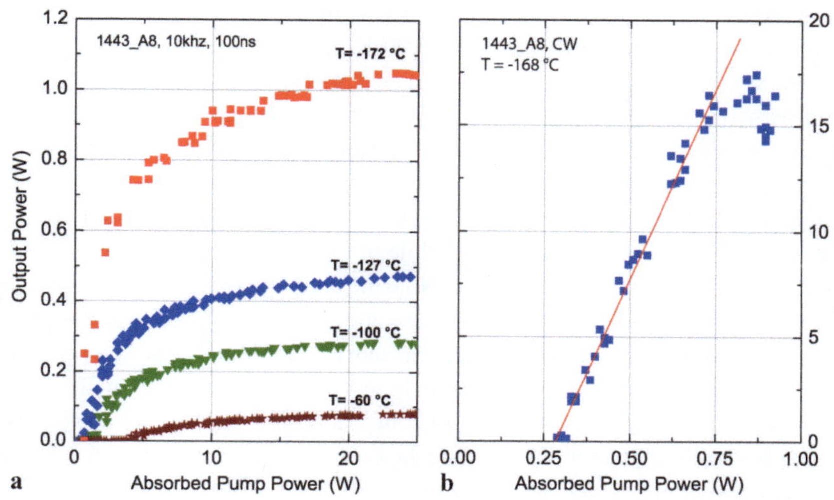

Fig. 6 Light in/light out characteristics at different temperatures for a PbTe-on-Si mid-IR VECSEL when pumped with a 1.55 µm wavelength laser (corrected for optical losses in the power measurement set-up). **a** Pulsed excitation with 100 ns pulses and 10 kHz repetition frequency, **b** CW excitation

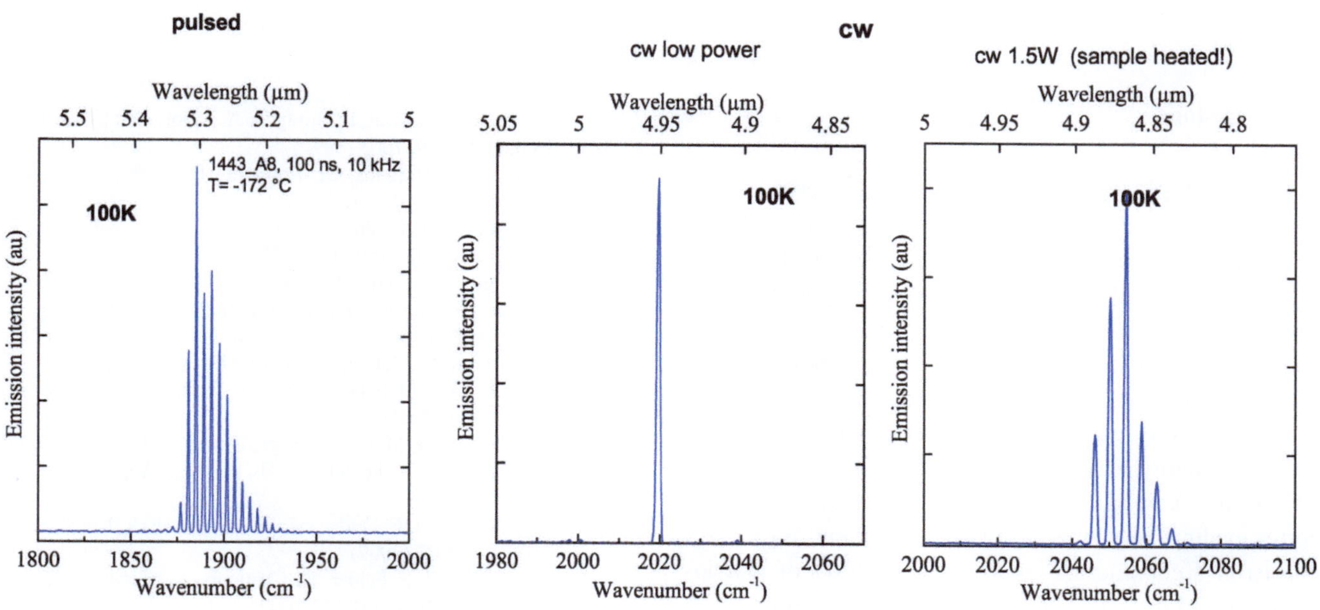

Fig. 7 Normalized laser spectra at 100 K for a PbTe-on-Si mid-IR VECSEL

standing electric wave (i.e. at half of the layer thickness). The structure is designed for 100 K operating temperature and grown on a BaF_2 substrate. Up to >600 mW_p output power is achieved at ~ 100 K, while threshold pump power is as low as 200 mW. Laser emission is at ~4.5 µm wavelength at this temperature. The corresponding wavelength for a VECSEL with "bulk" PbSe active layer at the same temperature is ~6.5 µm (Fig. 4), the blue-shift therefore amounts to ~1.7 µm. Threshold power is below ~0.2 W, and lasing occurs up to 170 K heat sink temperature [9].

2.3 PbTe VECSEL on Si-substrate with 1.1 W output power

A better heat removal, leading to higher output power, is expected when growing on a Si substrate, which has much better heat conductance than BaF_2. PbTe was applied as active layer with an optical thickness of one wavelength at 130 K operation temperature [10]. Indeed, at 100 K operating temperature, emission power is up to 1.1 W_p when illuminated with 100 ns pulses (Fig. 6a). In CW operation, up to 18 mW are observed before thermal rollover occurs (Fig. 6b). Spectra are multimode with high-power excitation, but monomode for low power. Figure 7 (left) shows a spectrum at 100 K with pulsed excitation, while in Fig. 7 (right) spectra with CW excitation are shown. Note the blue-shift due to the increased heating (positive temperature dependence of the bandgap) when switching from pulsed to low power CW and (still more pronounced) high-power CW operation. The structure lases up to 0°C. In a similar structure, but with a thinner PbTe active layer corresponding to one wavelength at *RT*, lasing occurs up to 25°C (Fig. 8).

3 Conclusions

The first VECSELs for the mid-IR range are described. They emit at 3.6–6.5 µm, CW up to 170 K, and above *RT* in pulsed-mode operation. The devices are fabricated employing lead chalcogenides (IV–VI narrow band gap semiconductors) rather than III–V materials.

Advantages of VECSELs are their good beam quality (circular emission cone with ~1° aperture angle), their very simple structure, easy and large wavelength tuning, and power scalability. This is in contrast to the well-known mid-IR quantum cascade lasers, which are edge emitters with corresponding very wide astigmatic emission angles and consist of a complicated arrangement of 100's of individual layers. Mid-IR VECSELs are therefore extremely well suited sources for spectroscopy.

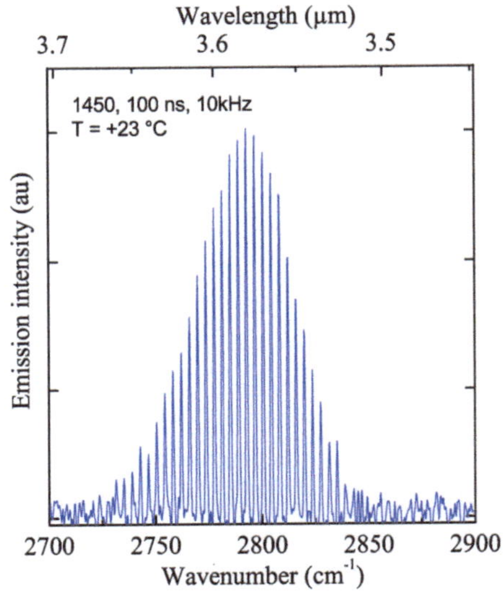

Fig. 8 Normalized laser spectra at *RT* for a PbTe-on-Si mid-IR VECSEL

RT CW operation is possible when employing improved heat spreaders like diamond. This technique is already developed for III–V based VECSELs with wavelength up to 2.x µm [2]. With the typical pump powers needed in the IV–VI samples shown above, a temperature rise of 20–40°C is calculated by stationary heat flow simulations when bonded to diamond substrates. Therefore, *RT* CW emission is expected when employing this technology.

Other wavelengths ranging from <3 µm up to >15 µm are accessible by using ternary active layers like $Pb_{1-z}X_zY$ (Y = Se or Te) where X = Eu or Sr for shorter, and X = Sn for longer wavelengths with respect to the binary compounds. As described above, QW blue-shift the emission and at the same time lead to still lower thresholds.

References

1. M. Kuznetsov, F. Hakimi, R. Sprague, A. Mooradian, IEEE Photon. Technol. Lett. **9**, 1063 (1997)
2. N. Schulz, J.-M. Hopkins, M. Rattunde, D. Burns, J. Wagner, Laser Photonics Rev. **2**, 160 (2008)
3. A.C. Tropper, H.D. Foreman, A. Garnache, K.G. Wilcox, S.H. Hoogland, J. Phys. D, Appl. Phys. **37**, R75 (2004)
4. W.W. Bewley, J.R. Lindle, C.S. Kim, M. Kim, C.L. Canedy, I. Vurgaftman, J.R. Meyer, Appl. Phys. Lett. **93**, 041118 (2008)
5. M. Beck, D. Hofstetter, T. Aellen, J. Faist, U. Oesterle, M. Ilegems, E. Gini, H. Melchior, Science **295**, 301 (2002)
6. M. Rahim, M. Arnold, F. Felder, K. Behfar, H. Zogg, Appl. Phys. Lett. **91**, 151102 (2007)
7. M. Rahim, F. Felder, M. Fill, H. Zogg, Opt. Lett. **33**, 3010 (2008)
8. M. Rahim, A. Khiar, F. Felder, M. Fill, H. Zogg, Appl. Phys. Lett. **94**, 201112 (2009)
9. H. Zogg, M. Rahim, F. Felder, M. Fill, D. Boye, A. Khiar, Proc. SPIE **7193**, 71931G (2009)
10. M. Rahim, M. Fill, F. Felder, D. Chappuis, M. Corda, H. Zogg, Appl. Phys. Lett. **95**, 241107 (2009)

Pulsed quantum-cascade-laser spectroscopy with intermediate-size pulses: application to NH_3 in the 10 μm region

B. Grouiez · V. Zeninari · L. Joly · B. Parvitte

Received: 10 September 2009 / Revised version: 4 March 2010 / Published online: 27 March 2010
© Springer-Verlag 2010

Abstract Pulsed quantum-cascade-laser spectrometers are usually used to detect atmospheric gases with either the interpulse technique (short pulses, typically 5–20 ns) or the intrapulse technique (long pulses, typically 500–800 ns). Each of these techniques has its drawbacks. Particularly the gas absorption spectra are generally distorted. We have previously developed another technique called intermediate-size pulses (typically 50–100 ns) technique for gas detection using pulsed QCL spectrometers. In this paper, infrared spectra of ammonia recorded with this technique in the 10 μm region are presented. For the NH_3 spectra recorded at low pressure (i.e. in the mbar range), the spectra show typical oscillations after the absorption. The Beer–Lambert law cannot explain these oscillations, termed the rapid-passage effect. Comparisons between experimental and calculated spectra will be realized. This phenomenon is not satisfactory from a spectroscopic point of view and spectra must be recorded at higher pressures. For the NH_3 spectra recorded at higher pressure (i.e. in the 50 mbar range), the oscillations disappear and the Beer–Lambert law could be reused. This paper will demonstrate that the intermediate-size technique gives reliable measurements for NH_3 detection. Moreover the typical apparatus function (0.003 cm^{-1} HWHM) is far lower from the typical apparatus function of the interpulse QCL spectrometers (0.015 cm^{-1} HWHM).

B. Grouiez · V. Zeninari (✉) · L. Joly · B. Parvitte
Groupe de Spectrométrie Moléculaire et Atmosphérique, UMR CNRS 6089, Université de Reims Champagne-Ardenne, UFR Sciences Exactes et Naturelles, Moulin de la Housse, BP 1039, 51687 Reims Cedex 2, France
e-mail: virginie.zeninari@univ-reims.fr
Fax: +33-03-26913147

Present address:
B. Grouiez
CEA Saclay, 91190 Gif-sur-Yvette cedex, France

1 Intermediate-size pulse quantum-cascade-laser spectrometry

Quantum-cascade lasers (QCLs) are ideal mid-IR sources for trace gas monitoring. They can be fabricated to operate at any of a very wide range of wavelengths from 3 to 30 μm where the absorption gas coefficients are stronger than in the near-infrared. Various techniques using continuous wave (cw) QCLs have been tested in the Reims laser group demonstrating very beautiful results from a spectroscopic point of view [1–8]. A comparison between the cw and the pulsed modes for a quantum-cascade laser has been realized in [9]. Despite the good results obtained with cw devices, pulsed mode operation has long been the main way for using QCLs [10] and remains the only available technique for several spectral ranges [11]. Two different ways of pulsing the laser have been demonstrated.

The first way is used by most spectrometers based on pulsed QCLs [12, 13] and is based on a short-duration current pulse applied to the QCL. This current generates a pulse in the spectral domain, which can be tuned through the absorption feature of interest by use of a slowly varying subthreshold current ramp or by use of a variation of the temperature of the laser chip but at much lower speed, owing to the high thermal capacity of the laser submount and laser base. The main drawback associated with this method is related to the laser line broadening that reduces spectral resolution and sensibility. The "line broadening" that is often attributed to the laser line width is in fact related to the detector's bandwidth [14].

The other method for pulsing QCL differs from the previous one. It uses the linear frequency downchirp that is induced when a long top-hat current pulse is applied to a QCL. The temporal profile of a generated pulse is recorded and used to infer the spectral nature of the pulse. The extent of the frequency window that is covered depends on the

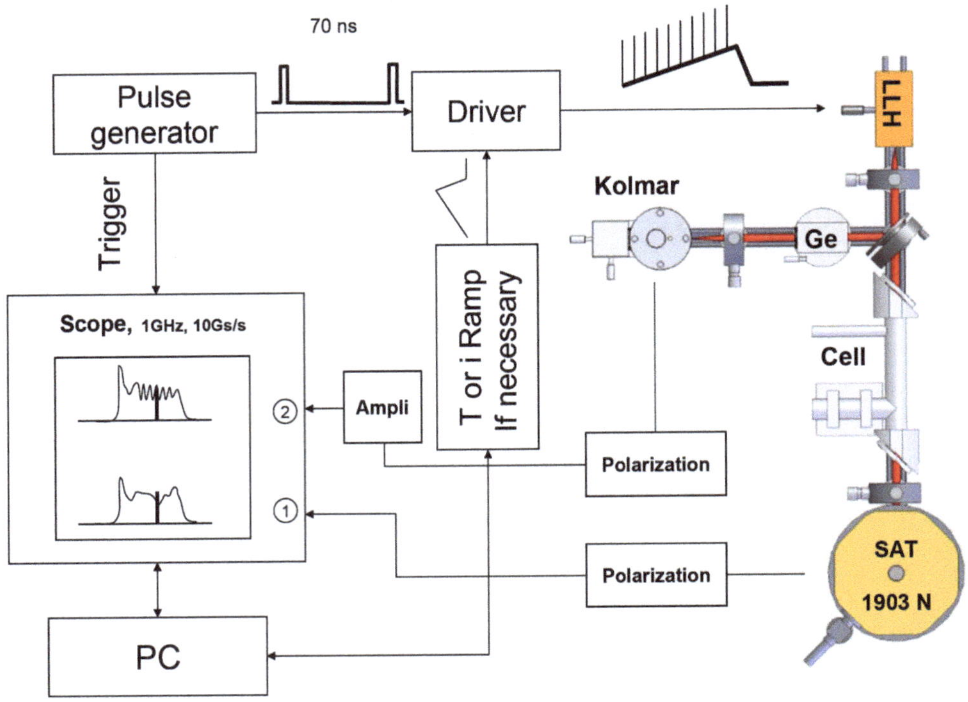

Fig. 1 Schematics of the experimental set-up

duration of the applied current pulse [15]. The spectral profiles of molecular gases can be recorded in real time [9, 16]. With this "long pulse" method, the resolution of the QCL spectrometer is limited not by the effective line width but by the temporal resolution of the detection system. Under some experimental conditions (low-pressure spectra, for example) the correlation between temporal and spectral profiles is not necessary straightforward. Particularly, if the duration of the interaction between the laser pulse and the molecular gas is short in comparison with the collisional lifetime, saturation and rapid-passage effects can occur as demonstrated in [17, 18].

Although this technique has numerous advantages when compared to the short-pulse technique, it suffers from several drawbacks: it requires much more specific instrumentation than the short-pulse technique, such as high-speed detectors and preamplifiers, a high-speed acquisition board, or a digital sampling oscilloscope. Moreover, the use of such long pulses is not recommended for most of the currently commercially available pulsed QCLs provided by Alpes Lasers [19]. Lasers allowing the use of long pulses seem to be available only for fewer wavelengths.

As none of those methods was completely satisfactory, a third way has been developed that uses intermediate length pulses, typically between 30 and 80 ns [14]. This interval of pulse duration was chosen to reach the linear frequency variation area that follows the transient at the beginning of the pulse without damaging the QC laser. We expected to obtain a narrower line width than with short pulses. The spectral range that is scanned inside so short pulses is usually small and we use either a subthreshold ramp (duty cycle of 0.3% with a frequency of 100 Hz) or a substrate temperature variation (duty cycle of 1% and ramping rate of 0.06 cm^{-1}/min) to increase the spectral range. These experiments were performed with a setup described in Fig. 1.

The pulse generator (Berkeley Nucleonics Corporation (BNC 565)) delivers 70 ns impulsions with a 1% duty cycle that are used in conjunction with a temperature or a subthreshold current ramp in order to obtain the laser wavelength variation. When using a temperature ramp, the laser temperature is scanned step by step, with typically 100 ms between successive steps. The series of impulsions are sent (using the driver from Alpes Laser starter kit which in part composed by a DEI PCO-7120 card) to the QC laser placed inside a Laser Laboratory Housing from Alpes Lasers. The laser beam passes through the absorption path cell containing ammonia ($L = 20$ cm) and is collected by SAT1903N high-speed MCT detector. The detector's signal is then sent to a digital sampling oscilloscope (Lecroy Waverunner104 Xi) so that each optical pulse is recorded. A part of the laser beam is sent through a Germanium etalon (with a FSR of 0.048 cm^{-1}) to a Kolmar detector in order to obtain the wavenumber calibration. Absorption features can be seen either inside each pulse (i.e. versus time) or at a constant delay inside successive pulses, versus subthreshold ramp/substrate temperature.

Each of spectra presented in this paper have been built using the intermediate-size method described in our previous paper [14]. Each pulse is recorded by the digital oscilloscope and a post-treatment permits one to rebuild the spectrum by selecting a constant delay inside pulses. The baseline due to the power variation of the laser during the

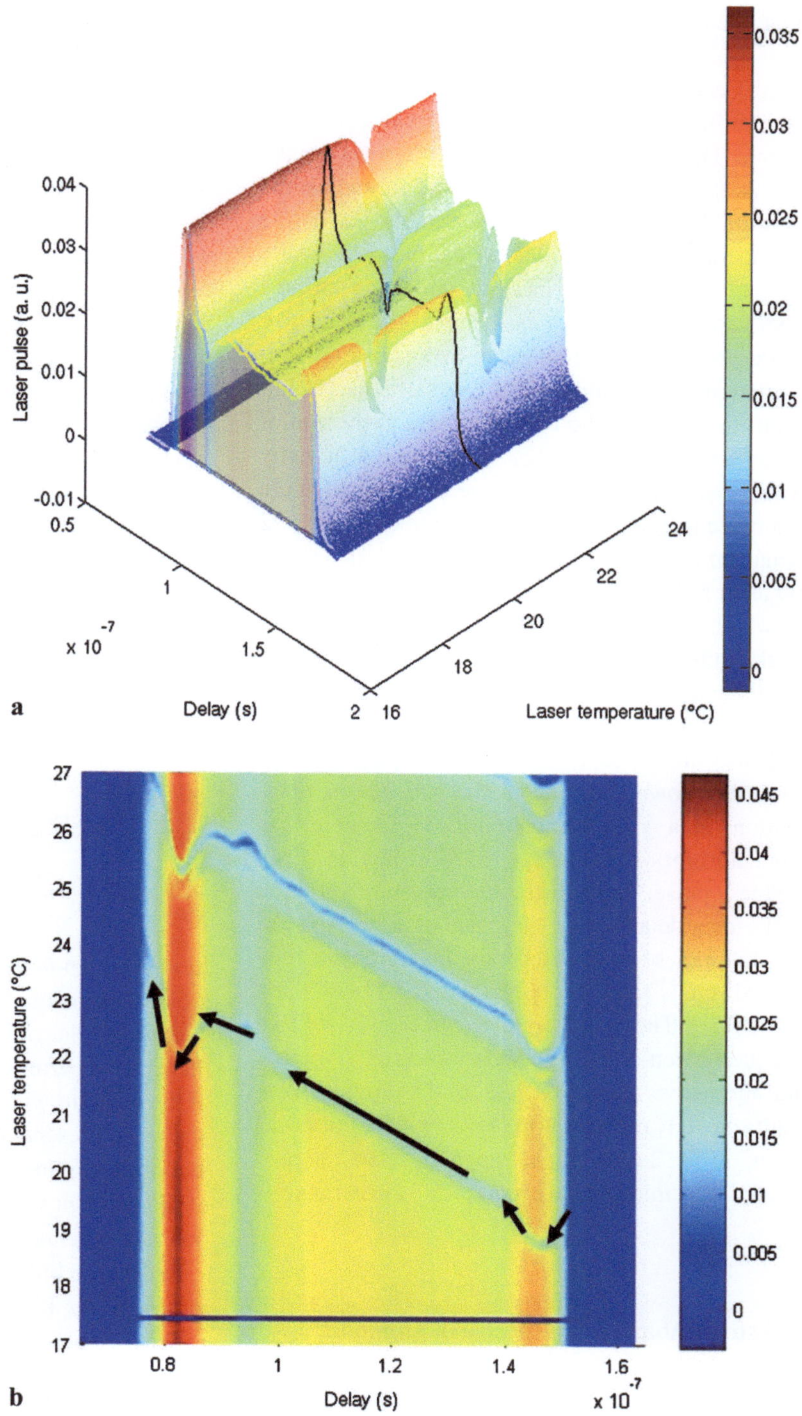

Fig. 2 (**a**) 3D view of successive experimental laser pulses (NH_3/air spectrum, $P = 60$ mbar, Concentration # 0.3%, Temperature # 296 K, Length = 20 cm). Successive laser pulses are presented versus the laser substrate temperature. The black trace corresponds to one pulse. (**b**) Top view of Fig. 2(a)

ramp is also taken into account during this post-treatment operation. Figure 2 presents the experimental data corresponding to 3000 successive measurements recorded during a substrate temperature ramp ranging from 17 to 27°C. Each measurement is in fact the average of 10 successive pulses and corresponds to one step of the temperature scanning. The total time for one measurement of about 7 μs is negligible when compared to the duration of one temperature step. Figure 2(a) is a 3D representation of the 3000 measurements made between 17 and 27°C. The signal amplitude (measured optical power) is coded using the color scale given on the right of the figure, from blue (no signal) to red (maximum of signal). The pulse recorded at $T = 21$°C is plotted in black. Figure 2(b) corresponds to a top view of Fig. 2(a). Each measurement corresponds to one horizontal line on Fig. 2(b). The horizontal scale represents the pulse length and the vertical scale gives the substrate temperature corresponding to each measurements. According to the color coding, the maximum of absorption appear as the thin blue lines.

Figure 2(b) can be used to follow the moving of the absorption line inside the pulse during the substrate temperature ramp. This is materialized by the arrows on the Fig. 2(b). This moving is an important clue to understand the variation of the instantaneous wavenumber versus the substrate temperature. In the middle of the pulse (typically between 1 and 1.4×10^{-7} s on Fig. 2(b)), the emitted wavenumber varies in a very regular manner: the frequency downchirp can be exactly compensated by a substrate temperature variation and the absorption line shifts nearly linearly. In both transients at the beginning and at the end of the pulse, the emitted wavenumber varies very differently at a given substrate temperature the same absorption feature can be seen twice in one pulse (see for example between 22 and 23°C and for a time delay about 0.8×10^{-7} s). The arrows allow us to follow the value of the instantaneous chirp frequency, which is directly linked to the variation in heating during the pulse. One can see that the sense of this one change rapidly and is also more important in transients than in the rest of the pulse.

Figure 2 also allows us to compare the spectral range due to frequency downchirp and to the substrate temperature variation: at a given temperature (21°C for example) only one absorption feature (at 967.997761 cm^{-1}) can be seen inside a pulse. The substrate temperature variation (for example for a delay inside the pulse of 1.2×10^{-7} s) allows one to scan several absorption features (one line at 967.997761 cm^{-1} and a doublet at 967.774745 and 967.738421 cm^{-1}). The first alternative can be used for narrow and isolated spectral lines (i.e. at low pressure) whereas the second alternative should be preferred at higher pressures. Results on NH_3 in the 10 µm region will be presented in this paper: in the low-pressure range (1 mbar range—Sect. 2) and in the mid-pressure range (50 mbar range—Sect. 3).

2 Low-pressure (mbar range) spectra of ammonia

An example of recorded pulse of ammonia at low pressure with the intermediate-size pulse technique and without added ramp is presented in Fig. 3. Figure 3(a) features a 70 ns pulse at a laser temperature without absorption and Fig. 3(b) features a recorded pulse at a laser temperature where ammonia absorbs. In both cases the first part of the pulse presents a transient and is followed by a regular frequency downchirp area. The last part of the pulse also presents a transient. The Fabry–Pérot etalon transmission shown on Fig. 3(b) allows us to determine the frequency downchirp tuning rate in this useful part of the pulse. At these low pressures the spectrum due to the chirp of the laser into the pulse let oscillations appears after the absorption. This phenomenon is named the rapid-passage effect and was observed with the intrapulse technique in [17].

The oscillations observed during the pulse cannot be explained by the Beer–Lambert law. The lifetimes of the levels that are linked to the Einstein coefficients and that are the basics of this law cannot take into account the individual comportment of the atoms. In this case there is no more equilibrium in the interaction between radiation and matter. A more realistic representation can be obtained thanks to a semi-classical model as was initially proposed by Duxbury team [17, 18, 20]. In this model the electromagnetic field is represented in a classical manner via the Maxwell equations and the atomic part is described by quantum mechanics.

This description is known as the Maxwell–Bloch equations. The Bloch differential equations are obtained by combining the coherent effect of the wave with the effect of a random phenomenon such as spontaneous emission and collisions. These equations can be presented by the following system [21]:

$$\frac{d\bar{u}}{dt} = -\frac{1}{\tau_2}\bar{u} + \delta\omega\bar{v}, \tag{1}$$

$$\frac{d\bar{v}}{dt} = -\delta\omega\bar{u} - \frac{1}{\tau_2}\bar{v} - \Omega_1\bar{m}, \tag{2}$$

$$\frac{d\bar{m}}{dt} = \Omega_1\bar{v} + \frac{m_0 - \bar{m}}{\tau_1}. \tag{3}$$

In this system $\bar{u}$ and $\bar{v}$ represent the mean values of the two crossed variables of the Bloch system. They are linked to the electric dipolar moment because the stationary solutions of this system permit to determine the dielectric susceptibility of the medium. $\bar{m}$ corresponds to the inversion of population and m_0 represents the initial value of the inversion of population. $\delta\omega$ represents the difference between the pulsation of the incident wave and the central pulsation of the transition and is related to the frequency downchirp tuning rate. τ_1 and τ_2 are linked to the lifetime of the levels. Finally Ω_1 represents the Rabi pulsation.

Analytical solutions of this system can be found when $\delta\omega$ remains constant during the experiment. Due to frequency downchirp $\delta\omega$ varies during each pulse and only numerical resolution of this system can be done. One important difference between the analytical solutions and the numerical solutions is that the oscillations in $\bar{u}$, $\bar{v}$ and $\bar{m}$ do not have a constant frequency. Moreover, as was demonstrated in [20], the resolution of this system do not allow to reproduce the observed signals and several effects must be included in the calculation:

- The velocity of the molecules affects the value of $\delta\omega$ due to Doppler effect. The system of coupled equation must be solved for several classes of velocities (200).

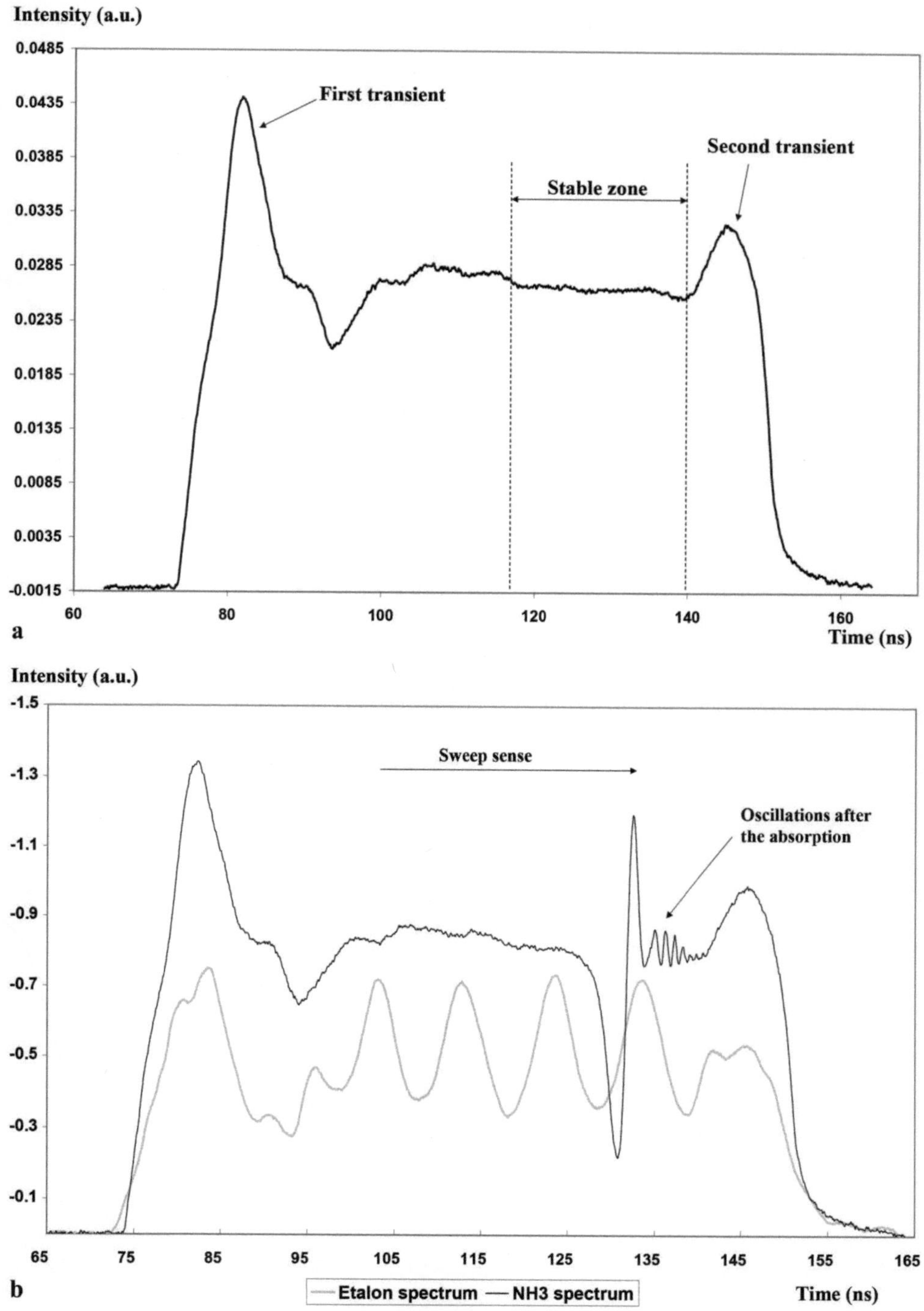

Fig. 3 (**a**) Record of a pulse with the intermediate-size pulse technique and (**b**) observation of rapid-passage effect in a 70 ns pulse with 0.37 mbar of pure ammonia

- The Rabi pulsation used in (2) and (3) is related to the transition dipole moment. The arbitrary orientation of the molecule versus the laser field must also be included in the calculation.

These two effects smooth the very fast oscillations observed in the solutions of (1), (2) and (3) and allow us to obtain a satisfactory agreement with observed signals. Some examples are presented on Fig. 4. Figure 4(a) presents some recorded signals at several pressures and Fig. 4(b) presents the results of the calculations. At this moment we do not obtain full agreement between measurements and calculations. As only numerical solutions can be obtained, they cannot be used in a typical inversion process like least-squares fitting. However, the overall agreement on the three pressure range is good.

The best solution to obtain spectra that are satisfactory from a spectroscopic point of view is to increase the pressure. In this case the lifetime τ_1 becomes very small and the second term of (3) $\frac{m_0-\bar{m}}{\tau_1}$ becomes larger than the first one $\Omega_1\bar{v}$; thus the oscillations disappear. Section 3 of this pa-

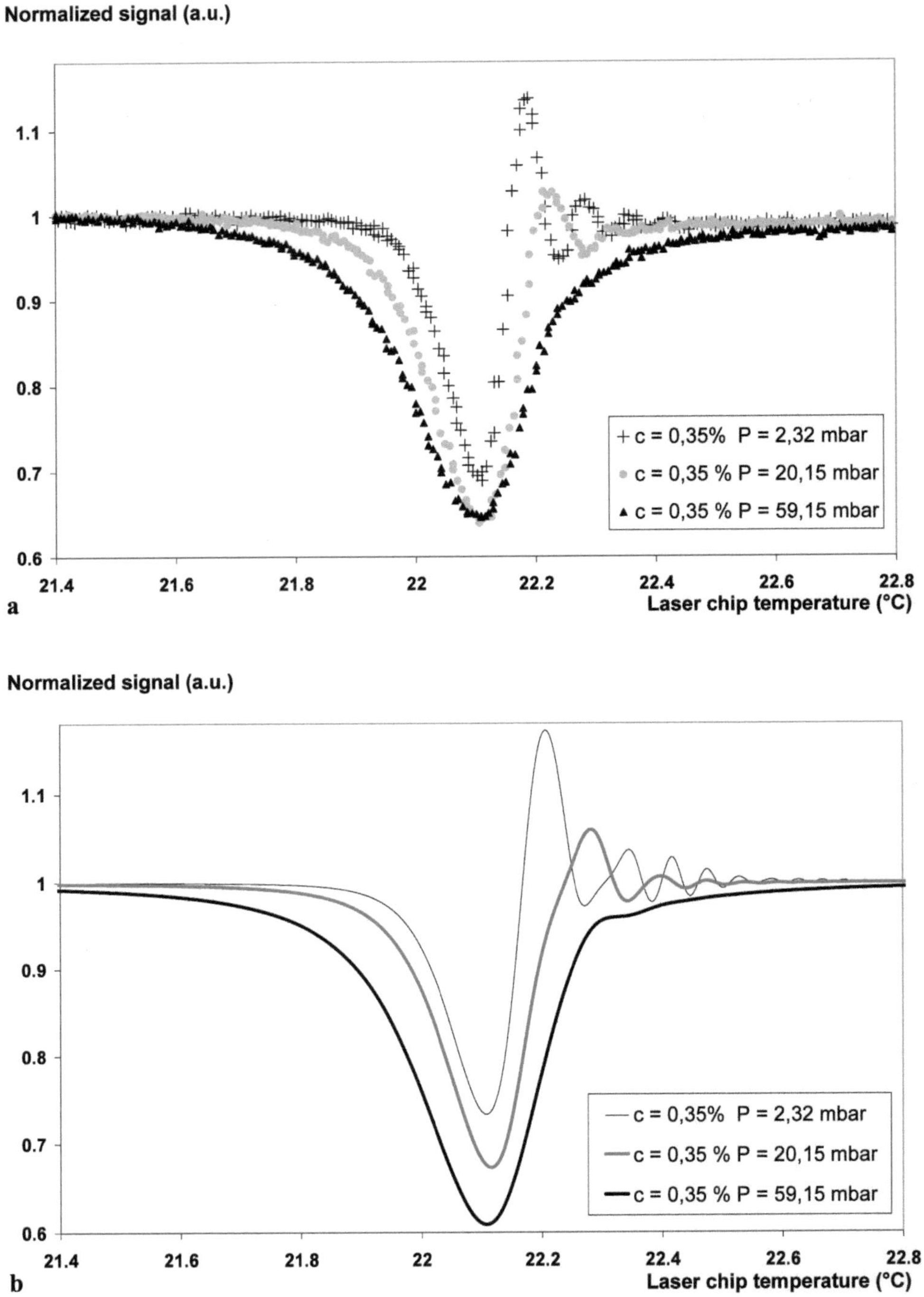

Fig. 4 (**a**) Recorded signals for different gas pressures. (**b**) Calculated signals for the same experimental conditions

per will present results obtained at higher pressures where rapid-passage effects do not exist anymore.

3 Mid-pressure (50 mbar range) spectra of ammonia

In this part either temperature or current ramps were added to the 70 ns length duration pulses. A spectrum corresponding to a 40 ns delay after the beginning of the pulse is presented in Fig. 5. This spectrum is very satisfactory from a spectroscopic point of view and can be directly compared to a calculation using Voigt profile and HITRAN parameters [22]. The fitted and experimental spectra are very similar. Numerous spectra were recorded at different pressures from 50 to 80 mbar. For all those spectra the collisional line width was fitted and compared to the HITRAN value. The fitting procedure has been described in detail in [23]. The experimental and HITRAN collisional line widths are plotted vs. pressure in Fig. 6. One can remark than the two lines are parallel; however, only the HITRAN curve passes through zero. This is the demonstration of the existence of an apparatus function.

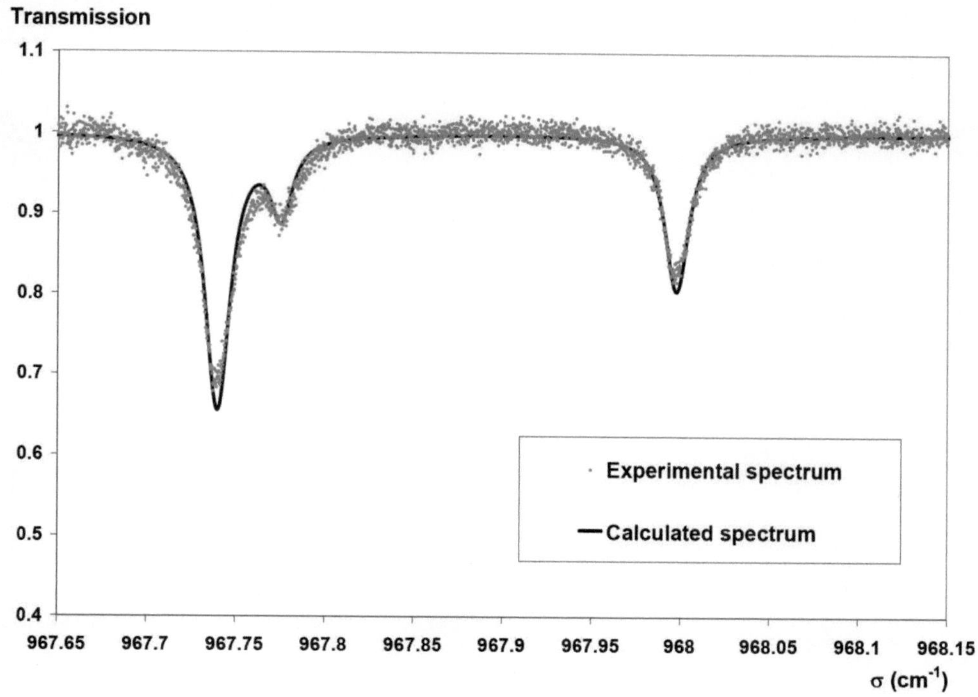

Fig. 5 Example of spectrum recorded with a temperature ramp of 10 K and comparison with a calculated spectrum using Voigt profile for a pressure = 71 mbar and concentration of NH_3 in air is 1.3%

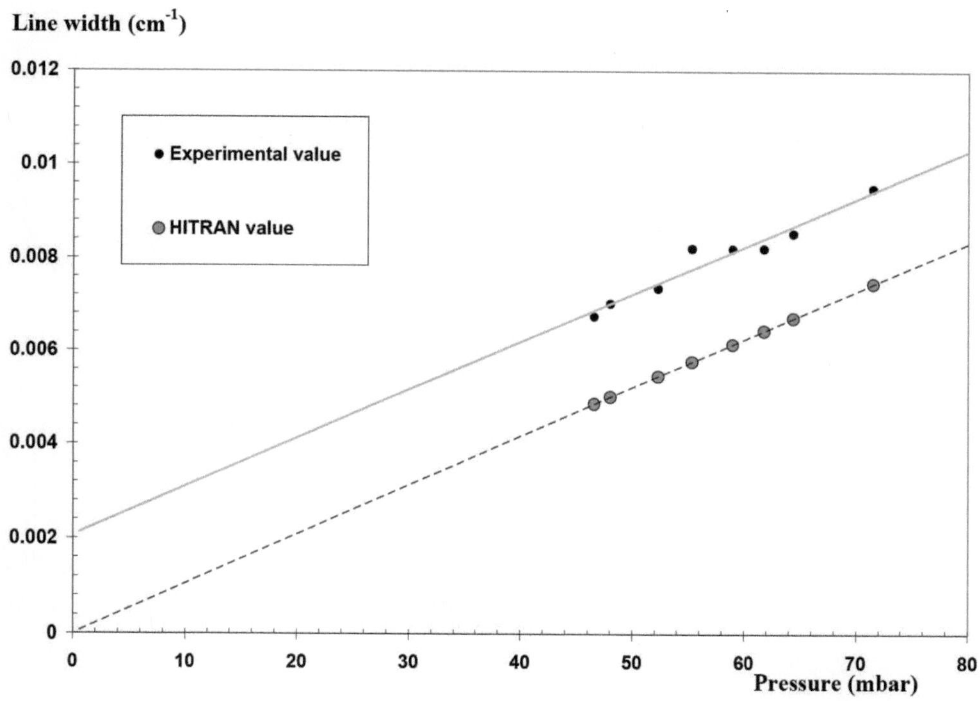

Fig. 6 Fitted line width vs. pressure for the experimental spectra. The HITRAN values are also displayed. See text for more details

The existence of an apparatus function is also demonstrated in Fig. 7. Figure 7(a) features the NH_3 absorption line using a temperature ramp and the comparison with the calculated spectrum using HITRAN data. Figure 7(b) shows the same line recording using a subthreshold current ramp. Note that the difference between the two techniques is that the temperature spectrum is longer to record (typ. 140 s) than the current one (typ. 10 ms). On the other side the current spectrum is less wide than the temperature one. For these pressures there is no problem but it may be a drawback for atmospheric pressure spectra.

Even if the experimental and HITRAN calculated spectra are similar, some differences remain. To confirm that these differences were due to the apparatus function, we calculate the HITRAN spectrum using Voigt profile and convoluted it by a Gaussian profile with $\Delta\nu = 0.003\ \text{cm}^{-1}$. These calculations are also displayed in Figs. 7(a) and 7(b). For these two figures one can clearly see that the experi-

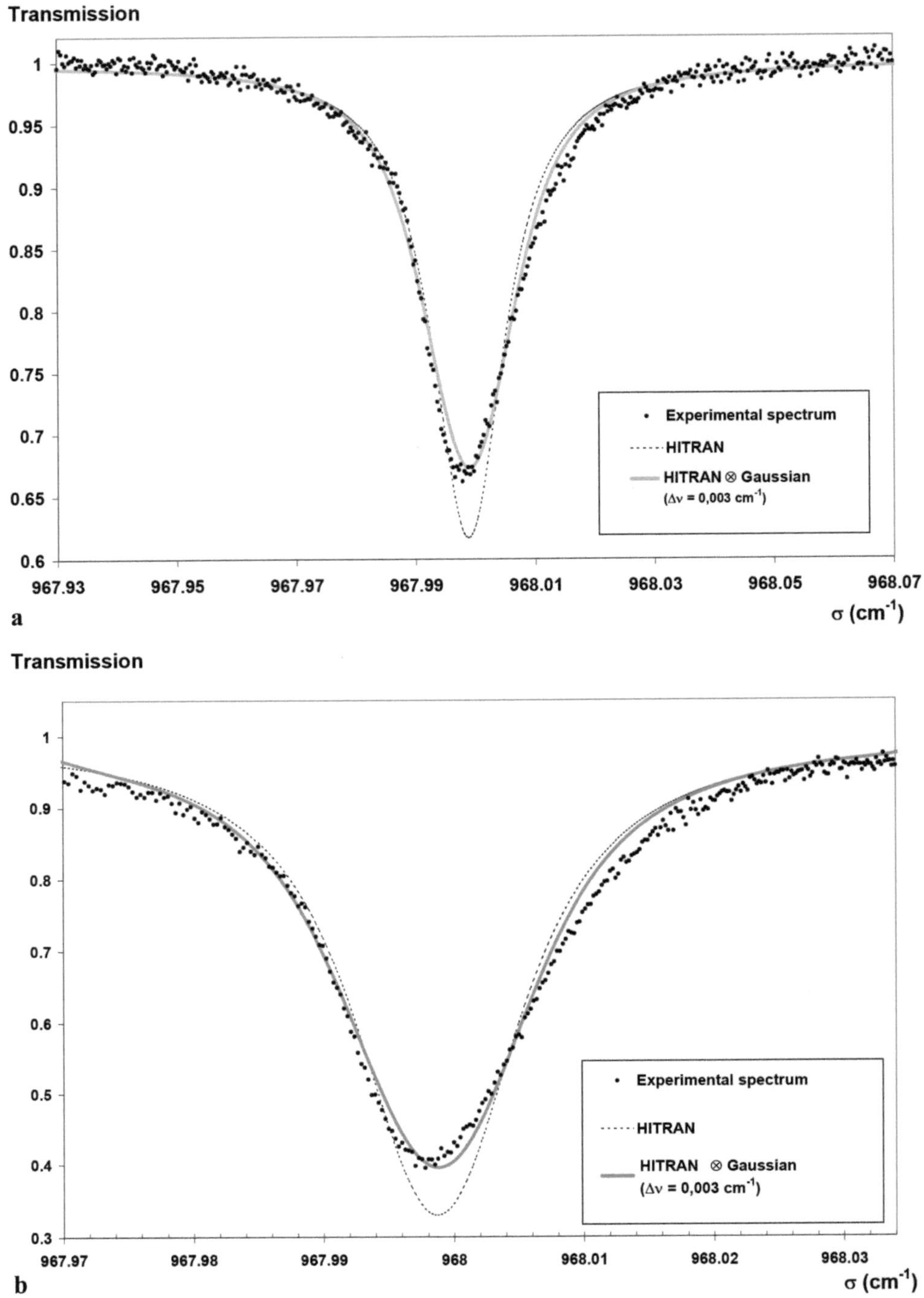

Fig. 7 Transmission experimental spectra with (**a**) a temperature ramp ($P = 58$ mbar and 0.28% of NH_3 in air) and (**b**) a current ramp ($P = 51$ mbar and 0.8% of NH_3 in air). On both parts of the figure are also displayed the calculated spectra using Voigt profile with the HITRAN data and another calculation introducing a convolution of the HITRAN spectra with a Gaussian lineshape (see text for more details)

mental spectra and the new calculation are very well fitted. The small difference remaining in the two spectra (on the right part of the wing) is probably due to a remaining part of the rapid-passage effect. The recorded and calculated spectra allows estimating the effective line width of the laser to be about 0.003 cm^{-1}, which is significantly shorter than typical values reported with shorter pulses (typically 0.015 cm^{-1} [12]). Thus the intermediate-size pulse technique is more satisfactory from a spectroscopic point of view.

4 Conclusion

Pulsed quantum-cascade-laser spectrometers are still used to detect atmospheric gases. We have previously demonstrated [14] that neither the interpulse nor the intrapulse technique was totally satisfactory to use the pulsed QCLs. The interpulse (short-pulse) technique gives distorted spectra and the intrapulse (long-pulse) technique cannot be applied to all QCLs. Thus we have developed the intermediate-size pulse technique for gas detection using pulsed QCL spectrometers. This technique was applied in this paper to record infrared spectra of ammonia in the 10 μm region. However, NH_3 spectra recorded in the mbar range show typical oscillations after the absorption due to the rapid-passage effect. This phenomenon had been previously observed with the intrapulse technique [17]. In this paper, the resolution of the Maxwell–Bloch equations has demonstrated a good agreement in comparison with experimental results at various pressures.

To realize correct spectroscopic measurements NH_3 spectra were recorded in the 50 mbar range, where the oscillations disappear. Thus the intermediate-size pulse technique gives reliable measurements for NH_3 detection using a temperature or a current ramp. Moreover, the typical apparatus function (0.003 cm^{-1} HWHM) is far lower from the typical apparatus function of the interpulse QCL spectrometers (0.015 cm^{-1} HWHM).

Acknowledgements The authors are indebted to the *Région Champagne-Ardenne* for financially supporting of the PhD of Bruno Grouiez.

References

1. L. Joly, V. Zéninari, B. Parvitte, D. Weidmann, D. Courtois, Y. Bonetti, T. Aellen, M. Beck, J. Faist, D. Hofstetter, Appl. Phys. B, Lasers Opt. **77**, 703 (2003)
2. B. Parvitte, L. Joly, V. Zéninari, D. Courtois, Spectrochim. Acta A **60**, 3285 (2004)
3. L. Joly, V. Zéninari, B. Parvitte, D. Courtois, G. Durry, Opt. Lett. **31**, 143 (2006)
4. A. Grossel, V. Zéninari, L. Joly, B. Parvitte, G. Durry, D. Courtois, Spectrochim. Acta A **63**, 1021 (2006)
5. A. Grossel, V. Zéninari, B. Parvitte, L. Joly, G. Durry, D. Courtois, Infrared Phys. Technol. **51**, 95 (2007)
6. A. Grossel, V. Zéninari, B. Parvitte, L. Joly, D. Courtois, Appl. Phys. B, Lasers Opt. **88**(3), 483 (2007)
7. L. Joly, C. Robert, B. Parvitte, V. Catoire, G. Durry, G. Richard, B. Nicoullaud, V. Zéninari, Appl. Opt. **47**, 1206 (2008)
8. A. Grossel, V. Zéninari, B. Parvitte, L. Joly, G. Durry, D. Courtois, J. Quant. Spectrosc. Radiat. Transfer **109**, 1845 (2008)
9. B. Grouiez, B. Parvitte, L. Joly, D. Courtois, V. Zéninari, Appl. Phys. B, Lasers Opt. **90**, 177 (2008)
10. J. Faist, F. Capasso, D.L. Sivco, C. Sirtori, A.L. Hutchinson, A.Y. Cho, Science **264**, 553 (1994)
11. J. Devenson, O. Cathabard, R. Teissier, A.N. Baranov, Appl. Phys. Lett. **91**, 141106 (2007)
12. K. Namjou, S. Cai, E.A. Whittaker, J. Faist, C. Gmachl, F. Capasso, D.L. Sivco, A.Y. Cho, Opt. Lett. **23**, 219 (1998)
13. A.A. Kosterev, F.K. Tittel, C. Gmachl, F. Capasso, D.L. Sivco, J.N. Baillargeon, A.L. Hutchinson, A.Y. Cho, Appl. Opt. **39**, 6866 (2000)
14. B. Grouiez, B. Parvitte, L. Joly, V. Zéninari, Opt. Lett. **34**, 181 (2009)
15. E. Normand, G. Duxbury, N. Langford, Opt. Commun. **197**, 115 (2001)
16. E. Normand, M. McCulloch, G. Duxbury, N. Langford, Opt. Lett. **28**, 16 (2003)
17. M. McCulloch, G. Duxbury, N. Langford, Mol. Phys. **104**, 2767 (2006)
18. M.T. McCulloch, E.L. Normand, N. Langford, G. Duxbury, D.A. Newnham, J. Opt. Soc. Am. B **20**, 1761 (2003)
19. www.alpeslasers.ch
20. G. Duxbury, N. Langford, M.T. McCulloch, S. Wright, Mol. Phys. **105**, 741 (2007)
21. L. Allen, J.H. Eberly, *Optical Resonance and Two-Level Atoms* (Dover, New York, 1975)
22. L.S. Rothman, D. Jacquemart, A. Barbe, D. Chris Benner, M. Birk, L.R. Brown, M.R. Carleer, C. Chackerian Jr., K. Chance, L.H. Coudert, V. Dana, V.M. Devi, J.-M. Flaud, R.R. Gamache, A. Goldman, J.-M. Hartmann, K.W. Jucks, A.G. Maki, J.-Y. Mandin, S.T. Massie, J. Orphal, A. Perrin, C.P. Rinsland, M.A.H. Smith, J. Tennyson, R.N. Tolchenov, R.A. Toth, J. Vander Auwera, P. Varanasi, G. Wagner, J. Quant. Spectrosc. Radiat. Transfer **96**, 139 (2005)
23. V. Zéninari, B. Parvitte, L. Joly, T. Le Barbu, N. Amarouche, G. Durry, Appl. Phys. B, Lasers Opt. **85**, 265 (2006)

Mid infrared interband cascade lasers for sensing applications

L. Nähle · P. Fuchs · M. Fischer · J. Koeth · A. Bauer · M. Dallner · F. Langer · S. Höfling · A. Forchel

Received: 7 October 2009 / Revised version: 16 December 2009 / Published online: 6 February 2010
© Springer-Verlag 2010

Abstract The growth of Interband Cascade Laser material to cover the wavelength range from 3–4 μm is presented along with the fabrication and characterization of Broad Area (BA) and Ridge Waveguide (RWG) devices based on this material. Pulsed operation slightly below room temperature is observed for both device types, and a strong reduction of threshold currents can be observed in the RWG lasers. Variation of the active Quantum Well width in the epitaxial structures enables laser emission in the 3–4 μm wavelength region.

A growing demand for mid infrared (IR) light sources (wavelength ≈2.5–20 μm) is developing in several fields of application. Compact and reliable laser sources in this wavelength regime exhibit excellent possibilities of use for example in the military (e.g. detection of explosives [1]) and commercial (e.g. industrial process control [2]) sector, partially owing to naturally occurring atmospheric transparency windows.

The wavelength range from 3–4 μm is particularly interesting for several sensing applications, such as the detection of hydrocarbons. Existing types of semiconductor laser sources have closed in on this wavelength area [3, 4]. However, decreasing hole confinement and increasing Auger recombination in GaSb-based type I interband diodes limit their usability towards higher wavelengths of ≈4 μm. Intersubband quantum cascade lasers (QCL) approach the wavelength range mentioned from above but suffer from fast phonon scattering losses. From both issues of the above described laser types emerges the need of a more efficient laser source in the 3–4 μm regime.

The discussed issues can be circumvented by using interband cascade lasers (ICL) [5, 6]. These technologically demanding laser sources utilize optical transitions between an electron state in the conduction band and a hole state in the valence band in a cascade of Sb-based type-II QW structures. A broken-gap band edge alignment enables the tailoring of the emission wavelength by altering the cascade structures.

In this letter we report on a collaborative work between the University of Würzburg and nanoplus GmbH presenting broad area (BA) and ridge waveguide (RWG) lasers based on ICL epitaxial material. Each sample was grown in a solid-source molecular beam epitaxy chamber equipped with cracking cells for both arsenic and antimony and standard group-III effusion cells. A second indium cell as well as a dual dopant cell for silicon/galliumtelluride provided a convenient degree of flexibility during growth. Initially, an about 300 nm thick GaSb:Te buffering layer was grown on top of an epi-ready (100) GaSb substrate. The active region consisted of 6 to 12 active cascades (see Fig. 1) and was sandwiched in-between two 200 nm thick GaSb:Te separate confinement layers and 400 repetitions of a 2.4/2.3 nm InAs:Si/AlSb super lattice cladding. The cascade design was similar to the one described in [7] using a partly Si-doped chirped InAs/AlSb super lattice as injector and adjacent W-QWs as optically active part. Their thicknesses were varied from 1.85–2.5 nm in order to cover the 3–4 μm emission wavelength range. A highly doped InAs:Si capping layer finalized the structures and ensured low ohmic resistances.

L. Nähle (✉) · P. Fuchs · M. Fischer · J. Koeth
Nanoplus GmbH, Oberer Kirschberg 4, 97218 Gerbrunn, Germany
e-mail: lars.naehle@nanoplus.com
Fax: +49-931-9082719

A. Bauer · M. Dallner · F. Langer · S. Höfling · A. Forchel
Technische Physik, Universität Würzburg, Am Hubland, 97074 Würzburg, Germany

BA devices were fabricated using a broad stripe gold p-contact as etch mask. Applying reactive ion etching, the structures were etched well through the capping layer. A bottom contact was evaporated.

RWG devices were realized using a chromium etch mask. After deep ion etching below the active cascades the structures were passivated by a several hundred nanometers thick passivation layer before evaporating the top and bottom contact. The passivation was necessary in order to avoid surface leakage currents due to unterminated dangling bonds forming interfacial traps on the etched semiconductor surface [8]. The deep etching through the active layers was performed to confine the current path through the structure and avoid lateral leakage currents. An electron micrograph of the dry etched ridge with very smooth sidewalls along with the depiction of a cleaved facet of a finished RWG structure are shown in Fig. 2.

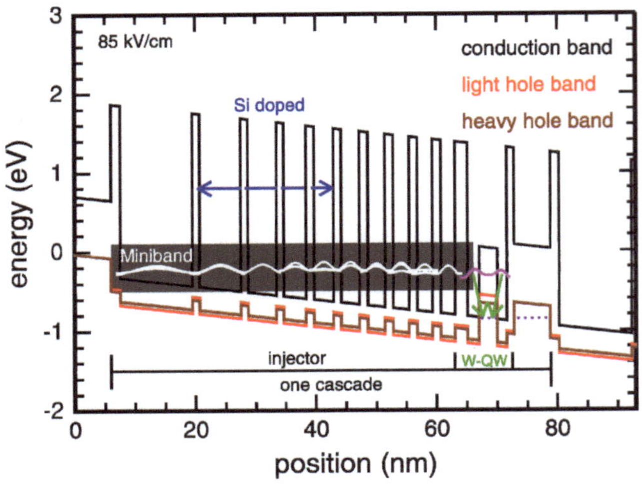

Fig. 1 Active cascade of the ICL structures used for the BA and RWG devices

Both BA and RWG devices were cleaved and mounted epi-side-up on copper heat sinks. Electro-optical characterization was conducted by installing the samples in a cryogenic cooler. Output power characteristics were investigated using a mercury cadmium telluride detector and spectral characteristics were studied employing an FTIR spectrometer. All measurements were conducted under pulsed conditions with 100 ns pulse width and 1 kHz repetition rate.

The results of these measurements for a BA device can be found in Fig. 3. P–I curves ranging from 100 to 290 K for a 2.0 mm × 0.1 mm device are shown on the left hand side. Threshold currents vary from 1.5 to 9.8 A at around room temperature. A representative U–I curve is integrated in the figure and shows an onset voltage of around 5.6 V. Spectral characterization shows coverage of the wavelength region

Fig. 2 Electron micrographs; *left*: dry etched ridge structure, *right*: cleaved facet of a RWG with passivation layers

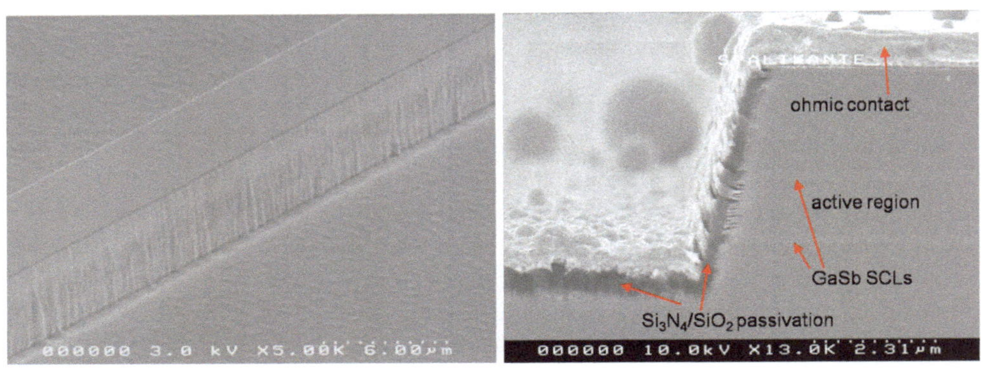

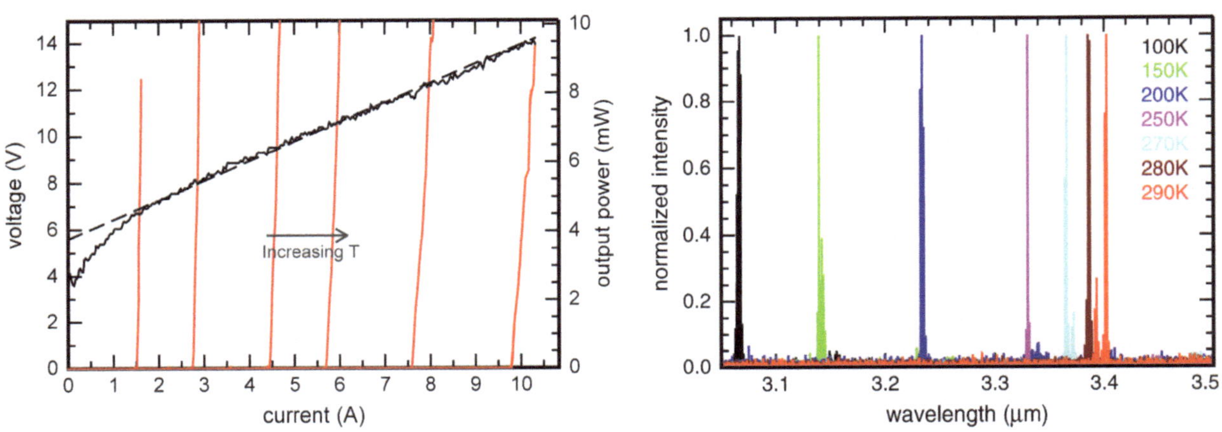

Fig. 3 Electro-optical characterization of a 2.0 mm × 0.1 mm BA device, *left*: P–I and U–I curves (U–I taken at 190 K), *right*: recorded spectra, both from 100–290 K

from 3.05 μm up to around 3.4 μm with varying temperature from 100 to 290 K. Selected spectra are presented on the right hand side of the graph.

In Fig. 4 results of comparative measurements on a 1 mm × 0.012 mm RWG structure can be observed. On the left hand side it can be clearly seen that the threshold currents of these devices are strongly reduced as compared to the BA devices, owing to the considerably smaller widths of the laser structures and the stronger confinement of the current path. Considering the temperature range from 100 to 280 K they vary from 0.15–0.72 A. Due to the smaller pumped laser volume the output powers are noticeably lower than in the BA devices. According to the U–I curve taken at 270 K the onset voltage in the RWG laser is slightly higher than in the BA laser, amounting to 6.0 V. This can most certainly be ascribed to a slightly better contacting of the BA lasers. Since the epitaxial structure is the same for both types of lasers the voltage drop along the active region should be the same in either case. On the right hand side of the figure the attained spectra for some selected temperatures are presented. They show basically the same behavior as in the BA devices. For comparison a wavenumber scale is included in the graph.

For both laser types the temperature shift of the gain of the ICL devices amounts to 1.7 nm/K or 1.7 cm^{-1}/K. The Fabry–Pérot mode spacing, which can be extracted

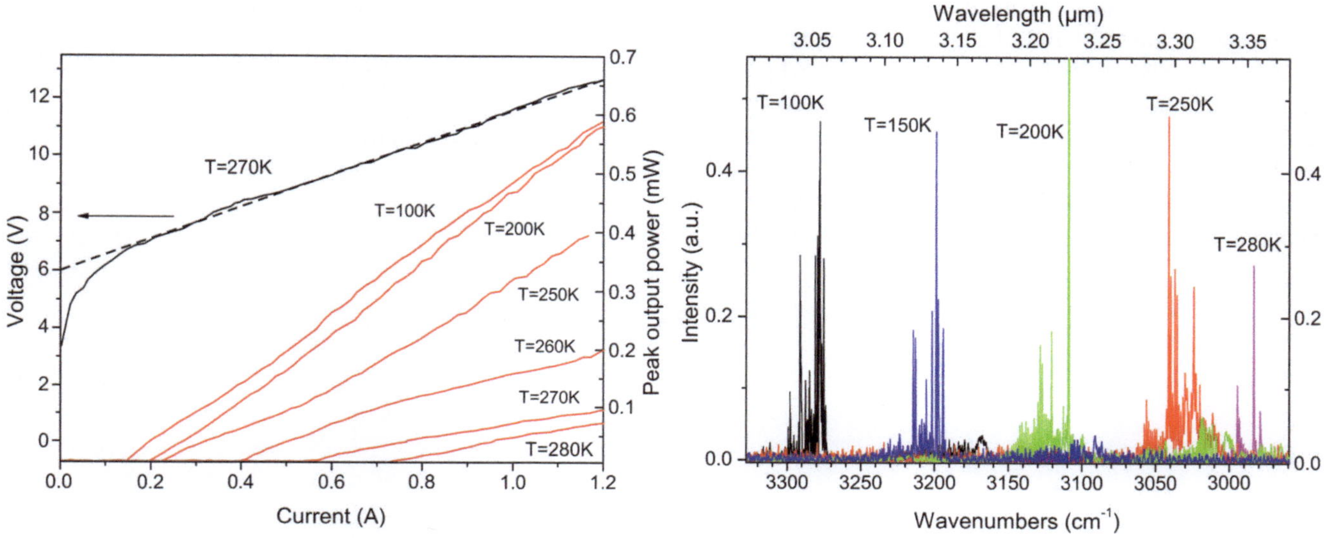

Fig. 4 Electro-optical characterization of a 1.0 mm × 0.012 mm RWG device, *left*: P–I and U–I curves, *right*: recorded spectra, both from 100–280 K

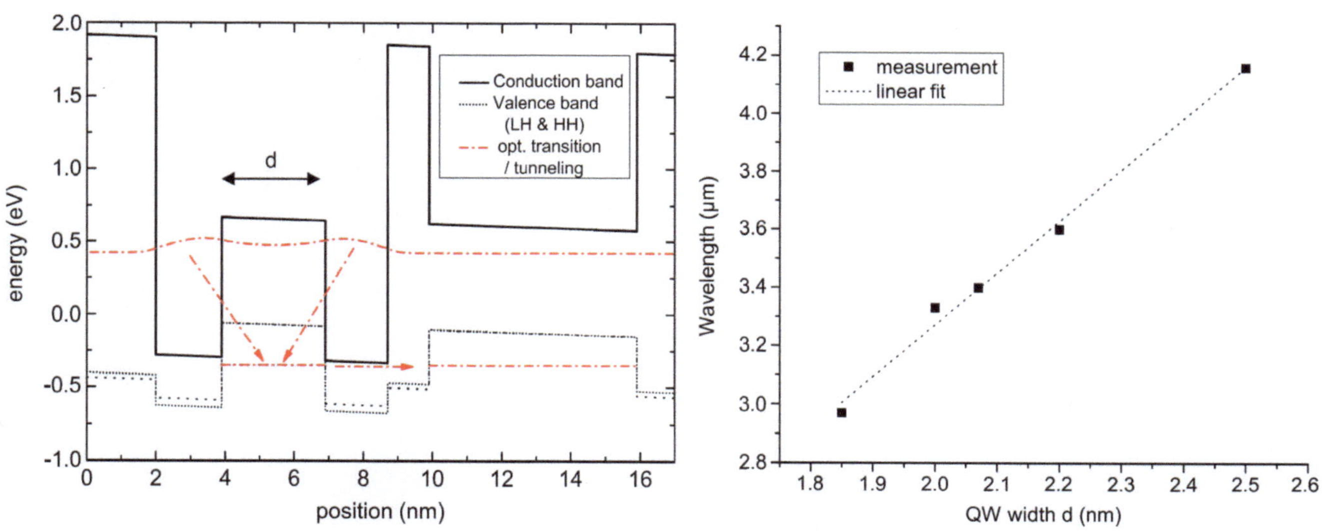

Fig. 5 *Left*: Depiction of a W-QW as used in the ICL active cascades along with the optical transition occurring. The QW width d was varied in order to attain an emission wavelength covering from 3 to 4 μm. *Right*: Measurement and fit of the emission wavelength of BA devices for a variation of the QW width

very accurately from the RWG measurements, is around $\Delta\lambda^{-1} = 1.7\ \text{cm}^{-1}$, which yields a refractive group index of $n_{\text{eff},g} = 3.65$, considering the 1 mm device length.

In order to cover the wavelength range from 3–4 µm different samples with varying QW widhts were grown. As depicted in Fig. 5, the width d of the active QW was varied from 1.85 to 2.5 nm. The emission wavelength at 77 K of BA devices manufactured on the different epitaxial materials is plotted against the QW width. The nearly linear dependence agrees very well with theoretical models that will be published elsewhere. The graph shows that, as intended, the wavelength range between 3 and 4 µm can be very well accessed using this approach.

Summarizing, the fabrication of BA and RWG lasers on ICL epitaxial material was presented along with characterization results for the two laser types. Both BA and RWG lasers operate up to slightly below room temperature and yield emission wavelengths varying from 3.05–3.4 µm with temperatures from 100–280/290 K. The threshold currents in the RWG devices are significantly reduced compared to the BA devices.

A study of the emission wavelength dependence on the variation of the active QW width in the active laser cascades showed very good results compared with theoretical models and enabled laser emission in the 3–4 µm wavelength region by tailoring the epitaxial structure.

References

1. C. Bauer, P. Geiser, J. Burgmeier, G. Holl, W. Schade, Appl. Phys. B **85**, 251 (2006)
2. M. Hümmer, K. Rößner, T. Lehnhardt, M. Müller, A. Forchel, R. Werner, M. Fischer, J. Koeth, Electron. Lett. **42**, 10 (2006)
3. L. Shterengas, G. Belenky, T. Hosoda, G. Kipshidze, S. Suchalkin, Appl. Phys. Lett. **93**, 011103 (2008)
4. S.Y. Zhang, D.G. Revin, J.W. Cockburn, K. Kennedy, A.B. Krysa, M. Hopkinson, Appl. Phys. Lett. **94**, 031106 (2009)
5. R.Q. Yang, Microelectron. J. **30**, 1043 (1999)
6. M. Kim, C.L. Canedy, W.W. Bewley, C.S. Kim, J.R. Lindle, J. Abell, I. Vurgaftman, J.R. Meyer, Appl. Phys. Lett. **92**, 191110 (2008)
7. C.L. Canedy, W.W. Bewley, J.R. Lindle, C.S. Kim, M. Kim, I. Vurgaftman, J.R. Meyer, Appl. Phys. Lett. **88**, 161103 (2006)
8. A. Gin, Y. Wei, J. Bae, A. Hood, J. Nah, M. Razeghi, Thin Solid Films **447–448**, 489 (2004)

Performance of a 12.49 meter folded path copper Herriott cell designed for temperatures between 296 and 20 K

M. Guinet · A.W. Mantz · D. Mondelain

Received: 7 October 2009 / Revised version: 18 February 2010 / Published online: 6 March 2010
© Springer-Verlag 2010

Abstract Using a set of gold-coated copper mirrors identical to those reported previously we were able to increase the useful absorption path from 5.9 meters to 12.49 meters. The low temperature performance of the new cell is similar to that of the previously described cell. The testing procedure and results are described here.

1 Introduction

Absorption line parameters, like pressure broadening, line mixing, collisional narrowing and pressure shifts are temperature dependent, reflecting the fact that molecular collision partners have different mean speeds at different temperatures. When modeling the earth atmosphere or the atmosphere of the outer planets, especially, it is important to have accurate experimental and calculated values for these parameters over the range of temperatures encountered in the atmosphere being studied.

Experimental determination of accurate values for the temperature dependent line parameters requires very careful measurements with absorption cells that may be characterized by very homogeneous and stable temperatures. Since many absorption features in the infrared are relatively weak, these absorption cells must also have long absorption path length. In order to keep volumes reasonable and masses to be cooled small, a folded optical absorption path is desirable. The Herriott design [1] is mechanically simpler than the White type cell [2] or the Chernin cell [3], so we decided to pursue the Herriott design for our cold cell applications. It is much more cumbersome to significantly change absorption path length with the Herriott design than with the other designs mentioned above, but the mechanical simplicity associated with the Herriott design means that optical alignment may be easily maintained, or reestablished using small adjustments to the input beam entrance angle, at different temperatures, and temperature uniformity is also easier to achieve with no external mechanical adjustments in the Herriott design.

We used spherical concave mirrors fabricated from oxygen free high purity copper, identical to the mirrors described previously [4], with a 15 cm radius of curvature and a mirror diameter of 50 millimeters. One of the mirrors has a 4-millimeter coupling hole drilled in it 16 millimeters from the center of the mirror to couple light into and out of the cell. We designed a Herriott cell that achieves 12.465 meters of folded absorption path using 41 images on the mirrors. The total path of 12.485 meters results from 0.02 meters of additional path traversed by the entrance and exit beams between the coupling mirror surface and the cell window. Geometrically more images are possible, however, signal to noise considerations practically limit the number of reflections to approximately 41 per mirror when using tunable diode lasers.

M. Guinet · A.W. Mantz · D. Mondelain (✉)
Laboratoire de Physique Moléculaire pour l'Atmosphère et l'Astrophysique, CNRS, UMR 7092, 75005, Paris, France
e-mail: mondelain@lpma.jussieu.fr
Fax: +33-01-44277033

M. Guinet · A.W. Mantz · D. Mondelain
Laboratoire de Physique Moléculaire pour l'Atmosphère et l'Astrophysique, UPMC Universite Paris 06, UMR 7092, 75005, Paris, France

A.W. Mantz
Department of Physics, Astronomy and Geophysics, Connecticut College, 270 Mohegan avenue, New London, CT 06320, USA

Fig. 1 A photograph of the alignment and spacing fixture shown in an arbitrary setting

Fig. 2 HeNe images on the entrance/exit mirror of a Herriott cell with wedged windows and set for approximately 50 images on the exit/entrance mirror. These are not the conditions used to construct the cell, and this set of conditions was used because the picture quality was clearer for this configuration

2 Alignment and testing of the copper Herriott cell

This new Herriott cell is basically as described in [4] with a few improvements resulting from an improved alignment and testing procedure.

In order to more thoroughly test the relationship between mirror spacing and the number of traversals for a given entrance angle, we designed and fabricated the test fixture shown in Fig. 1.

With reference to Fig. 1, the hardware consists of a fixed flange at the right of the fixture. The entrance/exit mirror with the coupling hole is mounted here, as are the three spacer rods that permit precision adjustment of the mirror spacing for the two mirrors that constitute the cavity. The vernier used to measure separation of the mirror support flange has a precision of 0.0025 centimeters. The second Herriott mirror is mounted on a sliding fixture, shown in Fig. 1 toward the left side of the photograph. The three spacer rods mentioned earlier support this sliding fixture. The left most flange supports the three spacer rods. This flange also provides a surface on which to mount one end of the fixture in order to adjust the beam input angle relative to the cell optical axis. A HeNe laser can be seen to the right of the fixture. This HeNe laser beam is used to measure beam angles relative to the cell optical axis, and HeNe images on either mirror are used to determine the cell path length, once the mirror spacing has been determined. We also adjust the spacing to reduce the likelihood of loosing signal resulting from image motion as the cell is cooled.

Introducing two wedged windows significantly alters the parameters that characterize an optimum mirror spacing and entrance beam angle. For this reason we find it imperative to use the wedged cell entrance and exit window as well as the wedged window that will be used on the outer vacuum box while performing the testing described above. Thus, the aluminum plate located to the right of the entrance flange in Fig. 1 replaces the end flange of the outer vacuum box, and the cell window (cannot be seen in this photograph).

Figure 2 shows images one may observe under a different set of conditions on one of the mirrors using the HeNe laser. The circular pattern of images is the set of images we would count in determining the absorption path. The elliptical set of images result from internal reflections in the wedged window, and the beam associated with this pattern leaves the cell at a different angle from the beam we are interested in measuring when properly adjusted. This pattern does not interfere with the beam associated with the circular pattern of images, the images that contribute to the absorption path length, when properly aligned. The entrance/exit coupling hole is located at the bottom of the mirror.

After the beam input parameters and cell tilt angle have been determined, we construct a test cell. This test cell is identical in all ways to the final Herriott cell, except for the fact that we remove a portion of the copper spacer, so we may make visible diagnostic tests to assure that the cell will perform as expected. The separation of the mirror support flanges inside this cell, as well as inside the final cell, is determined to a precision of 0.0025 centimeters. The test cell is seen in Fig. 3. The field mirror is shown to the left of the photo, and the back side of the entrance/exit mirror is shown to the right with a CaF_2 window installed in the test cell. The vacuum box for the system is behind the cell and to the left. The entrance window aperture is located on the upper surface of the box as it is oriented in the photograph. Refer to Fig. 3 in [4] to observe a completed cell.

3 Tests to determine the cell path length

The most complete theoretical description of the Herriott design may be found in [5, 6] where one finds the expression for the absorption path length, accurate through 4th order for the volume between the two spherical mirrors, given in (1),

$$L = Nd + \frac{Nr_0^4}{8f^2d} \tag{1}$$

where N is the total number of traversals, 82 (twice the number of images on a mirror), d is the distance between the two spherical mirror surfaces (here 15.1050 ± 0.0025 cm), r_0 is the radius of the circle (here 1.6 ± 0.1 cm) described by the images on each mirror (see Fig. 2 in [4]), and f is the focal length of the mirrors, in case of two identical mirrors, 7.5 ± 0.0025 centimeters.

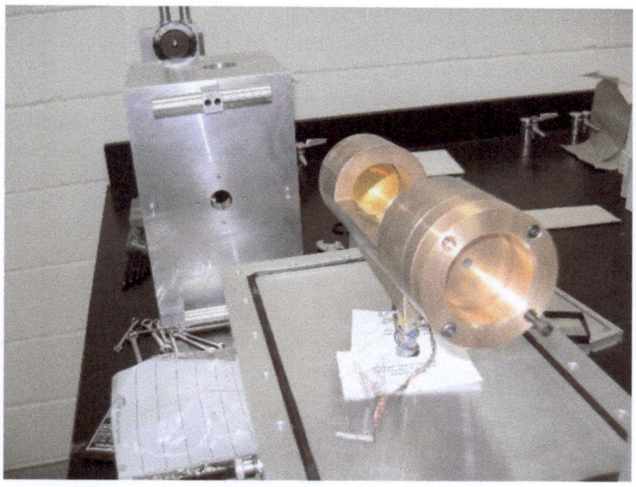

Fig. 3 A view of the copper cutaway cell from the entrance and exit window. Images on the field mirror are clearly visible for this test configuration, and a portion of the copper spacer is clearly removed. The tilt angle for the test cell is identical to the tilt angle for the final cell, because the same machined copper wedge is used on the refrigerator cold head for both assemblies. Also, the dimensions of the test cell and the final Herriott cell are identical to approximately 0.0001 inches as indicated on the digital position readout on the mill and lathe used to fabricate both cells

As mentioned above, we use the test fixture and test cell to determine d, r_0 and N described above. Once we have counted the images and determined the other necessary parameters, we use formula (1) to calculate absorption path length. The absorption path length inside the spherical cavity calculated using (1) is 12.465 ± 0.002 meters (total path is 12.485 ± 0.002 meters as explained earlier). The uncertainty is calculated using standard propagation of error methods. Experimentally, we verified the absorption path using a tunable diode laser. The laser we had available operated at 1132.800420 cm^{-1}. We measured the 02^00–00^00, P (44^e) rovibrational transition in $^{14}N_2{}^{16}O$. This high J transition precluded us from using a lower temperature than 150 K to test the cell, because the lower level population for this transition decreases rapidly at significantly lower temperatures. Additionally, the estimated error for this high J intensity is approximately 5%. From what we observed with our previous Herriott cell the distance between the two mirrors and thus the absorption path length are kept unchanged when decreasing the temperature even at low temperature.

The ratioed spectrum we recorded with the Herriott cell, stabilized at 149.6 K and coupled to the LPMAA diode laser spectrometer [7], is shown in Fig. 4 as square dots with experimental noise. Also shown in this figure is the calculated spectrum; the calculated spectrum is plotted with a solid line. To calculate the spectrum we used HITRAN 2008 data [8]. The absorption strength for this line is 4.758×10^{-22} $cm^{-1}/(\text{molecule}\,cm^{-2})$ with an accuracy of 5% for this weak line. The gas pressure is 1.829 Torr, or 2.438 hPa. The laser emission half-width (Voigt profile)

Fig. 4 A spectrum recorded at approximately 150 K for the P(44^e) rotational transition in the 02^00–00^00 band of $^{14}N_2{}^{16}O$ at 1132.8 cm^{-1}. The calculate spectrum as well as the residuals are also plotted in the figure as indicated

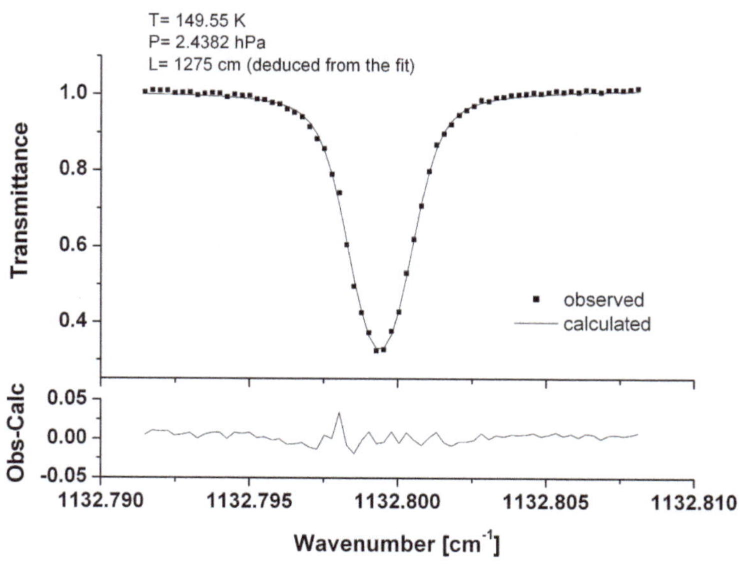

is approximately 6×10^{-4} cm^{-1}. Using these parameters, and assuming a Voigt absorption line shape, the absorption path length was calculated and adjusted until residuals were minimized. The best-fit path length deduced from the calculation was determined to be approximately 12.75 ± 0.64 meters (the uncertainty is mainly due to the uncertainty on the HITRAN intensity of 5%). The residuals (obs-calc) are plotted below the experimental and calculated spectra. This spectrum is noisier than we normally expect because this transition was also located at the cutoff edge of the CaF_2 windows used in the cell. In fact the windows and the cell transmitted only about 3% of the incident energy.

4 Conclusion

Using a set of gold-coated copper mirrors identical to those used in [4], and a more sophisticated test fixture designed and tested at Connecticut College, we were able to more than double the working absorption path length for an all copper Herriott cell designed to work at temperatures between 300 K and as low as 20 K. The path length at 150 K, based upon minimizing residuals was 12.75 ± 0.64 meters, based on HITRAN data, and the actual calculated length for 41 images was 12.485 ± 0.002 meters including the absorption path between the Herriott cell window and the entrance mirror reflecting surface.

Acknowledgements AWM gratefully acknowledges support for this work by NASA Grants NNX 08A078G and NNX 09AJ93G.

References

1. D.R. Herriott, H. Kogelnik, R. Kompfner, Appl. Opt. **3**, 523 (1964)
2. J.U. White, J. Opt. Soc. Am. **32**, 285 (1942)
3. S.M. Chernin, E.G. Barskaya, Appl. Opt. **30**, 51 (1991)
4. D. Mondelain, C. Camy-Peyret, A.W. Mantz, E. Tang, A. Valentin, J. Mol. Spectrosc. **241**, 18 (2007)
5. G. Moreau, C. Robert, J. Opt. (Paris) **16**, 177 (1985)
6. J. Altmann, R. Baumgart, C. Weitkamp, Appl. Opt. **20**, 995 (1981)
7. A. Valentin, Spectrochim. Acta A **52**, 823 (1996)
8. L.S. Rothman, I.E. Gordon, A. Barbe, D.C. Benner, P.F. Bernath, M. Birk, V. Boudon, L.R. Brown, A. Campargue, J.-P. Champion, K. Chance, L.H. Coudert, V. Dana, V.M. Devi, S. Fally, J.-M. Flaud, R.R. Gamache, A. Goldman, D. Jacquemart, I. Kleiner, N. Lacome, W.J. Lafferty, J.-Y. Mandin, S.T. Massie, S.N. Mikhailenko, C.E. Miller, N. Moazzen-Ahmadi, O.V. Naumenko, A.V. Nikitin, J. Orphal, V.I. Perevalov, A. Perrin, A. Predoi-Cross, C.P. Rinsland, M. Rotger, M. Šimečková, M.A.H. Smith, K. Sung, S.A. Tashkun, J. Tennyson, R.A. Toth, A.C. Vandaele, J. Vander Auwera, J. Quant. Spectrosc. Radiat. Transf. **110**, 533 (2009)

Miniaturized resonant photoacoustic cell of inclined geometry for trace-gas detection

A.V. Gorelik · A.L. Ulasevich · F.N. Nikonovich · M.P. Zakharich · V.A. Firago · N.S. Kazak · V.S. Starovoitov

Received: 15 September 2009 / Revised version: 8 December 2009 / Published online: 14 January 2010
© Springer-Verlag 2010

Abstract A photoacoustic cell intended for laser detection of trace gases is represented. The cell is adapted so as to enhance the gas-detection performance and, simultaneously, to reduce the cell size. The cell design provides an efficient cancellation of the window background (a parasite response due to absorption of laser beam in the cell windows) and acoustic isolation from the environment for an acoustic resonance of the cell. The useful photoacoustic response from a detected gas, window background and noise are analyzed in demonstration experiments as functions of the modulation frequency for a prototype photoacoustic cell with the internal volume $\sim$0.6 cm^3. The minimal detectable absorption for the prototype is estimated to be $\sim 1.2 \times 10^{-8}$ cm^{-1} W Hz$^{-1/2}$.

1 Introduction

Laser-based photoacoustic spectroscopy is one of the most sensitive techniques applied to local non-contact analysis of trace amounts of various chemical compounds in gas media [1–3]. The most appropriate light sources for this technique are lasers operated in the infrared wavelength region. They are high-power carbon dioxide (the oscillation wavelength region 9–11 μm) and carbon monoxide (2.8–4.1 or 5.1–8.0 μm) CW lasers [4, 5] , CW optical parametric oscillators (3–5 μm) [6], quantum cascade lasers (4–5 μm) [7, 8] and miniature laser diodes (0.7–2 μm) [9]. The photoacoustic technique is distinguished by high sensitivity at the (sub) ppb (parts per billion, 1:10^9) level with the minimal detectable absorption of 10^{-10} cm^{-1}, a short time resolution (down to a few seconds) and the capability to recognize reliably a large number of chemical compounds [2, 3]. The infrared-laser-based photoacoustic technique is generally recognized to be a reasonable approach to the inexpensive *in situ* multi-component analysis of atmospheric air in environmental pollution monitoring [2, 10, 11], exhaust gas monitoring [2, 12, 13], industrial process control [14, 15] and leak detection [16]. The technique has found successful exploitation in biology-related areas (plant physiology [3], crop and fruit storage [17], entomology [18] and medical diagnostics by means of exhalation analysis [19, 20]) for sensitive detection of gases being involved in the metabolic reactions and emitted by living systems.

The miniaturized spectroscopic hardware of trace-gas analysis is a promising line of development for the photoacoustic technique. The principle of the technique is based on measuring the amplitude and phase for the acoustic pressure oscillation (the so-called photoacoustic response) arising due to absorption of a modulated laser beam by molecules of gas inside a photoacoustic cell. The photoacoustic detection is realized with an enhanced sensitivity if the modulation frequency coincides with an acoustic resonance of the cell. Therefore, the modulation frequency should be correlated with the cell-resonator sizes: the frequency increases with reducing the sizes. The implementation of laser-beam modulation at ultrasonic frequencies allows reducing the cell size down to a few millimeters. Combining the high sensitivity inherent in the photoacoustic approach with an ability to probe the gas inside such a small-sized photoa-

A.V. Gorelik · A.L. Ulasevich · F.N. Nikonovich ·
M.P. Zakharich · N.S. Kazak · V.S. Starovoitov
B.I. Stepanov Institute of Physics, NASB, Nezavisimosti ave. 68, 220072 Minsk, Belarus

V.A. Firago · V.S. Starovoitov (✉)
Belarus State University, Nezavisimosti ave. 4, 220030 Minsk, Belarus
e-mail: vladstar@dragon.bas-net.by
Fax: +375-172-94-9002

coustic cell-resonator gives a possibility to analyze chemical compounds to be emitted by individual small-sized objects with an extremely low emission rate. A crude estimation shows that the application of the approach can provide detecting the gas leak emitted by an object with the rate down to $\sim 10^{-14}$ cm^3/s for C_2H_4 or $\sim 10^{-10}$ cm^3/s for CO_2. For comparison, in the aerobic reaction an individual cell of living organism emits CO_2 at a rate from 10^{-10} to 10^{-9} cm^3/s [21]. In the photosynthesis reaction, an individual cell of plant can absorb CO_2 at a rate higher than 10^{-8} cm^3/s [22]. The gas-leak-detection sensitivity will be increased 10^3–10^6 times compared to the non-miniaturized photoacoustic detectors, and more than 100 times in relation to the commercial mass-spectrometer-based leak-detection systems. In contrast to these systems, the ultrasonic photoacoustic leak detector needs no expensive vacuum chambers and can be applied to in situ localization of leak for a large number of substances to be emitted in atmospheric air. According to the theoretical estimations (see, for instance, [23–25]), the amplitude of photoacoustic response can be increased with reducing the cell sizes.

The cell performance is specified usually in terms of the signal-to-noise ratio. In order to obtain a high magnitude for the ratio, the cell design must satisfy some requirements. First of all, the design has to provide detection of the highest possible useful signal (that is, a photoacoustic response from the gas inside the cell) at a selected acoustic resonance. The cell must be reliably isolated from external acoustic noise and parasite electric pickups. And, finally, the background effect of the cell windows on the measurements should be reduced to a minimum. The background effect requires special attention for the miniaturized cells. This effect is associated usually with the absorption of laser beam in the cell windows resulting in local heating of the gas inside the cell. The measurement inaccuracy arisen from this effect is proportional both to the laser power instability and to the power absorbed by the windows. If a non-miniaturized photoacoustic cell is used, the effect can be critical for high-power laser applications (for instance, at an intracavity arrangement of the cell to a carbon dioxide or carbon monoxide laser operated with the intracavity power of a few tens of watts). The background effect is not very significant for non-miniaturized low-power laser systems (such as laser diodes or quantum cascade lasers operated at a power up to a few tens of milliwatts). But our experience shows that the size reduction for the photoacoustic cell can result in a dramatic increase in the background signal. This increase can be observed for a low-power laser beam. For instance, the window background signal is observed in our experiments for a small-sized photoacoustic cell at a laser beam power down to 10 mW. This observation does not contradict the theoretical conclusions [23–25].

The quartz-enhanced photoacoustic spectroscopy (QEPAS) is a well-developed approach to miniaturize the spectroscopic hardware with the help of the increased-frequency modulation of laser beam [26]. Instead of a gas-filled resonant acoustic cavity, the sound energy is accumulated in a high-Q quartz crystal frequency standard. Usually, the standard is a quartz tuning fork with an acoustic resonant frequency of 32 kHz in air. To this moment, great success is achieved in QEPAS-based gas detection with the help of compact cells (the internal cell volume reaches down to a few cubic centimeters) and different near- or mid-infrared laser systems (including diode and quantum cascade lasers, optical parametric oscillators) [27–29]. The experiments have proved that the high-sensitivity photoacoustic detection can be realized at ultrasonic modulation frequencies if the rates of intra- and intermolecular collisional vibration–vibration (VV) or vibration–translation (VT) energy redistribution for the detected species are high compared to the modulation frequency.

The goal of the work is to reveal the application potential for another possible approach to the miniaturized resonant photoacoustic cell. In contrast to the QEPAS technique, we support a traditional approach to photoacoustic detection: through a small hole in the cell shell, a condenser microphone registers the photoacoustic response to the laser beam modulated with the frequency of an acoustic resonance of the cells. According to the approach, an efficient way to miniaturize the resonant photoacoustic cell can be implemented with the help of a properly adapted cell design. We accept that the miniaturized photoacoustic cell must not be worse in performance compared to non-miniaturized one. Here we represent a simplified cell design and describe some helpful principles which allow optimizing the design for the best gas-detection performance. We demonstrate also how such an approach can be applied in practice. An experimental examination of a compact prototype cell is performed. In the examination, we estimate the measurement error associated with the noise in the absence of the photoacoustic laser beam and test the acoustic isolation of the cell from the environment. Then we analyze the photoacoustic signals (they are the useful photoacoustic signal and window background) to be initiated by a laser beam inside the cell. The performance of the cell is estimated in terms of the minimal detectable absorption.

2 Cell design

Here we describe a resonant photoacoustic cell of simple design. The cell is intended for detection of gases with the help of a collimated and linearly polarized laser beam. The cell consists of a shell and flat optical windows. Geometrically, the shell is an oblique section of a cylindrical tube. The face and end planes of the tube section are cut to the tube axis

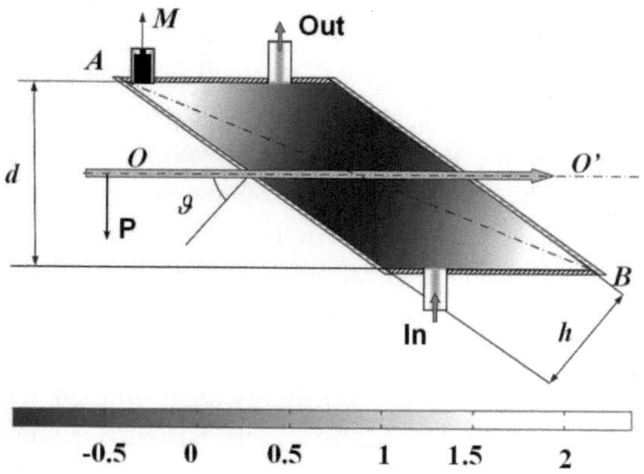

Fig. 1 The cell design: OO′, optical axis; **P**, electric polarization vector for the laser beam; M, microphone; In, gas inlet port; Out, gas outlet port. The wave pattern (grey-scale filling) inside the cell cavity demonstrates a spatial amplitude distribution for the acoustic standing wave of mode ν_2. The distribution corresponds to an optimal ratio between the diameter d and thickness h for a narrow collimated laser beam (the laser-beam diameter is equal to $d/16$). The anti-nodes of the distribution, which correspond to the positive or negative pressure oscillation, are depicted by *white* or *black* color of the filling

at an angle ϑ. The tube axis is accepted to be the optical axis OO′ for the laser beam inside the cell. The windows are attached to the face and end planes. The cell design is represented in Fig. 1 as a section of the cell by the plane formed by the optical axis OO′ and the electric polarization vector **P** for the laser beam. The internal cavity of the cell is specified in terms of the internal diameter d and thickness h of the tube section. In order to cancel the parasite reflection of the laser beam from the cell windows, the angle ϑ is accepted to be equal to the Brewster angle. Due to the inclined geometry, the acoustic eigenmodes for the cell cavity cannot be classified as longitudinal or transverse modes (for instance, as radial or azimuthal ones for the cylindrical resonator [23]). We identify the cell eigenmodes ν_n by the number of nodes n given by intersection between the relevant standing wave and the longest cell diagonal (the diagonal AB in the Fig. 1). The mode ν_1 (the mode with the lowest non-zero eigenfrequency f_1) is specified by one node on the diagonal. The mode ν_2 (the eigenfrequency f_2) gives two nodes on the diagonal. This mode is specified by 3 antinodes located at acute corners of the cell cavity (near the ends A and B of the line AB) and near the cell midpoint (the point given by intersection of the axis OO′ with the line AB).

The cell design is specified by the geometric shape for the cell cavity (the shape is described by the diameter d and thickness h), the microphone location and arrangement of inlet/outlet gas holes. These design parameters are chosen in such a way as to enhance the cell performance for an individual acoustic eigenmode. The parameters are chosen with the help of a numerical simulation. In the simulation, the spatial amplitude distribution for an acoustic standing wave, which corresponds to the selected mode, is determined. From this distribution, we find the useful signal (a photoacoustic response from the gas inside the cell) and window background (a signal caused by absorption of the laser beam in the cell windows). We estimate also the effect of external acoustic noise on the measurements. In general, the simulated signals depend on the quantities, which specify both the cell design and the laser beam. For simplicity, we accept fixed values for the laser-beam characteristics and analyze the signals as functions of the geometry parameters. The choice of an optimal ratio between the diameter d and thickness h leads to the mutual elimination of photoacoustic signals generated from the front and end windows and, as a result, to the total cancellation of window background. A proper arrangement of the inlet and outlet gas holes is intended to reduce the negative influence of the holes on the acoustic Q-factor and, simultaneously, to isolate the measurement from external acoustic noise for the selected mode. The analysis of the simulated data shows that the most efficient parameter optimization can be realized for the acoustic mode ν_2.

The cell design is optimized for the mode ν_2. In the optimization, we follow a simple procedure. First of all, we apply the numerical simulation in order to determine an optimal ratio between the diameter d and thickness h. Then, the position of the inlet and outlet gas holes is chosen near the nodes of the mode ν_2, the spatial amplitude distribution of which corresponds to the optimal values of d and h.

According to the numerical simulation, the maximal amplitude for the standing wave of mode ν_2 is realized at antinodes located near the acute corners of the cell cavity (see Fig. 1). The simulated wave amplitude in the corners is nearly twice as high as the absolute amplitude value in the cell midpoint. Therefore, the microphone is arranged near an acute corner of the cell. Such a position for the microphone allows also analyzing the acoustic properties for others eigenmodes.

We have applied the represented procedure to make a miniaturized version for the cell. The cell design is adapted to a collimated laser beam. The beam diameter and wavelength are accepted to be 0.5 mm and 9.2 μm, respectively. The cell windows are made of polycrystalline ZnSe. The angle ϑ ($\vartheta \approx 67.7°$) is the Brewster angle of ZnSe at $\lambda = 9.2$ μm. The diameter d and thickness h are found in the optimization procedure for the mode ν_2 and equal to 8 and 4.7 mm, respectively. The internal volume of the cell is 0.6 cm^3. The length of optical path inside the cell is ~12 mm. A miniature condenser Knowles FG-3629 microphone (sensitivity 22 mV/Pa at 10 kHz, noise ~54 $nV/Hz^{1/2}$ at 14 kHz) is applied to detect the acoustic signals. The microphone is connected to the internal cell cavity by a duct (with diameter 0.7 mm and height 1 mm). The spatial gap between the duct center and the nearest acute corner of the cell is 0.5 mm. The inlet and outlet holes are located on the plane formed by the optical axis OO′ and the electric polarization vector **P** of the laser beam. The distance from the hole to the nearest acute corner of the cell is 11 mm. The diameter for the inlet and outlet holes in the cell wall is 0.3 mm. Our numerical simulation predicts that in the frequency range from 0 to 20 kHz we can observe three acoustic resonances (near $f_1 = 8.1$ kHz, $f_2 = 14.2$ kHz and $f_3 = 19.2$ kHz) associated with the eigenmodes ν_1, ν_2 and ν_3. This prototype cell is applied in experiments where we analyze the acoustic properties of the cell and estimate the cell performance.

3 Noise analysis

We estimate the measurement error associated with the acoustic or electric noise registered by the microphone inside the prototype photoacoustic cell in the absence of laser beam. We assume that the cell cavity can be coupled acoustically with the environment through inlet and outlet gas ducts. Therefore, in the estimation we take into account a possible manifestation of acoustic disturbances produced outside the cell by a source.

The cell is inserted into a capsule, which provides a reliable sound isolation from the noise produced in our laboratory room (see Fig. 2). The capsule design allows the connection of the cell cavity with the environment through only the inlet and outlet flexible gas ducts of the inner diameter >1 mm. The compressed air blown through a nozzle is used as a source of external acoustic disturbances. These disturbances imitate the 'white' noise for the considered frequency range. The nozzle is placed near the suction hole of inlet gas duct. The conditioned laboratory air is applied as a gas medium filling the cell cavity.

In the experiment, we estimate the measurement error as a function of frequency f. The error is evaluated in terms of a measure of inaccuracy σ_f for the Fourier transform spectrum S_f of detected acoustic signals. The value of error σ_f is determined as a standard root-mean-square deviation for the Fourier transform S_f of acoustic signal:

$$\sigma_f = \left(\langle S_f S_f^* \rangle - \langle S_f \rangle \langle S_f^* \rangle\right)^{1/2}.$$

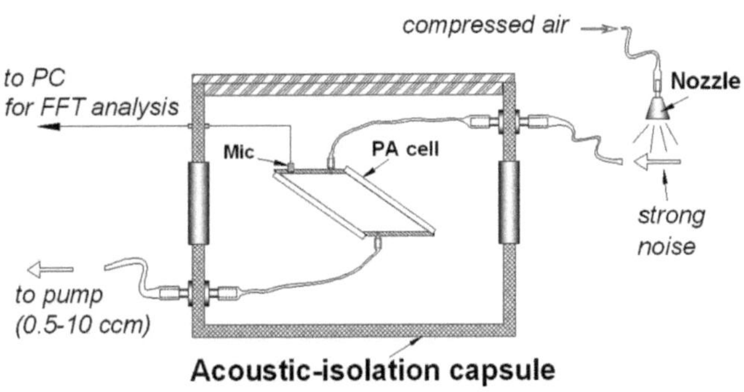

Fig. 2 Experimental arrangement for estimation of the measurement error associated with noise (Mic, microphone)

The quantity S_f is calculated with the help of the fast-Fourier-transform procedure performed for individual time-sample realizations of acoustic signal (each realization represents the microphone signal over a given time interval $\tau_{avr} \approx 0.13$ s). The symbol $\langle \ldots \rangle$ means the averaging over the ensemble of the realizations. In the experiment, the number of the signal-sample realizations is not less than 1000.

The estimation of the measurement error is performed for acoustic signals detected at different strengths of the external-noise effect. The negligibly small influence of the external noise on the measurements is realized when the source of external disturbances is switched off. The inlet and outlet gas ducts are closed and the cell cavity is acoustically isolated from the environment. A regime of strict silence is kept in the laboratory room. The strong effect of external noise on the measurements is produced if the source of external disturbances operates. The inlet and outlet gas ducts are opened and allow the gas flow through the cell at a flow rate ($\sim$30 cm^3/min), which provides a perfect gas renewal inside the cell for the time $\sim$1 s.

We analyze the deviation σ_f as a function of frequency f for the frequency range from 0 to 22 kHz. Fig. 3 demonstrates two typical frequency-dependences for σ_f obtained at different levels of external acoustic noise. When the external noise is negligibly small, the obtained dependence for σ_f corresponds to the minimal error attainable by the acoustic signal. The deviation σ_f is a slowly varying function of f. Slight resonant dependences are observed when the frequency is close to eigenfrequencies of the acoustic modes ν_1, ν_2 and ν_3. The magnitude of σ_f is in the interval of values from 10^{-7} to 4×10^{-7} V. We associate such a behavior of σ_f with the electric noise of the microphone and fluctuations of gas pressure inside the cell.

In the presence of a strong external noise, the frequency dependence of σ_f is essentially transformed. Frequency regions, which correspond to acoustic resonances of the cell, are strongly susceptible to the external disturbances. For these regions, the effect of external disturbances can result in a considerable increase for the measurement error. The disturbances lead to a substantial increase in σ_f for low-lying frequencies (the frequencies located near zero eigenfrequency [23] within the interval from 0 to 2 kHz), which correspond to the application region for the so-called non-resonant photoacoustic detection. The observed oscillation in ω-dependence of σ_f at low frequencies is associated with manifestation of acoustic resonant properties for the flexible inlet and outlet gas ducts. A considerable growth of σ_f with the disturbances is exhibited also for the frequencies located near the resonances with the modes ν_1 and ν_3. But, no increase of σ_f is observed for the frequencies resonant to the mode ν_2. Our experiment shows that for the frequencies $f \approx f_2$ the standard deviation σ_f is not affected by the external disturbances. Regardless of the strengths of applied

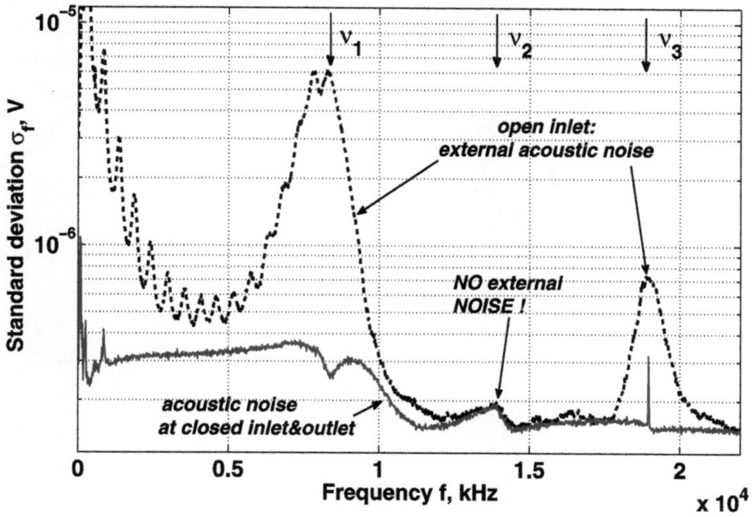

Fig. 3 Frequency dependence of the standard deviation σ_f for the Fourier transform of acoustic signals detected in the absence of laser beam at two different levels of external acoustic disturbances. The *solid dark grey line* gives the deviation σ_f obtained when the effect of external acoustic noise is negligibly small (the inlet and outlet gas ducts are closed). The *dashed black line* gives the deviation σ_f for the measurement at strong external acoustic disturbances (the inlet and outlet gas ducts are opened). The *arrows* point to the location of resonant frequencies for the acoustic modes ν_1, ν_2 and ν_3

disturbances, the quantity σ_f does not exceed 2×10^{-7} V. We associate this mode immunity to the disturbances with a proper arrangement of inlet and outlet gas holes drilled in the cell wall. Notice that such an acoustic isolation for an individual acoustic mode can be implemented when the external disturbances are applied to the cell through only the inlet and outlet gas ducts.

4 Analysis of useful and background signals

In order to study the laser-initiated photoacoustic signals for the prototype photoacoustic cell, we perform an experiment where a collimated beam of mid-infrared laser is applied. The laser beam is produced by a CW stabilized single-mode carbon dioxide laser tunable over the oscillation wavelengths in the range from 9 to 11 μm. The reliable selection of oscillation line and stable laser oscillation are provided with the help of a diffraction grating (150 grooves/mm) operated in the non-auto-collimating oscillation regime (the Littman scheme). The laser beam is modulated by a Germanium acousto-optic modulator (the efficiency of beam deflection is 80%). The beam deflected by the modulator is directed to a power meter. The non-deflected beam is directed from the modulator through a collimating two-lens system (the factor of beam reduction for the system is 4, the transmission factor is 76% at $\lambda \sim 9.2$ μm) to the photoacoustic cell. In the measurements, we accept the power of the deflected beam as an actual power of the probing beam. The aperture of the modulated laser beam near the photoacoustic cell is estimated to be 0.5 mm. The deflection of the laser beam by the acousto-optic modulator is switched with the help of a TTL-like reference signal supplied by a signal generator built in a digital oscilloscope Handyscope HS3. This digital oscilloscope is used also in order to register, store and transmit the electric signals (the signals supplied by the cell microphone and the reference signal) to the computer. The Fourier transform of the microphone signal obtained for a modulation frequency is accepted as the photoacoustic response to be detected. All measurements in the experiment are made for flows of ultra-high purity nitrogen (99.9995%), which contains ammonia at concentrations from 10 to 15 ppm. The flows are produced with the help of a nitrogen generator and a permeation-principle-based device. The rate of gas flow ($\sim$30 cm^3/min) to be blown through the cell is maintained automatically with the help of a flow controller in such a way to provide the perfect gas renewal inside the cell for the time $\sim$1 s. The pressure and temperature of the gas in the cell is accepted to be close to typical parameters for the laboratory room (740 torr and 22°C). Notice that, in the experiments, no capsule is applied for acoustic isolation of the cell. A more detailed description of the used experimental set-up is given in [30, 31].

In the experiment, we study the useful and background signals obtained for the cell from measurements performed on two individual laser-oscillation lines. They are lines distinguished by the strong or weak absorption ('*on-*' or '*off-line*', respectively) for ammonia. The oscillation line 9R(30) of carbon dioxide laser (the oscillation wavelength ≈ 9.22 μm) is applied for '*on-line*' measurements. According to the Hitran database (http://www.cfa.harvard.edu/hitran/), the absorption α_{on} on this line for ammonia is

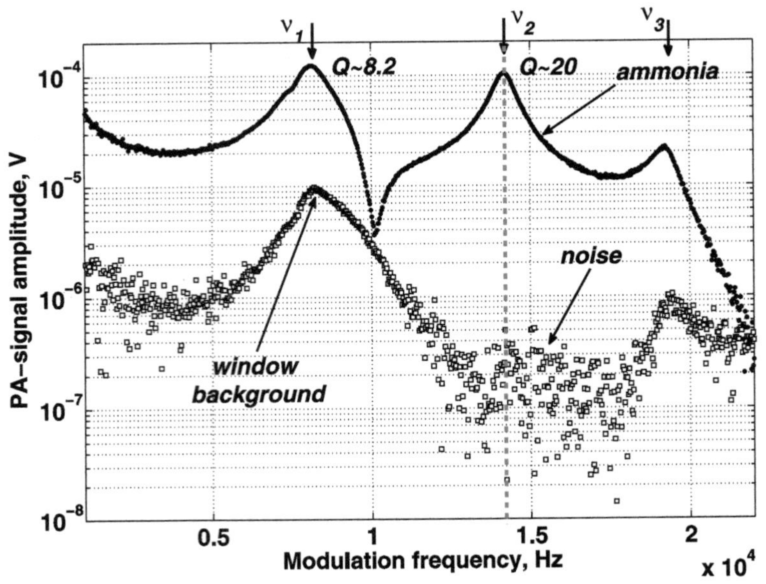

Fig. 4 Dependence of the amplitude of laser-initiated photoacoustic signals on the modulation frequency for a CO_2−laser beam. The measurements are made for a nitrogen flow admixed by ammonia (13.6 ppm) on two laser-oscillation lines distinguished by the strong (*'off-line'*, *black full circles*) or weak (*'off-line'*, *open squares*) absorption for ammonia. The *'on-line'* power (oscillation line 9R(30)) is 66 mW. The *'off-line'* power (line 9R(24)) is 130 mW. The time of averaging for every point is 0.13 s. The *arrows* point to the location of resonant frequencies for the acoustic modes ν_1, ν_2 and ν_3

73.9 $cm^{-1}\,atm^{-1}$. Such a strong absorption implies that the photoacoustic signal detected on the line can be accepted reliably as a useful response. The laser-oscillation line 9R(24) (9.25 μm) is selected as an *'off-line'*. According to the Hitran database, the absorption on this line for ammonia (0.068 $cm^{-1}\,atm^{-1}$) is negligible small in order to have any influence on the detected signal for the applied ammonia concentration. In addition, the oscillation line 9R(24) is located near to the *'on-line'* and specified by a stable single-mode operation. In the absence of ammonia in the gas flow, the photoacoustic *'off-'* and *'on-line'* signals are identical.

We analyze the *'on-'* and *'off-line'* signal amplitude and phase as functions of the modulation frequency f_m. The signals are detected in the presence of ammonia in the gas flow for the frequency range from 1 to 20 kHz. Figure 4 shows obtained frequency dependences for the amplitude in volts. The *'on-'* and *'off-line'* signals exhibit the different behavior on the frequency scale. Admixing ammonia to the flow at *'on-line'* measurements leads to an increase in the signal amplitude. The *'on-line'* signal clearly demonstrates resonant peaks when the modulation frequency is close to f_1, f_2 or f_3 predicted by the numerical simulation. The analysis of the obtained data shows that the Q-factor $Q_2 \approx 20$ ($Q_2 = f_2/\Delta f_2$, where the width Δf_2 is measured between the points where the amplitude is a $1/2^{1/2}$ value of the peak amplitude) for the resonant peak in the vicinity of f_2 is considerably higher compared to the relevant parameter $Q_1 \approx 8.2$ for the acoustic resonance near f_1. We explain such an improved performance for the acoustic mode ν_2 by a proper location of the inlet and outlet gas holes in the cell wall.

Whatever the ammonia concentration in the gas flow, the frequency dependence of the *'off-line'* signal amplitude remains invariant. Contrary to the *'on-line'* dependence, the *'off-line'* signal manifests only two resonant peaks. They are acoustic resonances near the frequencies f_1 and f_3. We associate the manifestation of these peaks with a background signal due to absorption of the laser beam in the cell windows. No resonance peak in the vicinity of f_2 is observed. Moreover, at $f_m \approx f_2$ the *'off-line'* signal is strongly noised and the signal amplitude attains a minimum, which is comparable with the standard deviation σ_f (the root-mean-square error) obtained for the detected acoustic signals in the absence of laser beam. The detailed analysis of the *'off-line'* signal measured at a modulation frequency $f_m \approx f_2$ shows that the signal phase is distributed in a quasi-uniform random manner over the range from 0 to 2π. It implies that at $f_m \approx f_2$ the magnitude of the window background is negligible small in comparison to the noise. We should note also a manifestation of the strong noise in measurements for the *'off-line'* amplitude at $f_m < 5$ kHz. This manifestation is explained by high susceptibility of the acoustically non-isolated cell to the external acoustic low-frequency disturbances, which are typical for the laboratory room.

Our study demonstrates that the optimization of the parameters, which specify the design of internal cavity for the photoacoustic cells of inclined geometry, leads both to an improvement of resonant properties and to a minimization of the window background for the acoustic mode ν_2. The minimal error of measurements is attained at modulation frequencies $f_m \approx f_2$ when the measurement error is associated with acoustic and electric noises of microphone.

5 Estimation of cell performance

Obviously, the gas-detection sensitivity for the cell is maximal when the modulation frequency f_m is close to f_2.

Therefore, we estimate the cell performance at this frequency. The performance is specified in terms of the normalized noise equivalent absorption (NNEA). We accept this quantity as the minimal detectable absorption α_{min}, which can be obtained if the signal-to-noise ratio is equal to 1 provided that the laser power is 1 W and the averaging time is 1 s. The coefficient α_{min} is found from

$$\alpha_{\mathrm{min}} = \alpha_{\mathrm{on}} C_{\mathrm{NH3}} \tau_{\mathrm{avr}}^{1/2} P_{\mathrm{on}} N_{\mathrm{off}} / S_{\mathrm{on}}.$$

Here the quantities C_{NH3} and τ_{avr} denote actual values for the ammonia concentration and averaging time. The quantity S_{on} gives the amplitude for the '*on-line*' photoacoustic signal at $f_m = f_2$. The power P_{on} is an '*on-line*' power for the modulated laser beam before the front cell window. The quantity N_{off} corresponds to the root-mean-square error for the acoustic signal obtained from the '*off-line*' measurements. According to our measurements, the signal-to-noise ratio $S_{\mathrm{on}}/N_{\mathrm{off}}$ is 1095 ($C_{\mathrm{NH3}} = 12.36$ ppm, $\tau_{\mathrm{avr}} = 0.13$ s, $P_{\mathrm{on}} = 49$ mW). The minimal detectable absorption α_{min} is estimated to be $\sim 1.2 \times 10^{-8}\ \mathrm{cm}^{-1}\ \mathrm{W\,Hz}^{-1/2}$.

6 Conclusion

Thus, we have presented a simple design of the photoacoustic cell. Principles which can help to optimize the design for the best gas-detection performance at an individual acoustic resonance of the cell and (simultaneously) to reduce the cell sizes were shortly described. These principles were applied in order to make a miniature prototype cell adapted to ammonia detection with the help of a CO_2-laser beam modulated at a frequency resonant with respect to the acoustic mode ν_2. The experiments demonstrated that the cell design provided an enhanced gas-detection performance for the acoustic mode ν_2 owing to the efficient cancellation of the window background and reliable acoustic isolation of the cell from the environment.

The represented design is specified by a few parameters which can be optimized in order, for instance, to adapt the cell to experiment needs and to enhance the cell performance. Certainly, the design simplicity makes the parameter optimization and cell manufacturing easy and extends the area of cell application in practice. The cell sizes can be reduced, hereafter, down to a few millimeters by a trivial scaling procedure. The obtained results demonstrate a great potential for the miniaturized photoacoustic cells of inclined geometry in creating compact ('pocket-sized') high-sensitivity laser sensors of chemical compounds. The miniature background-free and 'acoustically isolated' photoacoustic cell with Brewster windows placed inside the cavity of a high-power laser can find an application for in-situ detection of the smallest gas leaks.

Acknowledgements The authors acknowledge V.M. Stetsik and A.I. Chebotar for the help in performing the experiments. This work was performed in the framework of B-1252 project supported by International Science and Technology Center.

References

1. V.P. Zharov, V.S. Letokhov, *Laser Opto Acoustic Spectroscopy* (Springer, Berlin, 1986)
2. M.W. Sigrist (ed.), *Air Monitoring by Spectroscopic Techniques* (Wiley, New York, 1994)
3. F.J.M. Harren, G. . Cotti, J. Oomens, S.L. Hekkert, in *Encyclopedia of Analytical Chemistry*, ed. by R.A. Meyers (Wiley, Chichester, 2000)
4. W.J. Witteman, *The CO_2 Laser* (Springer, Berlin, 1987)
5. W. Urban, Rev. Infrared Phys. Technol. A **36**, 465 (1995)
6. M.M.J.W. van Herpen, S. Li, S.E. Bisson, F.J.M. Harren, Appl. Phys. Lett. A **81**, 1157 (2002)
7. B.A. Paldus, T.G. . Spence, R.N. Zare, J. Oomens, F.J.M. Harren, D.H. Parker, C. Gmachl, F. Capasso, D.L. Sivco, J.N. Baillargeon, A.L. Hutchinson, A.Y. Cho, Opt. Lett. A **24**, 178 (1999)
8. D. Hofstetter, M. Beck, J. Faist, M. Nagele, M.W. Sigrist, Opt. Lett. A **26**, 887 (2001)
9. R.F. Curl, F.K. Tittel, Annu. Rep. Prog. Chem., Sect. C. A **98**, 219 (2002)
10. R.A. Rooth, A.J.L. Verhage, L.W. Wouters, Appl. Opt. A **29**, 3643 (1990)
11. V.S. Starovoitov, S.A. Trushin, V.V. Churakov, J. Appl. Spectrosc. A **66**, 345 (1999)
12. A. Olafsson, M. Hammerich, J. Bulow, J. Henningsen, Appl. Phys. B **49**, 91 (1989)
13. S. Bernegger, M.W. Sigrist, Infrared Phys. A **30**, 375 (1990)
14. V.S. Starovoitov, S.A. Trushin, Pure Appl. Opt. A **2**, 505 (1993)
15. M.B. Pushkarsky, M.E. Webber, O. Baghdassarian, L.R. Narasimhan, C.K.N. Patel, Appl. Phys. B **75**, 391 (2002)
16. E. Huang, D.R. Dowling, T. Whelan, J.L. Spiesberger, J. Acoust. Soc. Am. A **114**, 1926 (2003)
17. J. Oomens, H. Zuckerman, S. Persijn, D.H. Parker, F.J.M. Harren, Appl. Phys. B **67**, 459 (1998)
18. F.G.C. Bijnen, J. Reuss, F.J.M. Harren, Rev. Sci. Instrum. **67**, 2914 (1996)
19. F.R.E. Gruijl, Radiat. Prot. Dosim. A **72**, 177 (1997)
20. L.R. Narasimhan, W. Goodman, C.K.N. Patel, Proc. Natl. Acad. Sci. A **98**, 4621 (2001)
21. P. Kestler, in *Comparative Biochemistry and Physiology of Insects*, ed. by K.H. Hoffman (Springer, Berlin, 1985)
22. R. Flindt, *Biologie in Zahlen* (Gustav, Stuttgart, 1988)
23. A. Miklos, P. Hess, Z. Bozoki, Rev. Sci. Instrum. A **72**, 1937 (2001)
24. S.L. Firebaugh, K.F. Jensen, M.A. Schmidt, J. Microelectron. Syst. A **10**, 232 (2001)
25. A.L. Palanetski, V.S. Starovoitov, Proc. SPIE, doi:10.1117/12.753310 (2007)
26. A.A. Kosterev, Y.A. Bakhirkin, R.F. Curl, F.K. Tittel, J. Opt. Lett. A **27**, 1902 (2001)
27. R. Lewicki, G. Wysocki, A.A. Kosterev, F.K. Tittel, J. Appl. Phys. B **87**, 157 (2007)
28. A.A. Kosterev, Y.A. Bakhirkin, F.K. Tittel, J. Appl. Phys. B **80**, 133 (2005)
29. A.K.Y. Ngai, S.T. Persijn, I.D. Lindsay, A.A. Kosterev, P. Groß, C.J. Lee, S.M. Cristescu, F.K. Tittel, K.-J. Boller, F.J.M. Harren, J. Appl. Phys. B **89**, 123 (2007)
30. A.V. Gorelik, V.S. Starovoitov, Opt. Spectrosc. A **107**, 890 (2009)
31. A.V. Gorelik, V.S. Starovoitov, 0810.0410

Gas cells for tunable diode laser absorption spectroscopy employing optical diffusers. Part 1: single and dual pass cells

J. Hodgkinson · D. Masiyano · R.P. Tatam

Received: 12 November 2009 / Revised version: 8 March 2010 / Published online: 20 April 2010
© Springer-Verlag 2010

Abstract New designs for gas cells are presented that incorporate transmissive or reflective optical diffusers. These components offer simple alignment and can disrupt the formation of optical etalons. We analyse the performance-limiting effects in these cells of random laser speckle (both objective and subjective speckle), interferometric speckle and self-mixing interference, and show how designs can be optimised. A simple, single pass transmissive gas cell has been studied using wavelength modulation spectroscopy to measure methane at 1651 nm. We have demonstrated a short-term noise equivalent absorbance (NEA, 1σ) of 2×10^{-5}, but longer term drift of up to 3×10^{-4} over 22 hours.

1 Introduction

Tunable diode laser spectroscopy (TDLS) is of great interest for gas detection in a variety of applications including health and safety, industrial process control, emissions monitoring and healthcare. The laser diode emission is scanned across a single absorption line of the target gas, and/or modulated in wavelength modulation spectroscopy (WMS) [1]. Many practical implementations of these techniques have shown that the detection sensitivity is limited by interference fringes and not by the theoretical limit given by detector noise [2, 3]. The interference fringes stem from Fabry–Pérot etalons between reflecting or scattering surfaces such as mirrors, lenses, optical fibre end faces, detector and laser head windows, semiconductor surfaces, and components of multipass cells [1], and also from low levels of optical feedback to the laser diode [4].

The etalons often exhibit a free spectral range similar to the linewidth of the absorbing species and appear as periodic spectral features with sufficient amplitude to obscure weak absorption signals [5]. Bomse et al. [6] suggested that gas cell windows are often the worst culprits. Narrow laser linewidths (tens of megahertz) used to resolve individual gas lines have correspondingly long coherence lengths (tens of metres). An etalon formed between the windows of a gas cell with a typical length of 10 cm will give rise to fringes with a spacing of approximately 1 GHz (0.033 cm^{-1}), which is comparable to gas absorption linewidths at atmospheric pressure.

A number of techniques have been used to reduce interference fringes in TDLS, which in general fall into three broad categories. In the first, a frequency jitter is applied to the laser diode and the signal is integrated, averaging out the fringes [7]. The second category involves post detection filtering using high pass filters [1], low pass filters [8] or Fourier domain analysis [9]. However, neither category can deliver an improvement in signal to noise ratios for fringes whose FSR is comparable to the gas absorption linewidth [1]. In the final category, the fringe spacing is mechanically modulated and the resultant signal is integrated to average zero, or to minimise the contribution of the fringes to the absorption signal. This has been implemented using longitudinal dithering of optical elements [6], an oscillating mirror [2, 10] and an oscillating Brewster-plate spoiler [11]. While effective, this approach limits the available detection bandwidth and adds to system complexity.

J. Hodgkinson (✉) · D. Masiyano · R.P. Tatam
Engineering Photonics Group, School of Engineering,
Cranfield University, Bedfordshire, MK43 0AL, UK
e-mail: j.hodgkinson@cranfield.ac.uk

Present address:
D. Masiyano
Two Trees Photonics, Garamonde Drive, Wymbush,
Milton Keynes, MK8 8LW, UK

All the above approaches benefit if etalon formation is minimised, by reducing and misaligning Fresnel reflections. Standard designs for TDLS gas cells therefore include wedging and angling of all windows, antireflection coating of windows and lens surfaces, and angle polishing optical fibre ends. This results in tight alignment tolerances that are difficult to fabricate and to maintain in the field [12]. We present the results of a new approach using optical diffusers in reflection or transmission. These components offer simple alignment and can disrupt the formation of optical etalons within the gas cell. However, their use introduces new issues to be considered that can also limit performance, namely laser speckle and the effects of optical feedback. In Part 1 of this paper we show how an understanding of these issues, and how they relate to TDLS, can be used to optimise the design of simple, single or dual pass cells. Both objective speckle (without a lens) and subjective speckle (in which a lens is used to collect light onto the photodetector) are considered. We report the results of a comparison of different cell geometries at the design stage, and experiments with a simple transmissive geometry. In Part 2 of this paper we also report a study of the same performance-limiting effects in integrating spheres—the optical diffuser analogue of multipass cells.

Interest in this approach is growing. Chen et al. have used a diffuse reflector manufactured from anodised aluminium in an industrial oxygen sensor [13]. A group at Lund University have used TDLS to explore porous and scattering media including pharmaceutical samples [14] and biological tissue [15]. Finally, the reflective cells that we consider here have much in common with the standard configuration for laser pointer style gas detectors, in which light from the laser is backscattered from an uncontrolled and typically diffusely reflecting background target [16–18].

2 Theory

2.1 Gas detection

Methane has a well-known $2\nu_3$ absorption band in the near infrared region of the spectrum, centred around 1.66 μm, shown in Fig. 1. We have used the R4 quadruplet line at 1.651 nm, however our results are generally applicable to TDLS based detection of other gases using different wavelengths of operation.

For monochromatic radiation, the level of light transmitted through a gas cell is given by the Beer-Lambert law;

$$\frac{I}{I_0} = \exp(-\alpha z) \qquad (1)$$

where I_0 is the light power transmitted in the absence of light absorption, α is the absorption coefficient of the measured in cm^{-1} (equal to the specific absorptivity of the gas ε

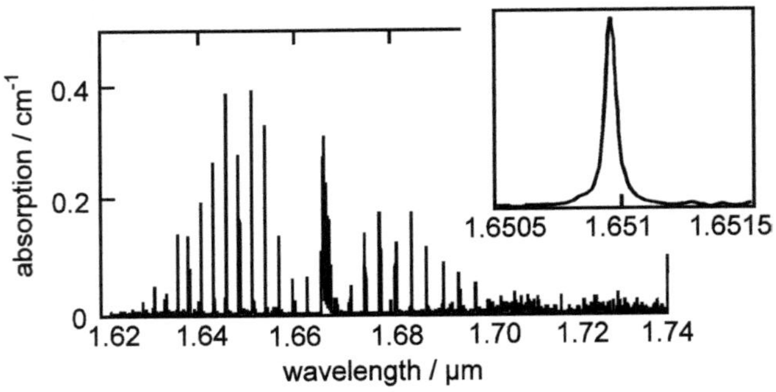

Fig. 1 Absorption spectrum for 100% methane in the near infra-red, calculated using data from the HITRAN database [19]

multiplied by the gas concentration) and z is the pathlength (here in cm). The gas absorptivity is often expressed in units of cm^{-1} per unit partial pressure of gas; at 1651 nm and at atmospheric pressure, methane has a value of ε at the line centre of 0.38 $cm^{-1}\,atm^{-1}$ [9]. At low values of αz, (1) approximates to the following linear relationship, for which the proportional change in detected signal is given by.

$$\frac{\Delta I}{I_0} \approx \alpha z \qquad (2)$$

We can therefore describe the uncertainty in the gas detection measurement as $\Delta I/I_0$, and translate this directly to a limit of detection for a given absorption coefficient. For wavelength modulation spectroscopy (WMS) using 2nd harmonic demodulation, a scale factor applies based on the modulation index used (the ratio of wavelength modulation amplitude to linewidth at half width half maximum). For an absorption line with a Lorentzian profile (typical for measurements at atmospheric pressure), the 2nd harmonic component I_2 is given by [20]

$$\frac{I_2}{I_0} = -2K\alpha z \qquad (3)$$

where the constant K can be optimised to a value of 0.343 by setting the modulation index to a value of 2.2 [21]. For methane at 1651 nm at atmospheric pressure, this corresponds to a dither of $\Delta\nu = \pm 2.8$ GHz or $\Delta\lambda = \pm 0.025$ nm.

2.2 Random laser speckle

Laser speckle patterns are formed when coherent light is backscattered from a randomly scattering surface whose surface roughness is comparable with the wavelength of illumination. The speckle is a superposition of random interferences and can be treated by assuming that the bright speckles are independent and obey Poisson statistics [22] such that the standard deviation of a single speckle is equal to the mean intensity. A detector placed within the speckle field integrates over a finite number of speckles; if N speckles are integrated, the uncertainty in the measured intensity is therefore $1/\sqrt{N}$. Random speckle noise therefore depends on the speckle size ε. We can identify two distinct forms of speckle as shown in Fig. 2, namely objective and subjective speckle.

For objective speckle, the speckle diameter ε_0 is given by [23]

$$\varepsilon_0 = 1.2\frac{\lambda z}{d} \tag{4}$$

where z is the distance from the optical diffuser to the detector and d is the illuminated extent of the diffuser. Note that this equation assumes that $d \ll z$. For subjective speckle, we have [23]

$$\varepsilon_s = 1.2\frac{\lambda L}{a} \tag{5}$$

where L is the distance from lens to detector, and a is the diameter of the lens aperture. If there are N independent speckles, the RMS deviation from a zero baseline for either form of speckle is given by

$$\frac{\Delta I}{I_0} = \frac{1}{\sqrt{N}} = \frac{\varepsilon}{D} \tag{6}$$

where D is diameter of the detector. In the worst case, if the deviations cannot be removed (e.g. by baseline subtraction), they translate to uncertainty in the measurement. We have evaluated the resulting uncertainty as $\Delta I/I_0$ for both types of speckle, making some assumptions about the cell geometry. The results are shown in Table 1.

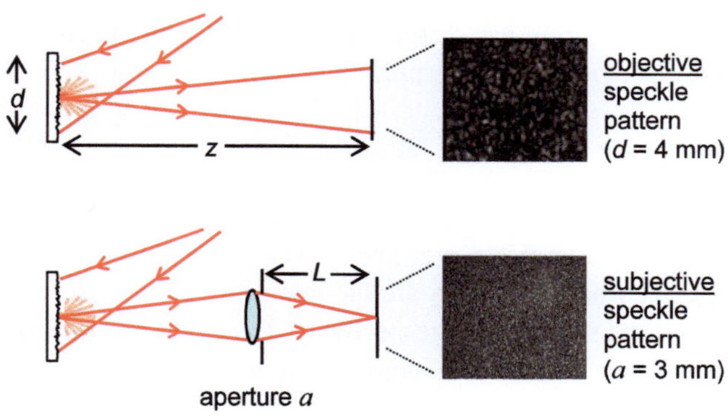

Fig. 2 Creation of objective and subjective speckle by laser illumination of a diffusely reflective surface

Now we consider the effect of wavelength modulation, for in many cases it is not simply the speckle that is of concern, rather how that speckle pattern changes with wavelength. For the small wavelength scans typical of TDLS, the speckles remain essentially in place, however each speckle moves in and out of phase. Fujii and Lit have calculated the mean deviation in the measured intensity at a single on-axis point for a change in wavelength $\Delta\lambda$ [24], as follows.

$$\frac{\Delta I_{\text{point}}}{I_0} = \left[1 - \exp\left(-h^2\Delta k^2\right)\right]^{\frac{1}{2}} \tag{7}$$

where Δk is the change in wavenumber, $k = 2\pi/\lambda$. This equation can be applied equally to direct spectroscopy, in which case $\Delta\lambda$ denotes the width of the wavelength scan or of the region used to perform a baseline analysis or fit, or to WMS, in which case $\Delta\lambda$ denotes the wavelength modulation. For small $\Delta\lambda$, this approximates to the following, which is linear with $\Delta\lambda$:

$$\frac{\Delta I_{\text{point}}}{I_0} \approx 2\pi h\frac{\Delta\lambda}{\lambda^2} \tag{8}$$

where h is a characteristic of the diffuse object and represents the RMS path difference in reflected or transmitted light paths from the diffuser. For a simple, single scattering reflector, this would be equivalent to twice the mean surface height deviation. If we assume that $h \sim 6$ µm, as for the experimental example given later in this paper, equation (7)'s deviation from linearity is less than 1% as long as $\Delta\lambda < 14$ nm, therefore (8) can be used with confidence. Equations (7) and (8) can be considered improvement factors for the resulting speckle noise; integrating over N independent speckles then yields the total RMS uncertainty in measured signal for small wavelength changes:

$$\frac{\Delta I}{I_0} = \frac{2\pi h}{\sqrt{N}}\frac{\Delta\lambda}{\lambda^2}. \tag{9}$$

Here, ΔI represents the uncertainty in DC intensity in a direct spectroscopy experiment; the corresponding uncertainty

Table 1 Estimated uncertainties in detected signals for random laser speckle

Type of speckle	Estimated uncertainty in $\Delta I/I_0$	Estimated change in $\Delta I/I_0$ under wavelength modulation[c]
Objective speckle (no lens)[a]	8×10^{-3}	$8 \times 10^{-3} - 8 \times 10^{-7}$
Subjective speckle (with lens)[b]	2×10^{-3}	$2 \times 10^{-3} - 2 \times 10^{-7}$

[a] $z = 100$ mm, $d = 25$ mm, $\lambda = 1651$ nm, $D = 1$ mm

[b] $z = 100$ mm, $a = 25$ mm, $L = 25$ mm, $\lambda = 1651$ nm, $D = 1$ mm

[c] $\Delta\lambda = 0.025$ nm, $h \geq \lambda$

in the 2nd harmonic component in a WMS experiment can be found by reference to (3). The resulting measured uncertainty is given in Table 1. The measured change in the speckle field therefore depends on the values of $\Delta\lambda$ and h. For $h \sim \lambda$ and $\Delta\lambda \sim 0.025$ nm, the level of uncertainty would be reduced by a factor of around 10^4. Thus, there is a significant benefit to operating over a narrow spectral range—the narrower, the better. If one were to work with low pressure gases, with narrower linewidths, the uncertainty $\Delta I/I_0$ would reduce, as indeed it does in conventional cells. For large values of h, (7) tends to unity and there is no difference in the measured uncertainty between a speckle field under a large phase change and the uncertainty associated with independent static speckle fields.

For a well-developed (high contrast) speckle field, the ideal value of h is considered to be of the order of the wavelength of light. As h decreases from this value, the surface becomes more similar to a smooth optical surface with no benefit offered by scattering. At much higher values of h, a single scattering surface will behave like a connected series of smooth surfaces, exhibiting interference with a higher uncertainty in the integrated intensity. We have also used multiply scattering materials such as Spectralon™, for which in reflective mode the light is believe to penetrate some distance (up to 10 mm) before re-emerging as backscattered light. These materials produce well-developed, high contrast speckle fields, but the effective value of h for such materials is unknown.

There is a final factor to consider, which is whether the speckle field moves during or between measurement integration periods (which in our experiments was defined by the time constant of our lock-in amplifier). In the case of stable speckle, creating spectral features that do not change from one measurement to the next, subtraction of the zero baseline would reduce the speckle uncertainty to nil, though subsequent long-term drift of the speckle would re-establish the speckle related uncertainty. In the case of rapidly changing speckle, averaging several speckle fields during a measurement integration period reduces the statistical error. Goodman has shown that the addition of M independent (uncorrelated) speckle fields reduces the speckle contrast by a factor of $\sqrt{M}$ [22], and we have confirmed this experimentally in a gas cell configuration, by statistical analysis of a single static speckle field [25].

2.3 Interferometric speckle

In speckle interferometry, a speckle field is deliberately mixed with a reference beam [26]. In a gas cell, a reference beam might be formed by the Fresnel reflection from a smooth cell window, as shown in Fig. 3.

As the wavelength is scanned, or alternatively as the value of z changes (for example because of thermal expansion), the resulting speckle patterns are correlated. By

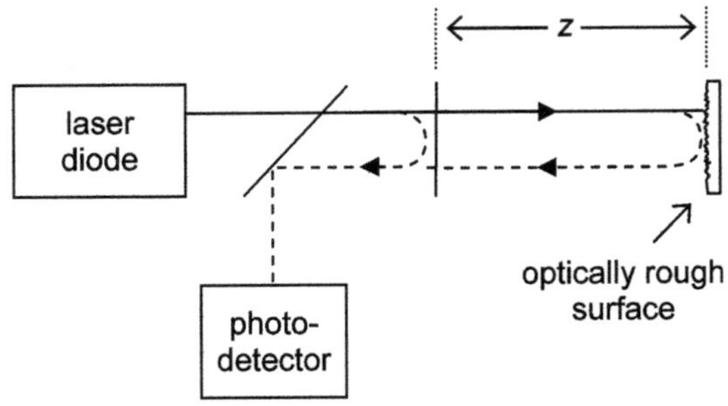

Fig. 3 Observation of interferometric speckle in an in-line configuration

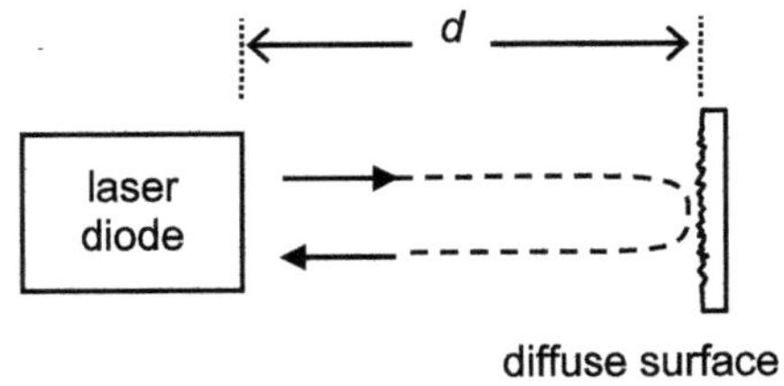

Fig. 4 Formation of self-mixing interference

subtracting subsequent speckle images from one another (or by performing a baseline subtraction on a detected signal), interference features are revealed. In speckle interferometry these are termed correlation fringes and represent an intensity modulation of the randomised speckle field. We have previously demonstrated such an intensity modulation arising from a gas cell that uses a diffusely reflective surface [25]. Sirohi has shown that the intensity modulation at the detector takes the form [26];

$$I = I_0(1 + A\cos\delta) \tag{10}$$

where δ is the optical phase change induced by a change of wavelength or distance z. The constant A is proportional to $\sqrt{I_1 I_2}$, where I_1 and I_2 are the intensities of the reference beam and speckle field respectively. Thus these interference fringes are the direct analogy of etalon-induced fringes for conventional gas cells.

2.4 Self-mixing interference

As an ideal diffuse surface scatters light in all directions, it can be hard to avoid a small proportion of this falling onto the laser diode. Figure 4 shows a simplified geometry.

Self-mixing or feedback interference results [27], and the laser output intensity takes the form;

$$I = I_0(1 + m\cos\delta) \tag{11}$$

where m is described as a modulation index (not to be confused with the modulation index used in TDLS described previously) and again δ is the optical phase change of the

Table 2 Summary of issues to be considered in TDLS gas cells using optical diffusers

Issue	Reduction strategy	Resultant level, as $\Delta I/I_0$
Random laser speckle	Use large apertures Move the sample Use wavelength modulation or baseline reduction	Fundamental uncertainty in the range $2 \times 10^{-3} - 2 \times 10^{-7}$
Interferometric speckle	AR coat and misalign cell window	Analogous to conventional etalon fringes $\leq 10^{-6}$
Self-mixing interference	Good quality isolation Balanced detection	$\leq 10^{-5} - 10^{-7}$

cavity. m is a function of the so-called feedback parameter C (dependent on the characteristics of the laser) and, for very weak feedback, proportional to $\sqrt{r}$ where r is the proportion of light reflected back into the laser diode. We have found that if steps are not taken to reduce feedback, the resulting interference fringes tend to dominate performance at low levels of r [4]. However, as the interference is a modulation of the total laser emission, the effect is reduced to first order by the use of balanced detection schemes [4, 28].

It is worth noting that at higher levels of optical feedback, other laser instabilities can also result including the appearance of additional modes or, in extreme cases, coherence collapse. However, we have not experienced these problems in our work with optical diffusers.

2.5 Summary

Table 2 summarises the particular issues to be considered when using optical diffusers in TDLS. It is worth noting that the theory relating to laser speckle has, in general, been developed for use in the field of speckle interferometry and not rigorously tested in its application to TDLS. We assume here that random laser speckle may not be removed by baseline subtraction, therefore it is considered to define the fundamental performance-limiting uncertainty of the systems we have studied. The other problems highlighted may be reduced by good design, at least in principle. Speckle uncertainty may be reduced by the use of a high NA lens and/or a large detector. The use of diffuse optics generally results in a loss of optical throughput because light is scattered in all directions and not collected by the detector. Fortunately, use of a high NA lens is also the logical response to such conditions. In a collimated system, use of an additional transmissive lens might introduce interference fringes, however in this case the light is uncollimated. We have not studied whether this contributes to a lower level of interference for reflections within the lens or between the lens and the detector. If such interference were troublesome, an off-axis reflective lens might be used instead.

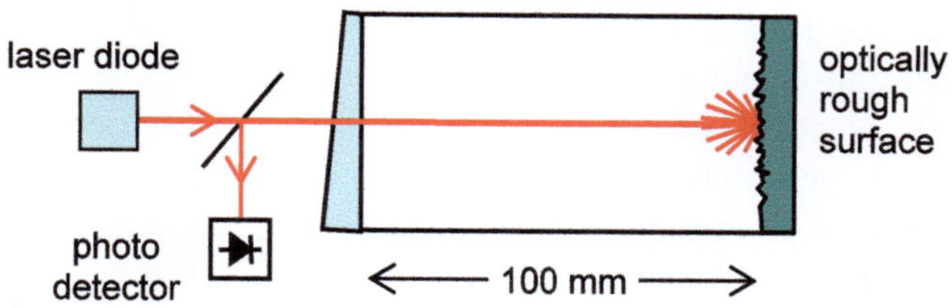

Fig. 5 Schematic diagram of a gas cell using a diffuse optical reflector

In this work, we chose to use WMS with 2nd harmonic demodulation. For our experimental purposes, compared to direct spectroscopy, this gave us the advantages of a zero baseline, a rapid response using our equipment and no dependence on spectral fitting algorithms. However, as the purpose of this paper is to investigate laser speckle in cells using optical diffusers, the principles should be capable of being translated to other spectroscopic techniques.

3 Gas cell design

We consider first a simple reflective gas cell and calculate the expected uncertainty in the measurement using the above theory. Figure 5 shows a schematic of the proposed cell geometry. Light from a laser diode passes through a beamsplitter, enters the cell via a wedged and AR coated window and strikes a diffuse reflector after making a pass though the cell. A small proportion of the light is backscattered along the incident axis and makes a second pass through the cell before being deflected by the beamsplitter towards a photodiode.

We estimate that the minimum practical value of z for this arrangement would be 150 mm, which for our other constraints ($d = 25$ mm, $D = 1$ mm, $\lambda = 1651$ nm) yields a proportional uncertainty of $\Delta I/I_0 = 10^{-2}$. This could be improved under wavelength modulation by up to $\Delta I/I_0 = 10^{-6}$ for an optimum value of h and $\Delta\lambda = 0.025$ nm. The cell has the advantage of using a double pass to improve detection limits. However, a number of potential disadvantages are also present. Firstly, there is a possibility of generating interferometric speckle with this arrangement. Although this would be avoided by misaligning a single window, the window is relatively close to the detector and the degree of misalignment required would be consequently greater. Secondly, the configuration has a relatively poor throughput as there is no lens to improve the collection efficiency at the detector, and the use of a beamsplitter exacerbates the problem. Finally, stray reflections associated with plate or cube beamsplitters may cause additional interference effects. Although pellicle beamsplitters avoid these problems, they are fragile and relatively expensive.

The same principles were applied to a range of different gas cell designs and the results are compared in Table 3. We have also compared the typical performance of a cell with a conventional configuration. In all cases we have used the same constraints, however depending on the geometry this might result in slightly different values of z, as in the case above. It was decided to implement the final transmissive design in our experiments for a number of reasons. The design is simple, and offers more convenient reduction of both interferometric speckle and self-mixing interference, the latter because the majority of the light is scattered in the forward direction, away from the laser diode.

In all the gas cell designs considered here, the estimated speckle related intensity uncertainty for a single static speckle pattern remained in the range 10^{-2} to 2×10^{-3}. This may be improved by increasing the detector size; a 5 mm diameter detector would give a 5-fold improvement in the signal: noise ratio. However, these levels are not sufficient to enable ppm level gas detection. Making the measurement over a narrow spectral range ($\Delta\lambda = \pm 0.025$ nm for WMS with methane at 1651 nm) would improve the signal: noise ratio by up to a factor of 10^4, depending on the surface properties.

To achieve this potential, it is therefore important that the diffuse surface is chosen carefully, and that the speckle field does not move substantially between, for example, intensity

Table 3 Summary of gas cell designs

	Cell geometry	Advantages	Disadvantages	Uncertainty $\Delta I/I_0$[a]
1	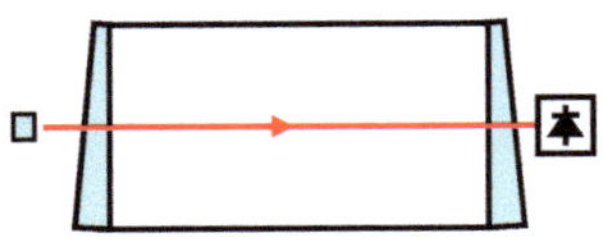	• etalon effects reduced by AR coated and wedged windows • high sensitivity can be achieved	• coatings can be expensive at non standard wavelengths • critical alignment	$10^{-5} - 10^{-6}$ [1]
2	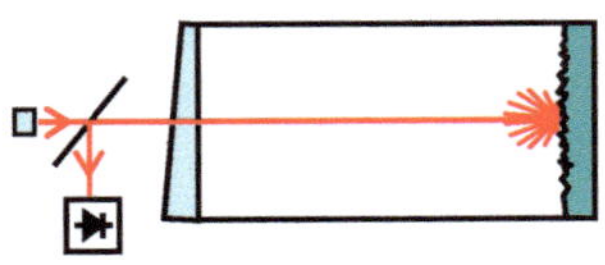	• double pass cell • simple geometry	• possibility of interferometric speckle • light inefficiency • requires 60 dB isolator to reduce self-mixing interference	$10^{-2} - 10^{-6}$
3	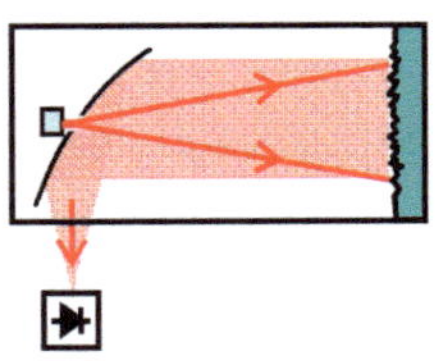	• double pass cell • simple and robust geometry • parabolic reflector improves light efficiency	• requires 60 dB isolator to reduce self-mixing interference	$2 \times 10^{-3} - 2 \times 10^{-7}$
4	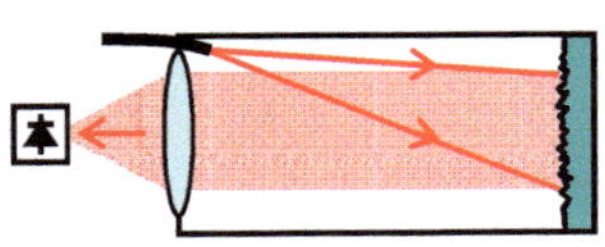	• double pass cell • simple and robust geometry • optical fibre delivery enables remote location	• interference effects associated with fibre coupling	$2 \times 10^{-3} - 2 \times 10^{-7}$
5	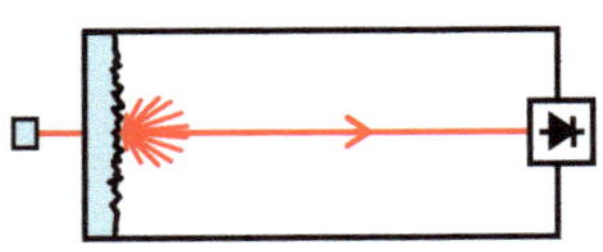	• simple and robust geometry • interferometric speckle and self-mixing interference reduced	• single pass cell • reflected specular component from 1st surface (can be reduced by AR coating/wedging)	$10^{-2} - 10^{-6}$

[a] 100 mm cell length, max 25 mm diameter optics, $\lambda = 1651$ nm, $D = 1$ mm, $\Delta\lambda = 0.025$ nm, $h \geq \lambda$

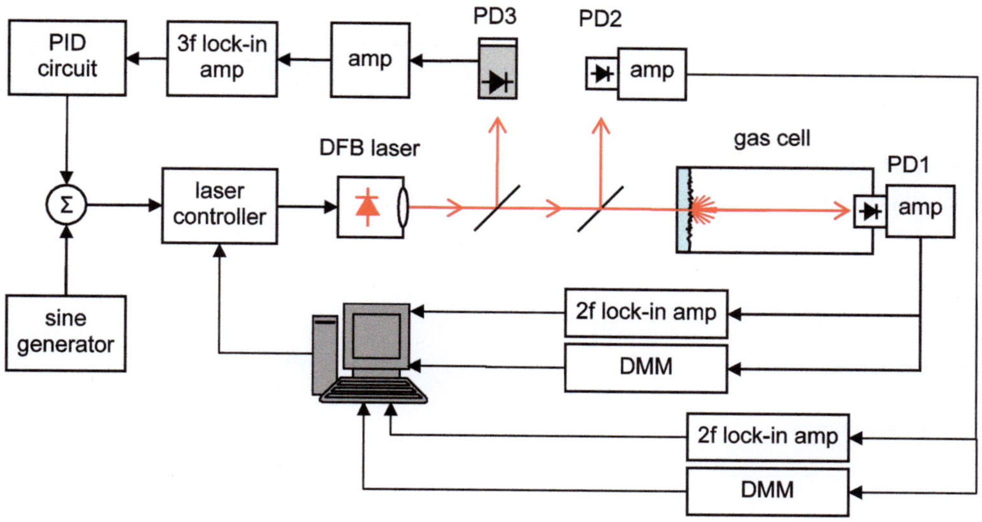

Fig. 6 Schematic experimental configuration for $2f$ wavelength modulation spectroscopy, using a $3f$ line lock

measurements taken at each side of the absorption line. The same issue exists for measurements in conventional cells; where vibrations are present, modulation frequencies should ideally be greater than the frequency of mechanical vibration so that the baseline remains stable during a single scan or wavelength modulation cycle.

Objective speckle uncertainty can be reduced by the use of a large detector and large illuminated area. When a lens is used to collect light (subjective speckle), the level of associated speckle uncertainty can be reduced using a high NA lens, which also improves the light collection efficiency. The speckle size is then independent of illumination geometry, which has advantages for measurements made in locations where the backscattering surface is difficult to control (for example, using backscatter laser pointers).

4 Experimental details

We employed WMS at $f = 6$ kHz with second harmonic ($2f$) detection at 12 kHz. Figure 6 shows a schematic diagram of our laser diode modulation and detection apparatus, in this case configured for use with our transmissive gas cell design. A sinusoidal signal at 6 kHz from a signal generator (Hewlett-Packard HP33120A) was applied to the laser controller (ILX Lightwave, ILX LDC-3722B) to give an amplitude at the laser diode of 24 mA (peak to peak). Our DFB laser diode package (Semelab Ltd.) incorporated a 1651 nm laser (NEL NLK1U5C1CA-TS) collimated with an aspheric lens (Lightpath 350230D). Gross wavelength tuning was achieved by controlling the diode temperature using a Peltier element within the package.

A proportion of the laser beam (~8%) was sampled by a pellicle beamsplitter (Thorlabs BP208) to a gas reference cell (PD3 in Fig. 6) consisting of a Ge detector at the bottom of a TO18 can containing 100% methane, giving an effective pathlength of 5.3 mm. We would wish to avoid the use of such beamsplitters in field applications, for example by using a partial reflection from a window or from the diffuser, or by using a proportion of the transmitted beam. However, for these experiments the use of beamsplitters permitted simpler comparison of different cell geometries. The signal from the photodiode was demodulated at $3f$ using a lock-in amplifier (Stanford SR850) to provide an error signal which was fed, via a PID (proportional, integrative, derivative) controller to the laser controller, controlling the dc current of the laser diode so as to lock the emission wavelength to the gas absorption line centre. In our standard WMS experiments, the laser emission was thus actively locked and we did not perform a simultaneous wavelength scan. However, in some experiments the lock was not used and a simultaneous slow wavelength scan through the gas absorption line was performed in order to observe the $2f$-demodulated spectrum.

The laser beam striking the optical diffuser had a diameter of approximately 8 mm. A proportion of the forward scattered light fell onto the detector/amplifier (PD1: Thorlabs, PDA50EC, 1 mm diameter) after making a single pass across the cell. A lock-in amplifier (Stanford Research Systems SR850, time constant $\tau = 1$ s) was used for $2f$ demodulation, its output voltage (X, in X, Y mode) sampled using a data acquisition card and transferred to a computer running Labview^{TM} software, through a data acquisition card (National Instruments PCI 6259). The DC signal from PD1 was sampled using a digital multimeter (DMM, Keithley 195A) before also being passed to the PC for data processing. Monitoring the DC level allowed us to normalise the $2f$-demodulated signals to compare results between different optical diffusers, each of which could project a different level of light onto PD1. A second detector amplifier (PD2 in Fig. 6) was used to measure a proportion (approximately 45%) of the incident beam intensity via a beamsplitter (Thorlabs BP245B3). This reference detector chan-

Table 4 Transmissive diffusers tested and their optical properties

Diffuser material	Characteristics	Relative transmission	Light pattern
No diffuser	open space	100%	collimated beam
Ground BK7 glass	220 grit	21%	strong specular reflection
(Thorlabs DG10)	600 grit	58%	well-developed speckle
	1500 grit	48%	part of beam not diffusely scattered, but displaced by angling surface
Zenith™ PTFE	100 µm thick	4.0%	well-developed speckle,
(Sphere Optics DF-100)	500 µm thick	2.2%	best used in reflection
Depolariser (Thorlabs DPU-25-C)	AR coated	99%	collimated beam, interference fringes
Holographic diffuser (Luminit)	0.5° circle pattern	>70%	Scattered light over specified cone angle

nel employed the same model of detector/amplifier, lock-in amplifier and DMM, with the same set-ups, as the main signal channel from PD1. Both the sample cell and the reference detector PD2 were angled slightly (around 10°) with respect to the optical path.

Test gases were fed to the gas cell from certified cylinders (Scott Specialty Gases), one containing hydrocarbon (HC) free air and the other containing either 1010 ppm or 50.1 ppm methane in HC free air. A bank of mass flow controllers (Teledyne Hastings HFC-302 with THPS-400 controller) was used to control flow rates from the two cylinders, with downstream mixing generating different concentrations in the range 0–1010 ppm.

5 Results

We tested the chosen geometry of Sect. 3, namely cell 5: a simple transmission cell design in which light from the laser diode passed through an optical diffusive element, which scattered the light mainly in the forward direction. A proportion of the light fell onto a 1 mm diameter photodiode after making a single pass across a 100 mm gas cell. For the purpose of comparison, we also tested a conventional transmission cell of the same length, employing two wedged and anti-reflection coated windows.

5.1 Choice of materials

Different types of diffusely transmitting materials were investigated. Their relative levels of transmission were compared for a 1 mm detector placed at a distance of 100 mm, and we observed the form of the scattered light pattern that they produced. The results are shown in Table 4.

We chose to use a 1500 grit ground glass diffuser because it combined a well-developed speckle pattern, good throughput and minimal specular reflection. The mean surface height deviation is uncertain but expected to be at or below 12.6 µm, as the ISO specification for 1500 grit refers to abrasion by particles with a mean diameter of 12.6 µm. For a material of refractive index 1.5, this gives a mean optical pathlength difference $h < 6$ m for a transmitter, and we would expect the improvement in noise level calculated using (8) to be less than or equal to a factor of 2.8×10^4.

5.2 Gas absorption measurements with a diffuse cell

The transmission cell was built using a 1500 grit ground glass window (Thorlabs DG10-1500) and 100 mm length, with a detector/amplifier (Thorlabs, PDA50EC) integrated into the end so as to avoid the use of additional windows. The only factor to be considered during alignment was to misalign the Fresnel reflection from the front (smooth) surface of the diffuser plate, therefore the process was very quick (<1 min). A separation of over 1 metre between the laser diode and the cell ensured that self-mixing interference effects were minimised. The output was recorded for 2nd harmonic WMS with a $3f$ line lock while a series of gas mixtures of different concentrations was passed through the cell. For each concentration point, zero data were recorded before and after (over a total time period of approximately 6 minutes) in order to remove any effects of drift. Figure 7 shows the results; a short-term noise equivalent absorbance (NEA) of 2×10^{-5} is estimated, corresponding to around 5 ppm methane (1σ).

5.3 Assessment of drift

The level of system drift was assessed by continuously recording data (with no gas in the cell) over a 22 hour period, for both the signal channel PD1 and the reference channel PD2. Allan variance plots (σ^2 as defined by Werle et al. [29]) were calculated for the recorded $2f$ measurements for each channel, using data normalised by division by the DC signals. These plots are shown in Fig. 8 and describe a system dominated by drift. Despite the use of normalised data

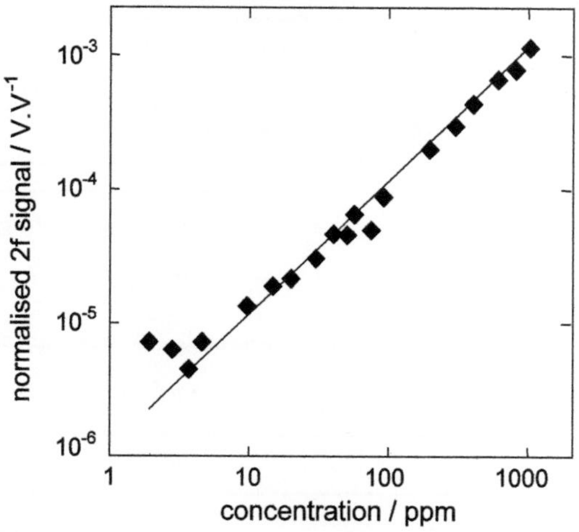

Fig. 7 $2f$-demodulated signals versus gas concentration for cell using a transmissive optical diffuser. $2f$ signals have been normalised by dividing by the measured DC signal, and the dashed line shows a linear response

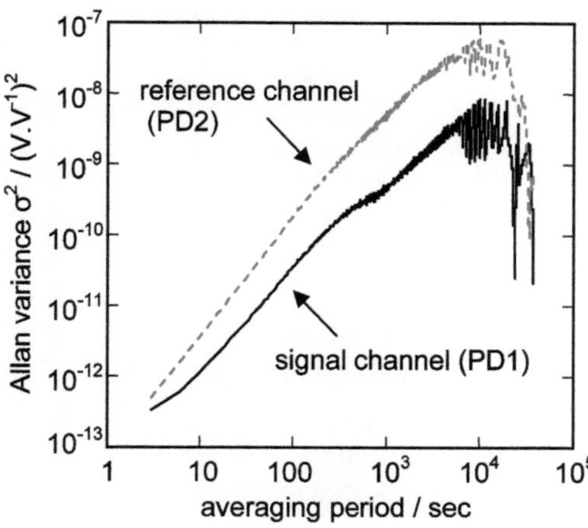

Fig. 8 Allan variance plot for normalised $2f$ time series data taken over a 22 hour period with an empty cell

for both channels, the Allan variance seems to be a factor of 3–10 times more pronounced on the reference channel than the signal channel. The drift on the signal seems to be bounded at a mean value ($\sqrt{(\sigma^2)}$) of an NEA of 3×10^{-4}, and at a value of 7×10^{-4} on the reference channel. Because of the high level of drift on the reference channel, we cannot associate the drift on the signal channel with the use of the diffuse optical cell alone, rather this data puts an upper limit on what the effect of the cell might be. The worst measured value of $\sqrt{(\sigma^2)}$ corresponds to a concentration of around 70 ppm, using Fig. 7.

The data is consistent with an interference effect; interference fringes have finite extent and would limit drift to a maximum value. We are aware of interference fringes in our laser diode output with a wide FSR, which cause instabilities similar to those seen on the reference channel. It is curious however that the Allan variance for the signal channel PD1 was consistently lower than that for the reference channel PD2, despite both these measurements being normalised ($2f$ measurement/DC measurement) and using identical apparatus. We were hoping to use the reference channel to correct for instabilities in the laser diode output, but this was not possible as the outputs from PD1 and PD2 were uncorrelated. Further work is needed to investigate the source of the drift in our system. Two additional effects may be present; (i) our pellicle beamsplitter may suffer from low frequency movement, which would affect the reflected reference beam more than the transmitted signal beam, and (ii) the signal channel may be measuring over a wider and more representative sample of the emitted laser beam than our reference channel.

5.4 Summary of results with transmissive cell

We consider speckle uncertainty on the detector to determine the fundamental limit of detection of this cell design. It would be improved however by the use of a more divergent illumination of the ground glass plate and/or the use of a high NA lens to collect light onto the photodiode. In our experiments, we have established a short-term (6 minute) limit of detection of around 5 ppm, in a system dominated by drift.

For a spectral scan of $\Delta\lambda = 0.025$ nm, beam diameter $d = 8$ mm, distance to detector $z = 100$ mm, detector diameter $D = 1$ mm and surface roughness deviation $h \leq 6$ μm, we would expect the level of speckle uncertainty to be $\Delta I/I_0 \sim 10^{-5}$, corresponding to an NEA for WMS of around 1.5×10^{-5}. Using (3), for a 100 mm pathlength cell and absorptivity of 0.38 cm^{-1} at the methane line centre, this translates to a detection limit of 4 ppm or below. Although this is consistent with our short-term results for gas concentration measurement (an NEA of 2×10^{-5}), it is not known to what extent the spectral effect of speckle was stable during these experiments.

The drift experiment puts an upper limit on the degree of instability of the speckle, and thereby the speckle uncertainty at an NEA of 3×10^{-4}. Unfortunately, our experimental platform was not itself stable and its performance may have dominated these results. Work is in progress to identify and remove the remaining causes of instability, which could be associated with feedback effects within the laser diode package [4], or by movement of our pellicle beamsplitter. The level of speckle uncertainty calculated in the absence of wavelength modulation (equation (6), Table 1 column 2) does not apply.

5.5 Speckle noise reduction

Two methods have been tested to suppress speckle noise by rotating or vibrating the diffuser. In these experiments, the

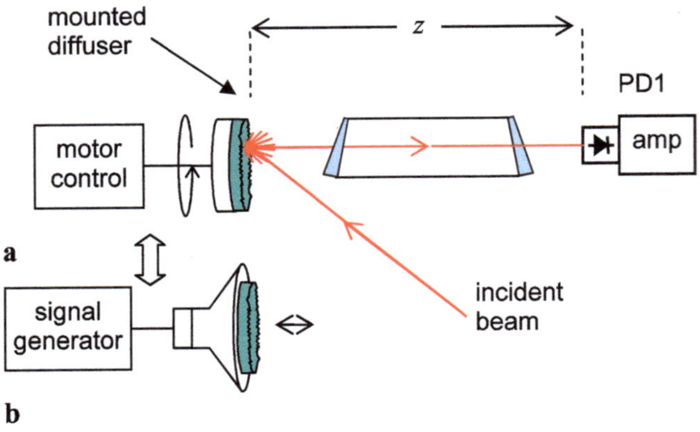

Fig. 9 Experimental configuration for speckle noise suppression using moving diffuse reflectors, by (**a**) rotation of a motor, and (**b**) vibration of a loudspeaker

speckle related intensity uncertainty was deliberately exaggerated, by using a small detection aperture, so that the effectiveness of the suppression could be quantified unambiguously. For simplicity the moving diffuser was placed outside a conventional gas cell, as shown in Fig. 9. In both cases, a 1 mm thick sample of Zenith™ (proprietary PTFE, Sphere Optics) was used as a reflective diffuser. We minimised the level of backscattered light returned to the laser diode by using a large angle of incidence and by placing the laser diode at a distance of over 1 m from the diffuser.

The speckle size was exaggerated by increasing the distance z between the diffuser and the detector to around 600 mm. The laser drive current was ramped over the range 35–70 mA in steps of 0.2 mA, while $2f$-demodulated detector output was recorded. In the first experiment we compared a stationary and rotating diffuser, the latter consisting of a 36 mm diameter, 1 mm thick disc of Spectralon™ rotated at a frequency of 1 kHz using the motor from an optical chopper. A lock-in time constant of $\tau = 100$ ms was used to allow scans to be completed within a reasonable time period. Scans were taken with the cell evacuated and also with the cell filled with 1.25% methane at atmospheric pressure. The resulting $2f$-demodulated gas signals are illustrated in Fig. 10 and show a large reduction in the received noise level when the diffuser was rotating.

Within one measurement integration period of 1 s, we expect to have averaged over the entire rotation of the diffuser, which for an 8 mm diameter beam striking the diffuser approximately 10 mm from the centre of rotation gives an equivalent of 8 independent speckle fields, which we would expect to yield an improvement in the noise level of $\sqrt{8} = 2.8$. A quantitative analysis of the results in Fig. 10 yields an improvement factor of 13, which is much greater. It is possible that additional speckle averaging was provided by imperfect alignment and diffuser wobble, giving a larger number of independent speckle patterns and a greater degree of suppression.

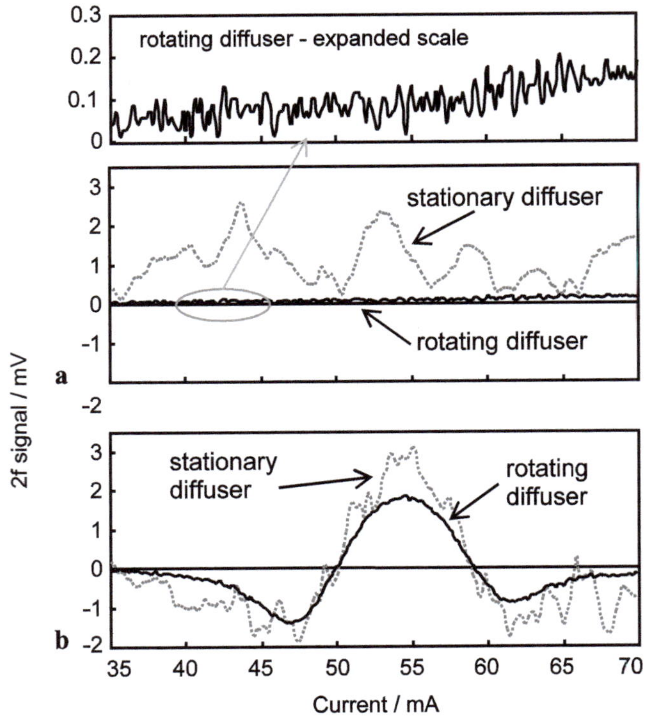

Fig. 10 Speckle noise reduction achieved by rotating a diffuser. $2f$-demodulated WMS scans of (**a**) HC free air, and (**b**) methane at a concentration of 1.25% vol. An exaggerated level of speckle noise was used to demonstrate the effect

In a second experiment, we glued the diffuser (36 mm diameter) to a loudspeaker diaphragm (Goldstar 541-861B, 3.2 Ω, 0.4 W). The loudspeaker was driven using an audio amplifier (Sanyo audio amp LA4597) with a 200 Hz sine wave, the amplitude adjusted manually using the amplifier's volume control (gain) to minimise ripple on the detector output. Vibrating the diffuser resulted in an observable reduction in speckle contrast, however the resulting amplitude of vibration is unknown. The results of a $2f$-demodulated methane line scan are shown in Fig. 11. Quantitative analysis shows that the speckle noise was reduced by a factor of 10 in this case.

6 Discussion and conclusions

We have analysed a number of factors affecting cell performance when optical diffusers are used instead of conventional, smooth windows or mirrors. Self-mixing or feedback interference can be performance limiting for both conventional and diffuse cells, but can be difficult to avoid in the latter, especially when diffuse reflectors are used. Our laser diode, which employed a relatively large (5 mm diameter) collimating lens for the purpose of our experiments, was particularly susceptible to this interference effect. To ensure that it did not affect our measurements, we used a large separation between the laser diode and the cell, but a more prac-

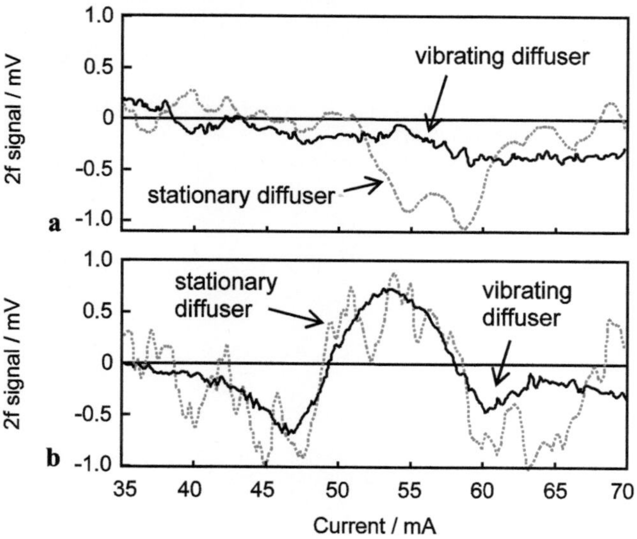

Fig. 11 Speckle noise reduction achieved by rotating a diffuser. $2f$-demodulated WMS scans of (**a**) HC free air, and (**b**) methane at a concentration of 1.25% vol. An exaggerated level of speckle noise was used to demonstrate the effect

tical solution would be needed for field instruments. This could include a smaller laser diode aperture, better optimised cell design and the use of an isolator.

Interferometric speckle in diffuse gas cells is the direct analogue of etalon-induced interference fringes in conventional cells and therefore needs to be removed. We have found in practice that it has not presented a problem and that its removal is simpler than for conventional cells, since the number of specular reflections from optical surfaces is reduced. Our chosen diffuse cell, with a transmission design, also minimises the possibility of generating a specular reference beam that can interfere with the speckle field.

Random laser speckle is believed to be a fundamental, performance-limiting effect for the diffuse optics that we have used. To gain the benefits of optical diffusers requires surface roughness that is comparable with the illumination wavelength, therefore speckle fields will be well-developed with high contrast. We have calculated the resulting speckle uncertainty for absolute measurements of light transmission, and for self-referenced measurements taken over spectral scans with a small $\Delta\lambda$, by baseline subtraction or using WMS. According to established speckle theory, the latter should offer levels of uncertainty compatible with ppm level gas detection, with $\Delta I/I_0$ in the range 10^{-2}–10^{-6} for our chosen 100 mm cell. The precise degree of uncertainty is dependent on the equivalent surface roughness of the optical diffuser. While this may be determined for simple, singly scattering materials such as the ground glass used in our chosen design, its value is unknown for multiply scattering bulk reflectors such as Spectralon™ and Zenith™ PTFE. It is also possible that multiple scattering in such materials will reduce uncertainty further through depolarisation [25].

Cells employing optical diffusers are simpler to align than standard cells. With careful design, the level of uncertainty determined by speckle on the detector was estimated for our system to give an NEA for WMS of around 1.5×10^{-5}. Our short-term experimental results are consistent with this, showing a short-term NEA (1σ) of 2×10^{-5}, but longer term drift of up to 3×10^{-4} over several hours. There may be additional sources of drift in our system that require further attention. We have also demonstrated two methods to reduce speckle uncertainty by a factor of 10 or more, based on vibration or rotation of the diffuser. These are analogous to the use of dithered optical window or spoiler plates in conventional cells, to reduce the visibility of interference fringes.

Acknowledgements This work was carried out under an EPSRC research grant (GR/T04601/01). Jane Hodgkinson is supported by an EPSRC Advanced Research Fellowship (GR/T04595/01). We would like to thank Dan Francis for his help with our data acquisition software.

References

1. H.I. Schiff, G.I. Mackay, J. Bechara, in *Air Monitoring by Spectroscopic Techniques*, ed. by M.W. Sigrist (Wiley, New York, 1994), Chap. 5
2. J.A. Silver, A.C. Stanton, Appl. Opt. **27**, 1914 (1988)
3. J. Reid, J. Shewchun, B.K. Garside, E.A. Ballik, Appl. Opt. **17**, 300 (1978)
4. D. Masiyano, J. Hodgkinson, S. Schilt, R.P. Tatam, Appl. Phys. B **96**, 863 (2009)
5. P. Werle, Spectrochim. Acta A **54**, 197 (1998)
6. D.S. Bomse, A.C. Stanton, J.A. Silver, Appl. Opt. **31**, 718 (1992)
7. J. Reid, M. El-Sherbiny, B.K. Garside, E.A. Ballik, Appl. Opt. **19**, 3349 (1980)
8. C.B. Carlisle, D.E. Cooper, H. Prier, Appl. Opt. **28**, 2567 (1989)
9. F. Capellani, G. Mellandrone, G. Restelli, in *Monitoring of Gaseous Pollutants by Tunable Diode Lasers*, ed. by R. Grisar, H. Preier, G. Schmidtke, G. Restelli (Kluwer Academic Publishers, London, 1987), pp. 51–60
10. S. Chou, D.S. Baer, R.K. Hanson, Appl. Opt. **36**, 3288 (1997)
11. C.R. Webster, J. Opt. Soc. Am. B **2**, 1464 (1985)
12. J. Hodgkinson, R.D. Pride, C. Tandy, D.G. Moodie, G. Stewart, in *Applications of Optical Fiber Sensors*, ed. by A.J. Rogers. Proc. SPIE, vol. 4074 (SPIE Press, Bellingham, 2000), pp. 90–98
13. J. Chen, A. Hangauer, R. Strzoda, M.-C. Amann, Laser spectroscopic oxygen sensor using diffuse reflector based optical cell and advanced signal processing, accepted for publication. Appl. Phys. B (2010). doi:10.107/s00340-010-3956-3 (this issue)
14. T. Svensson, M. Andersson, L. Rippe, S. Svanberg, S. Andersson-Engels, J. Johansson, S. Folestad, Appl. Phys. B **90**, 345 (2008)
15. M. Lewander, Z. Guan, K. Svanberg, S. Svanberg, T. Svensson, Opt. Express **17**, 10849 (2009)
16. T. Iseki, H. Tai, K. Kimura, Meas. Sci. Technol. **11**, 594 (2000)
17. R.T. Wainner, B.D. Green, M.G. Allen, M.A. White, J. Stafford-Evans, R. Naper, Appl. Phys. B **75**, 249 (2002)
18. B. van Well, S. Murray, J. Hodgkinson, R. Pride, R. Strzoda, G. Gibson, M. Padgett, J. Opt. A **7**, S420 (2005)
19. L.S. Rothman et al., J. Quant. Spectrosc. Radiat. Transf. **96**, 139 (2005)
20. W. Jin, Y.Z. Xu, M.S. Demokan, G. Stewart, Appl. Opt. **36**, 7239 (1997)

21. J.A. Silver, Appl. Opt. **31**, 707 (1992)
22. J.W. Goodman, in *Laser Speckle and Related Phenomena*, ed. by J.C. Dainty (Springer, Berlin, 1975), Chap. 2
23. A.E. Ennos, in *Laser Speckle and Related Phenomena*, ed. by J.C. Dainty (Springer, Berlin, 1975), Chap. 6
24. H. Fujii, J.M.Y. Lit, Opt. Commun. **22**, 231 (1977)
25. D. Masiyano, J. Hodgkinson, R.P. Tatam, Appl. Phys. B **90**, 279 (2008)
26. R.S. Sirohi, Contemp. Phys. **43**, 161 (2002)
27. G. Giuliani, M. Norgia, S. Donati, T. Bosch, J. Opt. A, Pure Appl. Opt. **4**, S283 (2002)
28. L. Persson, F. Andersson, M. Andersson, S. Svanberg, Appl. Phys. B **87**, 523 (2007)
29. P. Werle, R. Mücke, F. Slemr, Appl. Phys. B **57**, 131 (1993)

Gas cells for tunable diode laser absorption spectroscopy employing optical diffusers. Part 2: Integrating spheres

D. Masiyano · J. Hodgkinson · R.P. Tatam

Received: 12 November 2009 / Revised version: 8 March 2010 / Published online: 25 April 2010
© Springer-Verlag 2010

Abstract We have studied the effects of random laser speckle and self-mixing interference on TDLS based gas measurements made using integrating spheres. Details of the theory and TDLS apparatus are given in Part 1 of this paper and applied here to integrating spheres. Experiments have been performed using two commercial integrating spheres with diameters of 50 mm and 100 mm for the detection of methane at 1651 nm. We have calculated the expected levels of laser speckle related uncertainty, considered to be the fundamental limiting noise, and imaged subjective laser speckle in a sphere using different sized apertures. For wavelength modulation spectroscopy, noise equivalent absorbances (NEAs) of around 5×10^{-5} were demonstrated in both cases, corresponding to limits of detection of 1.2 ppm methane and 0.4 ppm methane respectively. Longer-term drift was found to be at an NEA of 4×10^{-4}. This lies within our broad range of expectations. For a direct spectral scan with no wavelength dither, a limit of detection of 75 ppm or fractional measured power uncertainty of 3×10^{-3} corresponded well with our prediction for the objective speckle uncertainty.

1 Introduction

Multi-pass cells are often used in tunable diode laser spectroscopy (TDLS) in order to increase the optical pathlength of the system and thereby improve limits of detection. Examples include designs by White [1], Herriott [2] and Chernin [3]. Essentially in each case a collimated beam of light is reflected a number of times between two mirrors to give a total pathlength that can be a considerable multiplier of the physical length of the cell. For example, a commercial Herriott cell design folds a 30 m beam into a physical length of 40.5 cm [4].

A second approach to multi-pass cells is to allow overlap of consecutive passes of the light beam across the cell and either integrate the total output (in the case of integrated cavity output spectroscopy, ICOS, or cavity enhanced absorption spectroscopy [5]) or to work in the time domain and measure the decay time of pulses in the cavity (in the case of cavity ringdown spectroscopy, CRDS [6]). Equivalent pathlengths for these systems are the longest currently available (e.g. 10 km path in a cavity of physical length 1 m [6]), resulting from the use of high quality dielectric mirrors. Confocal designs ease the alignment problem, such that these very long pathlengths may be aligned with care, but again this is a time-consuming process.

Each design has its advantages, and systems may be optimised for improved pathlengths, greater tolerance to cell bending during use, overlap between adjacent reflected beams and so on [7]. However, all suffer from the problem of difficult alignment, which is exacerbated by the need to also misalign potential optical paths through etalons and sources of feedback to the laser diode, in order to minimise interference fringes in the system. Such fringes are often the performance - limiting source of uncertainty in multi-pass cells (as indeed they can be in single pass cells) [8]. Multi-pass cells can be time-consuming to align, and mechanical tolerances for field use are tight.

It is for these reasons that interest has grown in the use of integrating spheres as gas cells. The internal walls of the

D. Masiyano · J. Hodgkinson (✉) · R.P. Tatam
Engineering Photonics Group, School of Engineering,
Cranfield University, Bedfordshire MK43 0AL, UK
e-mail: j.hodgkinson@cranfield.ac.uk

Present address:
D. Masiyano
Two Trees Photonics, Garamonde Drive, Wymbush,
Milton Keynes MK8 8LW, UK

sphere have very high levels of reflectance (up to 99%), but this is diffuse—the light is scattered in all directions. Good quality commercial spheres aim to use perfectly Lambertian reflectors, such that the radiance L (in $\mathrm{W\,m^{-2}\,sr^{-1}}$) from a given point is a constant in any direction. For a cavity with spherical geometry, it follows that the irradiance E (in $\mathrm{W\,m^{-2}}$) received from that point is then constant over the entire sphere [9, 10]. In the ideal case, after a single pass the irradiance at the sphere walls is perfectly uniform over the surface.

The advantages are as follows. Alignment tolerances are relaxed and absorption measurements may be made on turbid or irregular samples. Commercial integrating spheres are available as accessories for use with FTIR spectrometers [11]. For measurements using broadband emitters such as LEDs and filtered incandescent or discharge lamps, Elterman [10] has demonstrated measurement of the absorption of glass samples, Fry et al. [12] have measured visible absorption in aqueous solutions, Fecht and Johnson [13] have adapted a sphere for measurements on falling aqueous sample streams, and Hodgkinson et al. [14] have adapted a closed photothermal cell by inserting an integrating cavity. Work at the University of Limerick has resulted in broadband spectroscopic sensors for CO_2 (at 1.57 μm and 2 μm) [15] and NO_2 and SO_2 (in the UV/visible) [16].

The use of integrating spheres in TDLS has long been of interest. To our knowledge, Venkatesh et al. [17] were first to use an integrating sphere, monitoring carbon monoxide around 5 μm. Abdullin and Lebedev [18] have used a copper sphere with a CO_2 laser to measure ammonia, and Tranchart et al. [19] have used a 10 cm diameter sphere at 830 nm and 1.2 μm to detect water vapour and butane respectively. These studies confirmed that integrating spheres offer increased pathlengths within a relatively small volume without the need for precise alignment. However, the performance limitations of these systems have not been established. In Part 1 of this paper, we applied our methodology for the study of diffuse reflections in TDLS [20] to the use of optical diffusers in simple, single and dual pass cells. In Part 2, we now apply this understanding to the use of integrating spheres as multi-pass cells, with particular attention to the behaviour of random laser speckle and self-mixing interference.

2 Theory

A complete description of the speckle theory used here is given in Part 1 of this paper. Here, we concentrate on the specifics of the use of integrating spheres. Figure 1 shows a simplified model of the integrating sphere used in our experiments.

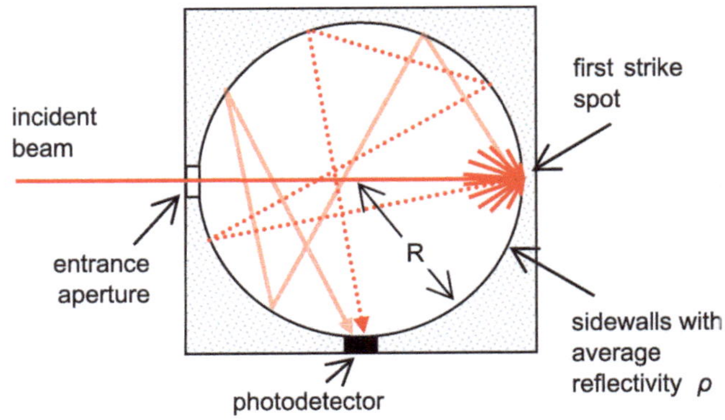

Fig. 1 Simplified model of an integrating sphere, showing a collimated beam making a first pass across the cell to the first strike spot. Two examples are shown of light beams making subsequent random multiple passes across the cell

Tranchart et al. [19] have shown that for an integrating sphere of internal radius R and high values of mean surface reflectivity ρ, the radiant flux (in Watts) Φ for low $\alpha z (\alpha z \ll 1)$ and high $\rho((1-\rho) \ll 1)$ is approximated by

$$\frac{\Delta\Phi}{\Phi_0} \approx \frac{\alpha z_0}{(1-\rho)}, \tag{1}$$

where Φ_0 is the radiant flux in the absence of gas, $\Delta\Phi$ is the change in radiant flux $(\Phi - \Phi_0)$ and the mean pathlength for a single pass is $z_0 = \frac{4}{3}R$. They then derive the effective total pathlength for low or zero absorbance α within the sphere, Z_{eff};

$$Z_{\mathrm{eff}} \approx \frac{4}{3}\frac{R}{(1-\rho)}. \tag{2}$$

We have shown that in the general case of higher values of α and lower ρ, the following expression applies [21]:

$$\frac{\Phi}{\Phi_0} = \frac{\ln\rho}{\ln\rho - \alpha z_0}. \tag{3}$$

The throughput of the sphere is defined as the ratio of light collected at the detector to that entering the sphere, and is given by [19]

$$\frac{\Phi_{\mathrm{exit}}}{\Phi_{\mathrm{in}}} = \frac{\rho f_{\mathrm{exit}}}{1-\rho(1-f_{\mathrm{total}})}, \tag{4}$$

where Φ_{exit} is total flux exiting a given port (Watts); Φ_{in} is total incident flux (Watts), f_{exit} is exit port fraction (area of exit port /surface area of sphere), f_{total} is total port fraction (sum of all port areas/surface area of sphere), and ρ is the sphere wall reflectance $(0 \le \rho \le 1)$.

Note that both the equivalent pathlength and the throughput are sensitive functions of the mean surface reflectivity, ρ. This parameter is governed by the material reflectivity at the wavelength of operation and by the proportion of non-reflective area taken up by, for example, the entrance aperture, photodetector and housing, plus gas inlet and outlet ports. It is not usually sufficient to calculate ρ by using manufacturer's reflectivity data and knowledge of the cell geometry; instead, cell calibration is required using (1) or (3). This calibration needs to be completed whenever there is a change to the optical set-up inside the sphere, or if in field use, it would need to be completed with sufficient frequency to prevent any build-up of dirt inside the sphere from causing the pathlength to drop by enough to render the instrument specification invalid.

2.1 Random laser speckle

Using (4)–(6) of Part 1, we can evaluate the uncertainty resulting from random laser speckle as $\Delta\Phi/\Phi_0$ for both objective and subjective speckle, by making some assumptions about the cell geometry. For our integrating sphere, the lensless detector has a limited field of view ($\pm 30°$) that defines the illumination geometry of (4), Part 1. If we were to use a high NA lens in front of the detector, we would evaluate (5), Part 1 assuming $L/a \sim 1$. Note that, for each type of speckle, the results are independent of the diameter of the sphere. The results are shown in Table 1, and in this case show the same expected level of uncertainty in the measurement of $\Delta\Phi/\Phi_0$.

Now consider the effect of wavelength modulation using (9) of Part 1. For $h \sim \lambda$ and $\delta\lambda \sim 0.025$ nm, the uncertainty is reduced by a factor of 10^4, and for higher values of h the improvement is between this value and unity (no improvement). The resulting expected uncertainty is also given in Table 1.

Integrating spheres developed for the mid IR use roughened surfaces coated with gold. For single scattering from this surface, the value of h would be equal to twice the RMS deviation of the roughened surface. However, spheres designed for use in the visible and near IR typically use proprietary sintered poly [tetrafluoroethylene] (PFTE) materials known as Spectralon™ or Zenith™. These are multiple scatterers, in which light penetrates some distance into the bulk (up to 10 mm) before re-emerging in the backscattered direction. These materials produce well-developed, high contrast speckle fields, but the effective value of h is unknown. Therefore we cannot predict the expected speckle-related uncertainty with any precision; the range of expectation in Table 1 is very broad, namely 4 orders of magnitude.

It should also be noted that the theory above has not taken into account the effects of depolarisation of the light. Multiply scattered light from an integrating sphere is expected to be fully depolarised, and light backscattered from a bulk scattering material such as Spectralon or Zenith may be partially depolarised, depending on the material thickness. Full depolarisation is expected to result in the creation of two independent speckle fields at orthogonal polarisation states [22], which would further reduce the speckle uncertainty by a factor of $\sqrt{2}$.

2.2 Other interferometric effects

Interferometric speckle is described in Part 1, Sect. 2.3. However, for an integrating sphere geometry as shown in Fig. 1 we have disrupted any possible return path of a reference beam. As long as there is no direct line of sight from the entrance aperture window to detector, we do not expect to encounter interferometric speckle effects when using integrating spheres.

Self-mixing or feedback interference fringes can arise when low levels of light are backscattered to the laser diode, as described in Part 1, Sect. 2.4. The first strike spot is the main source of feedback to the laser diode. Our experiments (see below) have been designed to minimise this feedback, which is particularly troublesome when using backscattering materials.

3 Experimental details

The details of our experiments are similar to those described in Part 1, with the exception of the integrating sphere itself. Figure 2 shows a schematic diagram.

Two integrating spheres were used, both incorporating Zenith™ as the reflective material. The first (Thorlabs IS200-4, manufactured by Sphere Optics) had an internal diameter of 50.8 mm; for simplicity we term this the "50 mm

Table 1 Estimated uncertainties in detected signals for random laser speckle

Type of speckle	Estimated uncertainty in $\Delta\Phi/\Phi_0$	Estimated change in $\Delta\Phi/\Phi_0$ under wavelength modulation[c]
Objective speckle (no lens)[a]	2×10^{-3}	2×10^{-3}–2×10^{-7}
Subjective speckle (with lens)[b]	2×10^{-3}	2×10^{-3}–2×10^{-7}

[a] $\pm 30°$ field of view, $\lambda = 1651$ nm, $D = 1$ mm

[b] $L/a \sim 1$, $\lambda = 1651$ nm, $D = 1$ mm

[c] $\Delta\lambda = 0.025$ nm, $h \geq \lambda$

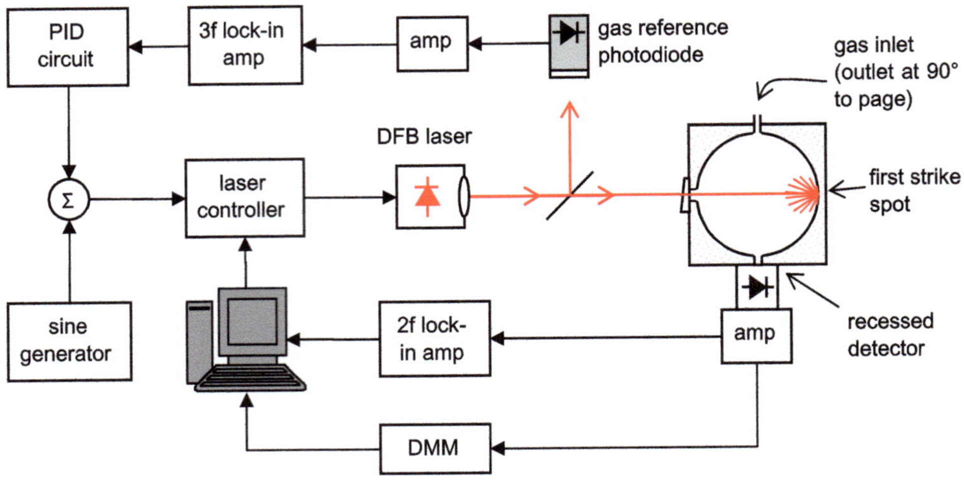

Fig. 2 Schematic experimental configuration for $2f$ wavelength modulation spectroscopy, using a $3f$ line lock

sphere". Light from the laser diode entered the integrating sphere via a 6.35 mm diameter entry port covered by an angled and antireflection coated window (Thorlabs PS812-C). The detector (12.5 mm diameter port, Thorlabs PDA400) was recessed to prevent a direct line of sight light from the first strike spot, therefore no baffle was necessary. The detector's field of view was restricted to ±32°. Two additional ports (1.5 mm diameter) allowed entry and exit of test gas mixtures. The second integrating sphere (Sphere Optics SPH-4Z-4) had an internal diameter of 101.6 mm (termed the "100 mm sphere") and a similar port arrangement, except that in this case the detector was not recessed and an internal baffle was present. We employed 6 mm diameter ports for the inlet aperture and detector, and 1.5 mm diameter ports for the gas inlet and outlet. In a previous publication, we have shown that the equivalent pathlength of the 50 mm and 100 mm spheres, in the absence of gas, was 1.2 m and 3.4 m respectively [21].

4 Results

4.1 Speckle behaviour

To confirm the behaviour of random laser speckle inside the 50 mm diameter sphere, we took images through an open sphere detector port using a silicon CCD camera (Pearpoint P176) and illumination via an entrance port at 90° from a HeNe laser. As shown in Fig. 3, images were taken using a lens placed at the exit port, with the CCD chip at a further distance of 50 mm; this geometry decreased the NA and should therefore exaggerate the speckle size.

Figure 4 shows the resulting speckle images taken with different sized lens apertures. Also shown in Fig. 4 are the speckle sizes predicted using (5), Part 1 for $L = 50$ mm and $\lambda = 633$ nm. Limited resolution of the images means that only the speckles in Fig. 4(d) are fully resolved, therefore images (a)–(c) suffer some loss of contrast to differing degrees. The dark vertical line in the images corresponds to a construction feature of the integrating sphere—the join between two hemispheres.

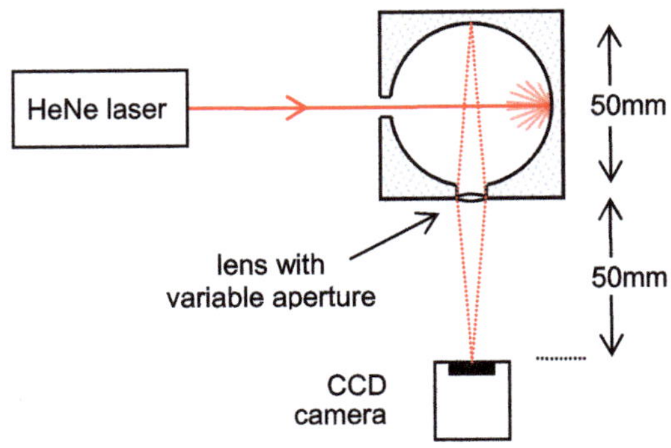

Fig. 3 Experiment to image subjective speckle in an integrating sphere

To understand how the speckle uncertainty could affect gas measurements, we replaced the camera with a photodetector (Thorlabs PD400) at a distance of 50 mm from the lens and illuminated the integrating sphere using our 1651 nm DFB laser. $2f$-demodulated signals were recorded while the laser injection current was ramped in the range 40–70 mA, corresponding to a sweep of approximately 21 GHz or 0.2 nm. Spectra were recorded for a 10 mm and 1 mm lens aperture. The results in Fig. 5 have been normalised to account for differences in the mean recorded light intensity for each aperture. The standard deviation in the baseline was recorded as 0.13×10^{-3} and 0.68×10^{-3} respectively, showing a reduction in speckle noise by a factor of 5 when the aperture was increased by a factor of 10. This is smaller than the 10-fold improvement that we would have expected from (5) of Part 1, but nevertheless shows a considerable reduction.

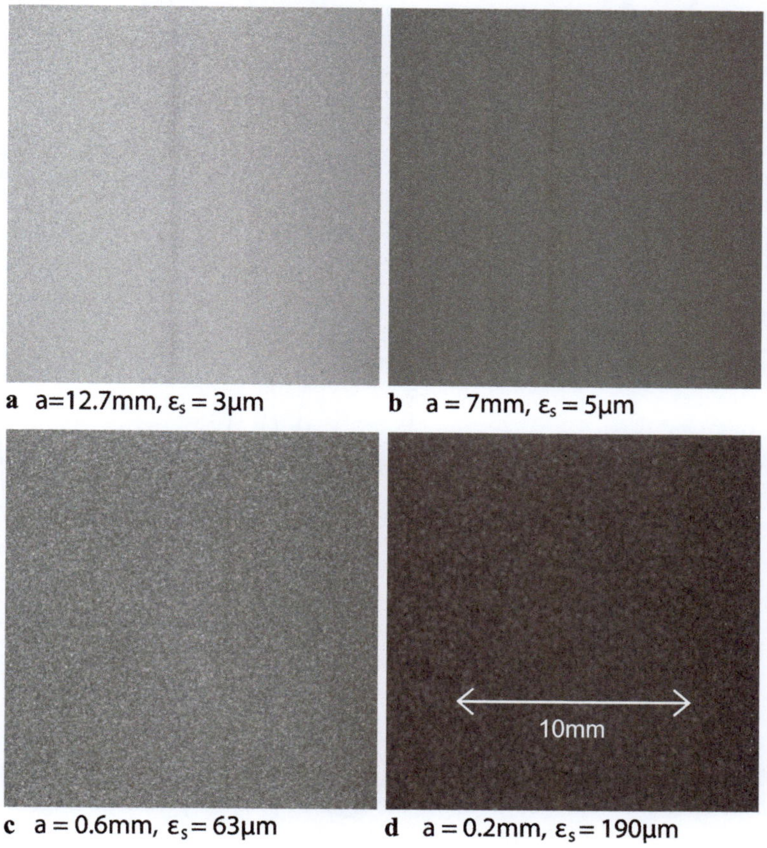

Fig. 4 Images of subjective speckle taken through an open port in a 50 mm diameter integrating sphere using a 633 nm HeNe laser and a CCD camera. A construction feature (the join between two hemispheres) is visible

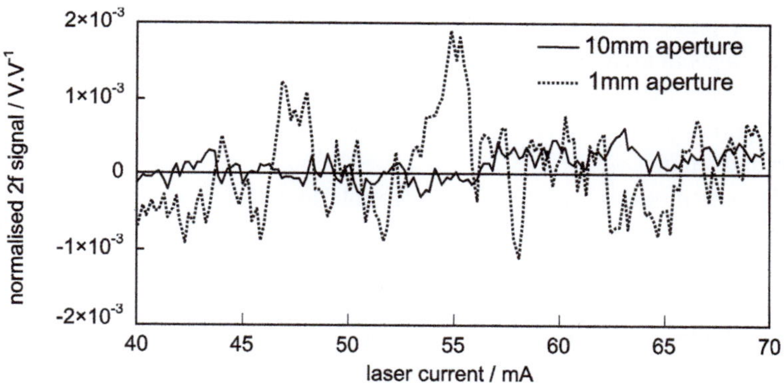

Fig. 5 $2f$-demodulated signals, normalised for mean recorded intensity, showing a reduction in subjective speckle noise following the use of a larger lens aperture

4.2 Self-mixing interference

The laser diode was configured with a monitor photodiode at the rear facet in order to record the total laser diode output; we have previously found that this is a robust diagnostic test for self-mixing interference fringes as opposed to etalon-generated fringes [23]. Using the 50 mm diameter sphere, the exit aperture of the laser diode package was located 100 mm from the sphere first strike spot and for simplicity the window at the entrance aperture was removed, as shown in Fig. 6.

$2f$-demodulated signals from the monitor diode were recorded while the DC laser diode current was ramped from 35 to 70 mA. The results in Fig. 7 show the generation of a significant, detectable interference fringe whose free spectral range is consistent with the round-trip cavity length

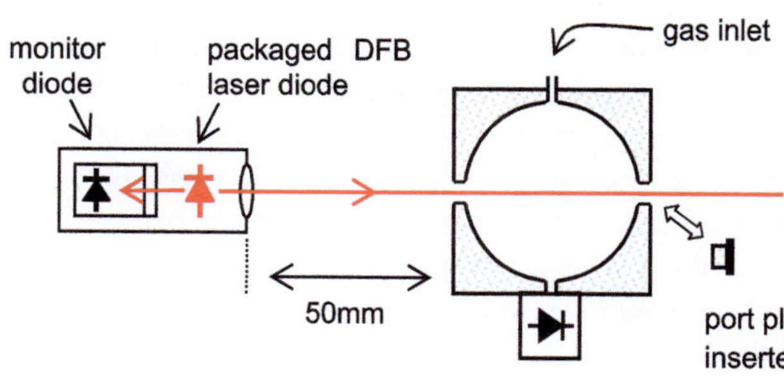

Fig. 6 Experiment to measure the effects of feedback from the first strike spot to the laser diode

Fig. 7 Signals from a monitor photodiode located at the laser diode rear facet to identify self-mixing interference. (**a**) Results of experiment with and without feedback from a 50 mm integrating sphere. (**b**) Subtracted signals showing self-mixing fringes caused by feedback from the sphere

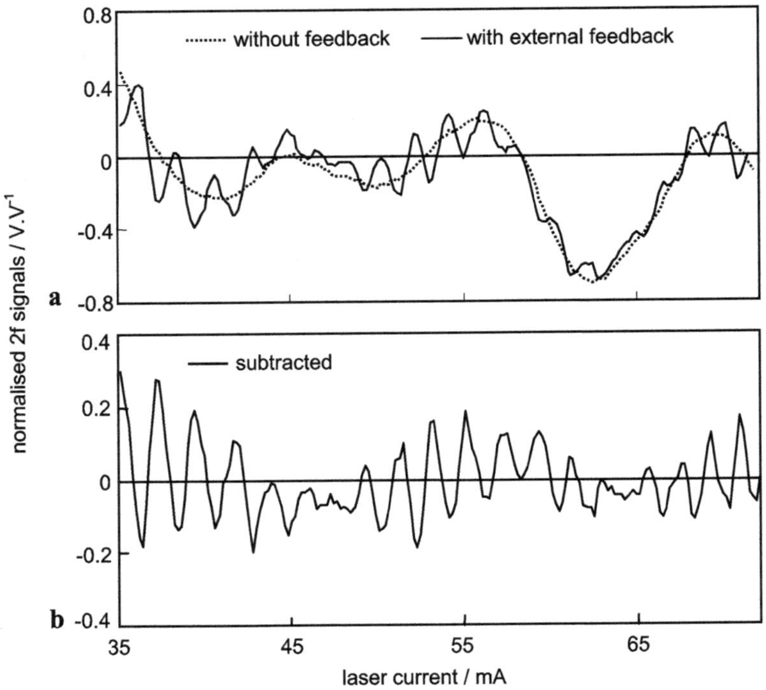

from laser to sphere surface. In a control test, a port plug at the first strike spot was removed to allow the light to pass through the sphere. The results of this test, also shown in Fig. 7, show a broader background fringe caused by the reference monitor photodiode itself; we have previously observed such fringes in more extensive self-mixing tests [23]. All our other experiments were conducted without the use of a monitor photodiode at the rear facet, in order to reduce this background fringe.

Our DFB laser diode is particularly susceptible to self-mixing interference because of its use of a large collimating lens, which acts to collect a greater proportion of backscattered light from the diffusely reflecting surface and deliver it to the laser diode. It is therefore possible that a reduced fringe amplitude would result from a better laser design, and from the use of balanced detection. With our apparatus, for the purpose of establishing limits of detection in the absence of self-mixing, we chose to simply place the laser diode a considerable distance from the sphere (1 m) in subsequent experiments. This reduces the fringes to a negligible level because of two effects: (i) a decrease in the intensity of backscattered light reaching the laser diode, and (ii) a reduction in the free spectral range of the fringes, which in a WMS experiment reduces their apparent magnitude [23].

4.3 Results for gas detection

Firstly, for the 50 mm diameter sphere, we recorded a direct scan of a gas line without implementing WMS. Because of reduced signal to noise ratios with the direct scan, a higher concentration of 5770 ppm was used in this experiment. Figure 8 shows the resulting background scan and gas line scan. A baseline analysis yields a deviation from a flat baseline (1σ) at any single point in the scan of 70 ppm or an uncertainty in $\Delta\Phi/\Phi_0$ of around 3×10^{-3}, which is close to our predicted value of 2×10^{-3} in Table 1. A similar level of performance might also be expected using direct spectroscopy with a conventional transmission cell.

Secondly, we implemented WMS and repeated the spectral scan. Using the 50 mm diameter sphere, $2f$-demodulated signals were recorded while the dc current was ramped from 40 to 75 mA, giving a scan across the methane line of width 22 GHz or 0.2 nm. The lock-in amplifier time constant was set to 1 s to enable comparison with later results, and data points were recorded every 2 s. This resulted in some slight loss of spectral resolution, but enabled scans to be completed within a few minutes. The results are shown in Fig. 9. Analysis of the baseline yields an RMS deviation (1σ) noise equivalent absorbance (NEA) of around 10^{-4}(1.6 ppm for methane). For a WMS dither of 0.025 nm, we would expect the intensity uncertainty $\Delta\Phi/\Phi_0$ to be between 2×10^{-3} and 2×10^{-7}, with a corresponding NEA (via (3) in Part 1) of between 3×10^{-3} and 3×10^{-7}.

In a final experiment the laser emission was locked to the peak of the absorption line. The $2f$-demodulated signal was recorded for a series of different concentrations of methane in air using the lock-in amplifier with a time constant of 1 s. The results are shown for both the 50 mm and 100 mm spheres in Fig 10. Their effective pathlengths at

Fig. 8 Direct spectral scan through a methane line at 1651 nm, using a 50 mm diameter integrating sphere. (**a**) Raw data, (**b**) calculated transmission

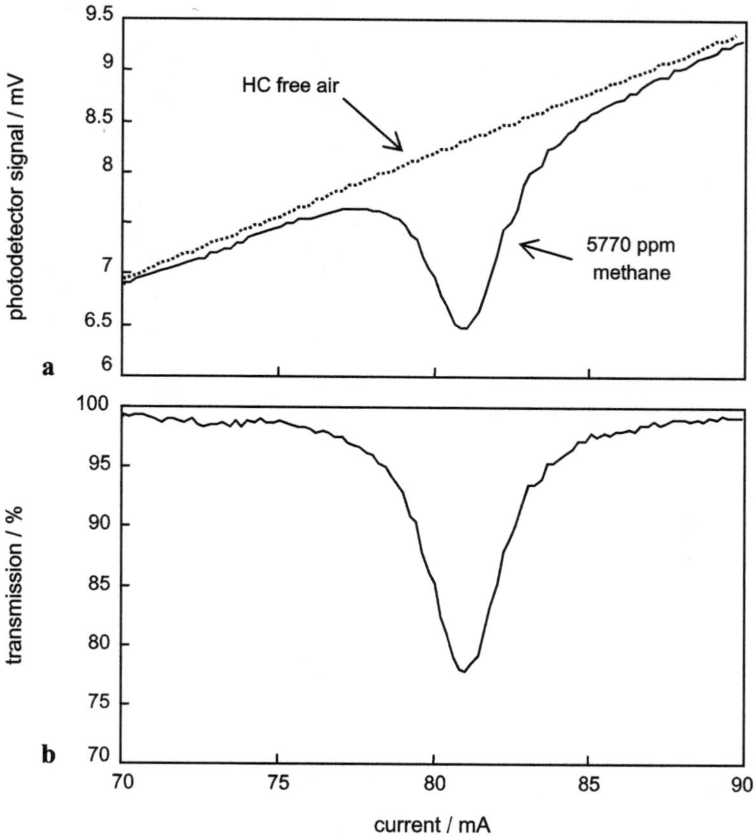

Fig. 9 $2f$-demodulated signals taken using a 50 mm diameter sphere containing 50 ppm methane and hydrocarbon free air

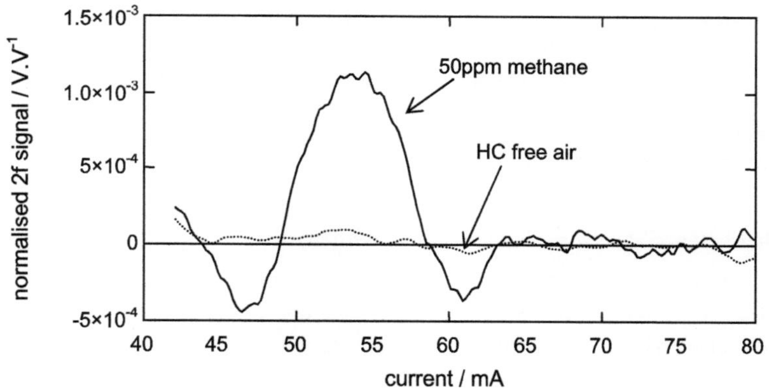

low levels of absorption were established to be 1.2 m and 3.4 m respectively. This calibration followed the methodology established in reference [21], using (3) to establish the value of ρ with high precision, then applying (2) which is valid at low absorption. Normalised $2f$ signals from the 100 mm sphere were approximately $3\times$ those from the 50 mm sphere, in line with the pathlength ratio of 2.8.

Limits of detection were established by taking the RMS deviation of a baseline measurement (1σ) over several periods of 1 min, and normalising to a standard gas measurement. This yielded limits of detection (1σ) of 1 ppm for the 50 mm sphere and 0.4 ppm for the 100 mm sphere, translating to NEAs of around 5×10^{-5} for both cases. We also measured an optical throughput of 0.3% for the 50 mm sphere (corresponding to the ratio of detected power to input power). Despite this low figure, the loss of light was not performance limiting.

4.4 Assessment of drift

The apparatus in Fig. 2 was adapted along the lines of Fig. 6 in Part 1, so as to include a second beamsplitter taking a proportion (approximately 45%) of the light to a reference detector. This reference channel employed the same model of detector as the main signal channel and the same lock-in am-

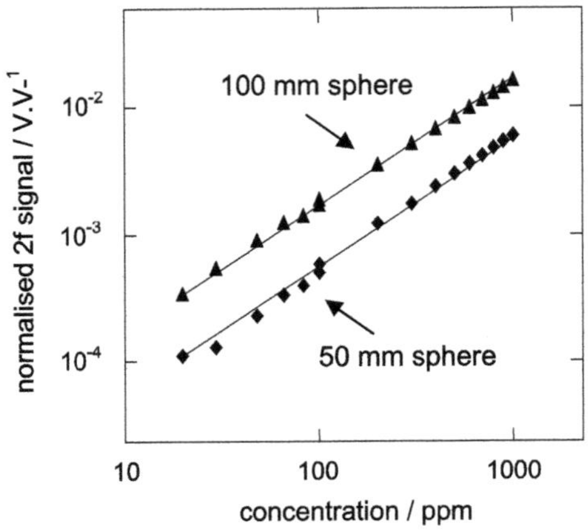

Fig. 10 Normalised $2f$-demodulated signals taken for different methane concentrations measurements in a 50 mm diameter and a 100 mm diameter sphere. The dashed lines show linear responses

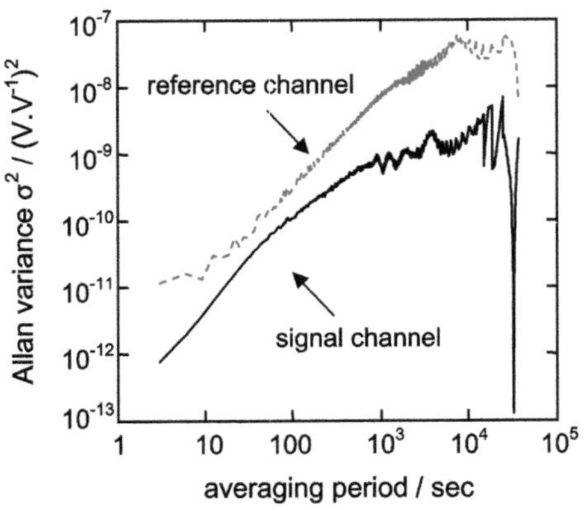

Fig. 11 Allan variance plot for normalised $2f$ time series data taken over a 21 hour period with an empty cell

plifier, with an identical set-up. The level of system drift was assessed by continuously recording data (with no gas in the cell) over a 21 hour period, for both the signal channel and the reference channel. Allan variance plots (σ^2 as defined by Werle et al. [24]) were calculated for the recorded $2f$ measurements for each channel, using data normalised by division by the DC signals. These plots are shown in Fig. 11 and, as in Part 1, describe a system dominated by drift.

The system drift appears to reach a maximum level of $\sqrt{\langle\sigma^2\rangle}$ corresponding to an NEA of 4×10^{-4} for the integrating sphere and 1×10^{-3} for the reference channel. As we found in Part 1, the reference channel consistently drifts by more than the signal channel (this effect has been found to be repeatable). It may be that our signal channel has a larger and more representative sample of the laser diode output, and that the additional drift in the reference channel is cause by it sampling only a small portion of the beam. Or, our pellicle beamsplitter may be unstable over long timescales, with small movements affecting the reflected reference beam more than the transmitted signal beam.

5 Discussion and conclusions

We expect that random laser speckle is the fundamental performance limiting effect for integrating spheres. For direct spectral scans, our results have deviations from a flat baseline of $\Delta\Phi/\Phi_0 = 3 \times 10^{-3}$, which is of a similar magnitude to the level predicted by our theory (2×10^{-3}). For WMS, our results yield short-term limits of detection (1σ) of 1 ppm for the 50 mm sphere and 0.4 ppm for the 100 mm sphere, or NEAs of around 5×10^{-5} in both cases. Longer-term drift was found to be at an NEA of 4×10^{-4} for the 50 mm sphere. These figures fall within our expected range, however that range is too broad to act as confirmation of the theory laid out in Part 1. We expected that speckle-related uncertainty in $\Delta\Phi/\Phi_0$ would be identical for spheres of different diameter.

The main uncertainty in the theory lies in the value of h, the mean optical path deviation for the surface, which is determined by characteristics of the surface material and not quantified for the Spectralon or Zenith PTFE scatterers used in our integrating spheres. Assuming our longer-term drift figure of 4×10^{-4} is at or above the level set by (9) in Part 1, we can set an upper limit for h at around 10 mm. This is rather large compared to the figure used for the diffuse optic in Part 1 (6 μm), but is on a scale commensurate with the penetration depth of light into the reflective material. Further work is needed to investigate this parameter and to test the theory rigorously in this application, as well as to determine whether there are other sources of drift in our system that could be dominating the results.

The question then arises, whether there is scope to improve on the limits of detection achieved here. Firstly, for the purpose of our experiments we required integrating spheres that could be flexibly reconfigured, and used apertures large enough to permit capture of the subjective speckle field. The equivalent port area was therefore larger than could be achieved through careful design, and in particular the detector port area and the entrance aperture could be reduced in size. It may be possible to reduce the detector port to 2 mm diameter and the entrance aperture to 3 mm diameter. This would have the effect of increasing the mean reflectivity of the inner surface of the sphere, by reducing the non-reflective area (f_{total} in (4)), and thereby increasing the pathlength. Using (2), we anticipate that mean pathlengths could then be improved from 1.2 and 3.4 m to approximately 2.8 m and 6.5 m respectively for the 50 mm and 10 mm diameter spheres, improving limits of detection by a factor of approximately 2 in both cases. These pathlength multipliers do not compete with the best alternative multi-pass cells, but nevertheless offer a considerable improvement over single pass cell.

Secondly, we believe that residual drift effects may persist in our laser diode package, even after removal of the internal reference diode used in Sect. 4.2. Previous work has shown that the use of a simple balanced detection scheme can improve the resulting uncertainties by a factor of 10 or more [23]. It is not certain whether this would yield the same improvement to the overall limit of detection, or reveal underlying random speckle noise that we expect to act as the fundamental limit. For an integrating sphere geometry, the noise levels for objective and subjective speckle are both expected to be similar. In contrast with the simple transmission cells of Part 1, the detection aperture for objective speckle is very large, reducing the associated speckle noise to a level comparable with subjective speckle recorded using a high NA lens.

Thirdly, we know that commercial integrating spheres are designed to give perfectly even irradiance over the surface of the sphere, by using materials that exhibit near ideal Lambertian scattering. For gas detection, our need for simple alignment translates to even irradiance at the surface local to the detector, and constant scattering behaviour at the first strike spot. This may offer the potential for designs better suited to gas detection. There may also be scope to better understand the value and role of the parameter h borrowed from speckle theory, and to reduce h while maintaining high levels of diffuse scatter.

We have confirmed the findings of other workers in this field, namely that integrating spheres are very simple to align, and that etalons are disrupted within the cell as long as there is no direct line of sight from the entrance aperture to the detector. We have also found it straightforward to avoid etalons between the entrance aperture window and other "upstream" optics. However, we have discovered that self-mixing or feedback interference can be performance limiting for lasers placed close to the integrating sphere, and optical feedback must therefore be reduced to realise low gas detection limits. Finally, although integrating spheres have very low levels of optical throughput (~0.3% for the 50 mm sphere), this has not limited performance for the near-infrared InGaAs detectors that we used.

We can compare the performance of single pass cells and integrating spheres. In Part 1, using a single pass cell with a diffuse transmitter, we demonstrated a short-term NEA (1σ) of 2×10^{-5}, and longer-term drift of up to 3×10^{-4} over 22 hours. Here, our short-term NEA was around 5×10^{-5} and our long-term drift around 4×10^{-4}. For the singly scattering diffuser used in Part 1, the value of h was easier to estimate, and indeed it would be possible to confirm this by direct measurement, whereas for the multiply scattering material used in the integrating sphere its value is unknown. For subjective speckle with the same detection aperture and the same material properties, we would expect the level of speckle uncertainty to be similar for (i) a single pass cell of length L, and (ii) an integrating sphere of diameter L, with the latter simply having a further $\sqrt{2}$ improvement as a result of depolarisation. For objective speckle, the latter has a wider field of view, reducing the expected speckle uncertainty compared to the single pass cell. For both objective and subjective speckle, the limit of detection for the integrating sphere then benefits from its longer pathlength, by a factor of between 20 and 50 depending on how the sphere has been configured (24 and 34 in our experiments, for the 50 mm and 100 mm spheres respectively).

To conclude, we have studied the effects of random laser speckle and self-mixing interference on TDLS based gas measurements made using integrating spheres. Experiments were conducted at 1651 nm for detection of methane, but the results can be translated to other wavelengths in the near-infrared. For a direct scan using a 50 mm diameter commercial integrating sphere, the estimated short-term NEA (1σ) was 3×10^{-3} or 75 ppm methane. Using wavelength modulation spectroscopy reduced the short-term NEA to around 5×10^{-5}, corresponding to 1.2 ppm and 0.4 ppm respectively for sphere diameters of 50 mm and 100 mm. Long-term drift was estimated to be at an NEA of 4×10^{-4} for the 50 mm sphere, using WMS. We believe there may be scope to improve on these limits in future work.

Acknowledgements This work was carried out under an EPSRC research grant (GR/T04601/01). Jane Hodgkinson is supported by an EPSRC Advanced Research Fellowship (GR/T04595/01). We would like to thank Dan Francis for his help with our data acquisition software.

References

1. J.U. White, J. Opt. Soc. Am. **32**, 285 (1942)
2. D.R. Herriott, H. Kogelnik, R. Kompfner, Appl. Opt. **3**, 523 (1964)
3. S.M. Chernin, E.G. Barskaya, Appl. Opt. **30**(1), 51 (1991)
4. Toptica GmbH, *Product specification. Compact Herriott cell for absorption spectroscopy: CMP-30*. Available at www.toptica.com (2009)
5. R. Engeln, G. Berden, R. Peeters, G. Meijer, Rev. Sci. Instrum. **69**, 3763 (1998)
6. A. O'Keefe, D.A.G. Deacon, Rev. Sci. Instrum. **59**, 2544 (1988)
7. A.G. Berezin, S.M. Chernin, D.B. Stavrovskii, in *Proc. 7th International Conference on Tunable Diode Laser Spectroscopy—* TDLS09, Paper E2 (2009)
8. H.I. Schiff, G.I. Mackay, J. Bechara, in *Air Monitoring by Spectroscopic Techniques*, ed. by M.W. Sigrist (Wiley, New York, 1994). Chapter 5
9. Labsphere Inc. *A Guide to Integrating Sphere Theory and Applications* (Labsphere, North Sutton, 1998)
10. P. Elterman, Appl. Opt. **9**(9), 2140 (1970)
11. L.M. Hanssen, K.A. Snail, in *Handbook of Vibrational Spectroscopy*, vol. 2, ed. by J.M. Chalmers, P.R. Griffiths (Wiley, Chichester, 2002), p. 1175
12. E.S. Fry, G.W. Kattawar, R.M. Pope, Appl. Opt. **31**(12), 2055 (1992)
13. I. Fecht, M. Johnson, Meas. Sci. Technol. **10**, 612 (1999)
14. J. Hodgkinson, M. Johnson, J.P. Dakin, Appl. Opt. **44**, 4360 (2005)
15. E. Hawe, P. Chambers, C. Fitzpatrick, E. Lewis, Meas. Sci. Technol. **18**, 3187 (2007)
16. E. Hawe, C. Fitzpatrick, P. Chambers, G. Dooly, E. Lewis, Sensor. Actuat. A **141**, 414 (2008)

17. C.G. Venkatesh, R.S. Eng, A.W. Mantz, Appl. Opt. **19**(10), 1704 (1980)
18. R.M. Abdullin, A.V. Lebedev, Sov. J. Opt. Technol. **55**(3), 139 (1988)
19. S. Tranchart, I.H. Bachir, J.-L. Destombes, Appl. Opt. **35**(36), 7070 (1996)
20. D. Masiyano, J. Hodgkinson, R.P. Tatam, Appl. Phys. B **90**, 279 (2008)
21. J. Hodgkinson, D. Masiyano, R.P. Tatam, Appl. Opt. **48**(30), 5748 (2009)
22. A. Bozeit, J. Burke, H. Helmers, H. Sagehorn, R. Schuh, Opt. Laser Technol. **30**, 325 (1998)
23. D. Masiyano, J. Hodgkinson, S. Schilt, R.P. Tatam, Appl. Phys. B **96**(4), 863 (2009)
24. P. Werle, R. Mücke, F. Slemr, Appl. Phys. B **57**, 131 (1993)

Mid-infrared quantum cascade detectors for applications in spectroscopy and pyrometry

D. Hofstetter · F.R. Giorgetta · E. Baumann · Q. Yang · C. Manz · K. Köhler

Received: 20 August 2009 / Revised version: 3 February 2010 / Published online: 17 March 2010
© Springer-Verlag 2010

Abstract In this paper, we give an overview of quantum cascade detector technology for the near- and mid-infrared wavelength range. Thanks to their photovoltaic operating principle, the most advanced quantum cascade detectors offer great opportunities in terms of high detection speed, reliable room temperature operation, and excellent Johnson noise limited detectivity. Besides some important features dealing with their fabrication and their general characteristics, we will also briefly present some possibilities for performance improvement. Elementary theoretical considerations adopted from photoconductive detectors confirm that optimization of such devices always involves various tradeoffs.

1 Introduction

Although intersubband (ISB) photodetectors were first demonstrated already more than twenty years ago, they remain a topic of high scientific and practical interest. These early devices were photoconductive quantum well infrared photodetectors (QWIPs) [1]. It was soon discovered that QWIPs have many unique properties, especially regarding high frequency operation. As ISB detectors are unipolar devices, their fundamental speed limit is the ISB scattering time of electrons ($\tau_{\text{scatter}} \approx 1$ ps) rather than the parasitic capacitance of the p-n-junction or the finite transit time through the intrinsic region [2]. This typically results in cutoff frequencies on the order of 100 GHz [3]. As a further advantage, ISB detectors can be designed for a wide range of wavelengths using a single material system just by choosing adequate semiconductor layer thicknesses. This is in sharp contrast to interband devices, where the bandgap determines to a large extent the detectable photon energy.

D. Hofstetter (✉)
University of Neuchatel, Avenue de Bellevaux 51,
2009 Neuchatel, Switzerland
e-mail: Daniel.Hofstetter@unine.ch
Fax: +41-32-7182901

F.R. Giorgetta · E. Baumann
Optoelectronics Division, National Institute of Standards and Technology, 325 Broadway, Boulder, CO 80305, USA

Q. Yang · C. Manz · K. Köhler
Fraunhofer Institute for Applied Solid State Physics,
Tullastrasse 72, 79108 Freiburg i. Brsg., Germany

An excellent overview on the different working principles of ISB infrared photodetectors is given in a book by Schneider and Liu [4]. In very basic terms, one can distinguish between photoconductive and photovoltaic detectors. In the class of photoconductive devices, the most common detector is the QWIP, where the change of device resistance under illumination is measured. There are several types of QWIPs; they differ mainly by the kind of ISB transition exploited. According to the position of the upper detector level, the optical transition can be either bound-to-quasi-bound [5] or bound-to-miniband [6]. In the widely used bound-to-quasi-bound QWIP, which was also the first ISB detector to be demonstrated in 1987, the detection energy depends on the conduction band discontinuity between the quantum well (QW) and the barrier material. The QW thickness and conduction band discontinuity (through the material composition) are chosen such that the second quantized electronic level is close to resonance with the barrier's conduction band edge; this measure ensures a good carrier extraction efficiency under application of an appropriate bias voltage. By using the bound-to-miniband design, the QWIP detection wavelength can be partly decoupled from the conduction band discontinuity. Nevertheless, a resonance condition between the upper detector state and the miniband states must be fulfilled in order to obtain a good responsivity. Today, QWIP focal plane arrays sensitive at wavelengths between 8 μm and 10 μm have reached commercial maturity. As a particularly advanced example of such a QWIP-based camera, Gunapala et al. presented a high performance 1024 × 1024 pixel dual band focal plane array based on GaAs/AlGaAs with cutoff wavelengths of 5.1 μm and 8.4 μm [7]. The big advantages of such photoconductive ISB detectors are a high responsivity and a good detectivity, especially at low operating temperatures. However, their quite considerable dark current leads to a non-negligible dark current noise which dominates the noise budget at higher temperatures. This drawback is only partly compensated by the relatively high responsivity. In contrast, a photovoltaic ISB detector profits from the fact that there is no dark current and thus no dark current noise. Unfortunately, such a device has no photoconductive gain and therefore a lower responsivity, but the better noise behavior can at least partly make up for these shortcomings.

It is therefore somewhat astonishing that over the years, the development of photovoltaic ISB photodetectors has seen a slower overall progress. Already back in 1991, Schneider et al. [8, 9] observed pronounced photovoltaic effects in an asymmetric multi QW structure. The potential asymmetry was obtained via a sheet of delta-doping close to the active QW. As shown by Schneider as well, optimized photovoltaic QWIPs have superior noise properties, the capability to operate at higher photon fluxes, and an improved dynamical range in comparison to a photoconductive QWIP [10]. Especially with respect to the improved noise behavior, it is thus no surprise that the first QW-based infrared detector for THz radiation was a photovoltaic device rather than a photoconductive QWIP [11, 12]. A similar ISB detector for mid-infrared wavelengths was later presented by Gendron et al. [13]; in analogy to the functioning of a quantum cascade laser (QCL), this device was named quantum cascade detector (QCD). In such a QCD, the asymmetric potential used for unilateral carrier transport is formed by a series of QWs with increasing thicknesses. Interesting enough, a QCD-like semiconductor structure was invented already in 1987 under the name 'optical charge pump' [14]. Like the bound-to-miniband QWIP, this device offers more design freedom for a given material composition. A very schematic comparison between the band structures of a photoconductive QWIP and a photovoltaic QCD is shown in Fig. 1.

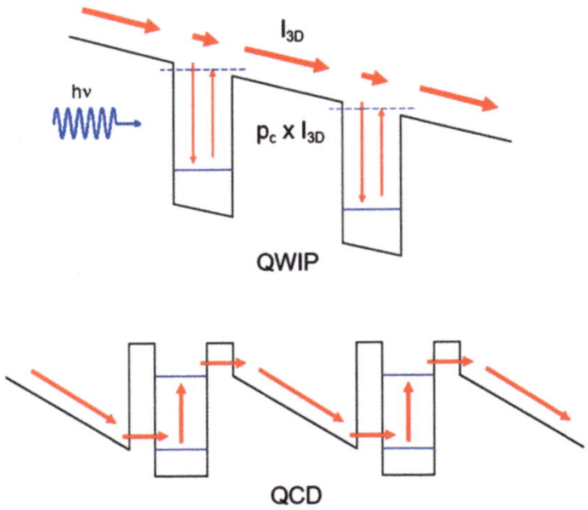

Fig. 1 Schematic conduction band diagram of a QWIP and a QCD. In the QWIP, electron transport is accomplished by an external voltage bias whereas in a QCD, an internal potential ramp ensures the carrier transport

Compared to a QWIP, the absence of dark current noise is not the only advantage of a QCD. The missing dark current also prevents capacitance saturation in the read-out circuit and thus allows longer integration times. Finally, the thermal load of the detector is strongly reduced, which is of interest if the available cooling is limited, for example in space-born or hand-held terrestrial staring systems. As a final point, the somewhat narrower linewidth of QCDs compared to QWIPs results in a reduced background photon noise. Although QCDs have already progressed a lot [15], they do not yet take full advantage of their potentially low Johnson noise; mainly because of an insufficiently high room temperature device resistance. Our most recent results, which will be presented in this paper, show nevertheless that QCDs with competitive performance can be achieved using improved designs.

We will discuss the design principles of QCDs as well as the choice of material system for the different detection wavelengths. Experimental results of QCDs at various wavelengths are reviewed. First, QCDs based on InGaAs/InAlAs lattice matched to InP detecting between 4.7 μm and 10 μm are presented. As lattice matched InGaAs QCDs can only detect wavelengths above ∼4 μm due to the conduction band discontinuity of 520 meV, an alternative approach is presented for shorter wavelengths: strained InGaAs/InAlAs. Finally, a broadband QCD covering the wavelength region from 4.7 to 7.4 μm will be described. We will also discuss some of the trade-offs to be made when designing improved QCDs; they will lead to certain guidelines for detectors with particular applications.

2 QCD theory

The electro-optical behavior of QCDs can be described using exactly the same theory as in QWIPs. The basic optical process for mid-infrared detectors such as QWIPs or QCDs is the ISB transition [16]. In contrast to interband transitions, ISB transitions result in relatively narrow and peaked absorption features, whereas interband transitions are characterized by spectrally large absorption features with a cut-off

Table 1 General behavior of important QCD parameters as a function of the sheet carrier density, n_s, and the number of active QWs, N_{QW}

	η	g_p	$R_p \propto \eta \cdot g_p$	$D^*_{BLIP} \propto R_p/\sqrt{g_p^2 \eta}$	$R_0 A$	$D^*_J \propto R_p \cdot \sqrt{R_0 A}$	T_{BLIP}
n_s small	n_s	const	n_s	$\sqrt{n_s}$	$1/n_s$	$\sqrt{n_s}$	const
n_s large	1	const	const	const	$1/n_s$	$1/\sqrt{n_s}$	$\downarrow$
N_{QW} small	N_{QW}	$1/N_{QW}$	const	$\sqrt{N_{QW}}$	N_{QW}	$\sqrt{N_{QW}}$	const
N_{QW} large	1	$1/N_{QW}$	$1/N_{QW}$	const	N_{QW}	$1/\sqrt{N_{QW}}$	$\downarrow$

wavelength. The main reason for this striking difference is the opposite curvature of the electron and hole dispersion curves in the interband case and the parallel curvature of the electronic subbands in the ISB case. The current responsivity of a QCD, R_p, is defined as detector output current I_S per unit of input signal power P_S and is given by

$$R_p = \frac{I_S}{P_S} = \frac{q}{h\upsilon}\eta g_p = \frac{\lambda q}{hc}\eta g_p; \quad g_p = \frac{p_e}{N_{QW} p_c} \tag{1}$$

where $\nu = c/\lambda$ is the signal frequency, λ the signal wavelength, c the vacuum speed of light, q the elementary charge, h Planck's constant, η the absorption efficiency, and g_p the photodetector gain. Furthermore, p_e is the escape probability of an excited electron in the active QW, p_c its capture probability into the active QW ground state, and N_{QW} the number of active QWs. Optimization of R_P is thus accomplished through improvement of both absorption efficiency and photodetector gain. For sufficiently low absorptions, the total absorption efficiency is proportional to the absorption efficiency of a single QW. However, since the detector gain is inversely proportional to the number of QWs, the current responsivity is in first approximation independent of the number of periods.

Another common figure of merit for photodetectors is the detectivity $D^* = R_p/i_n\sqrt{A\Delta f}$, which is the ratio between peak responsivity R_p and mean noise current i_n (or the inverse noise equivalent power) normalized by the detector area A and the measurement bandwidth Δf. Its units are cm $Hz^{0.5}$/W, also known as Jones. For any given infrared detector, the detectivity D^* reveals two distinct temperature regimes. At low temperatures, D^* is dominated by photon noise due to the 300 K blackbody radiation seen by the device. This is the background limited operating (BLIP) regime. Above a certain temperature, T_{BLIP}, other noise mechanisms become dominant. For a QWIP, D^* at $T > T_{BLIP}$ is determined by dark current noise, whereas the detectivity of a QCD is dominated by Johnson noise. Taking into account these different facts, we get for the detectivity of a QCD

$$\begin{aligned} D^*_J &= R_p\sqrt{\frac{R_0 A}{4k_B T}} \\ D^*_{BLIP} &= \frac{R_p}{\sqrt{2q^2 g_p \int g_p \eta(\nu')\frac{d\Phi_{BG}(\nu')}{d\nu'}d\nu'}} \end{aligned} \tag{2}$$

where R_0 is the differential device resistance around 0 V, T the device temperature, and $d\Phi_{BG}(\nu')/d\nu'$ the spectral background photon flux density. For a QCD, D^* is maximized by ensuring a high device resistance (thus a low Johnson noise) without lowering the escape probability and thus the detector gain. The most important design parameters are the layer thicknesses (determining the band profile as discussed below), the doping density n_s of the active QW, and the number of periods N_{QW}.

Let us first have a look at the dependency on the doping density (sheet carrier density) n_s (see also Table 1). As long as the absorbance is small, $N_{QW}\alpha_{2D} \ll 1$, so that $\eta = 1 - \exp(-N_{QW}\alpha_{2D}) \approx N_{QW}\alpha_{2D}$, the absorption efficiency η and thus the responsivity R are proportional to α_{2D} and thus to the doping density n_s [15]. Looking at (2), it follows that D^*_{BLIP} is proportional to $\sqrt{n_s}$. As the device resistance decreases proportionally in $1/n_s$, D^*_J is proportional to $\sqrt{n_s}$ as well. In this case, T_{BLIP} remains constant. This does, however, not imply that n_s should be as high as possible. The absorption efficiency η cannot become larger than 1 and therefore saturates for too high doping concentrations. For high n_s, R and D^*_{BLIP} therefore saturate, whereas $D^*_J \propto 1/\sqrt{n_S}$ decreases with a further increase of n_s. This leads to a decrease in T_{BLIP} as well.

Similar considerations can be made regarding the dependency on the number of periods N_{QW}. Assuming again that $N_{QW}\alpha_{2D} \ll 1$ (small absorption) and thus $\eta \propto N_{QW}$, R is independent of the number of periods N_{QW} and D^*_{BLIP} increases with $\sqrt{N_{QW}}$. As the device resistance grows linearly with N_{QW}, the Johnson noise dependent detectivity D^*_J shows the same behavior as D^*_{BLIP}: $D^*_J \propto \sqrt{N_{QW}}$. This means again that T_{BLIP} does not change. For large N_{QW}, the aforementioned assumption of small absorption is not true anymore, η and D^*_{BLIP} saturate, R is proportional to $1/N_{QW}$, and D^*_J is proportional to $1/\sqrt{N_{QW}}$: T_{BLIP} decreases. As a conclusion, we notice that for any given doping density, an ideal number of periods must be found. This is also shown in Fig. 2 where a two-dimensional plot of D^*_J as a function of the number of periods and the doping density is shown. If—for instance—the doping density is increased (at constant number of periods), the responsivity increases as well. However, the resistance goes down, leading to a maximum in D^*_J. The same effect can be seen when increasing the number of periods (at constant doping density). Although the resistance

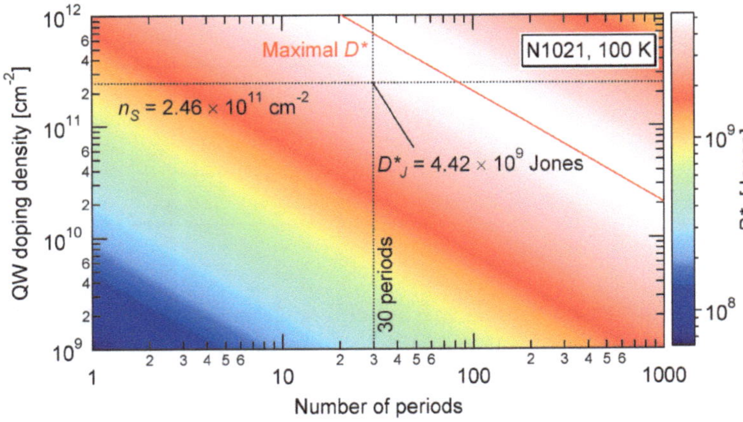

Fig. 2 Two-dimensional representation of the Johnson-noise limited detectivity of a QCD. The simulation is based on (1) and (2) and was done for a 7.5 µm device at 100 K

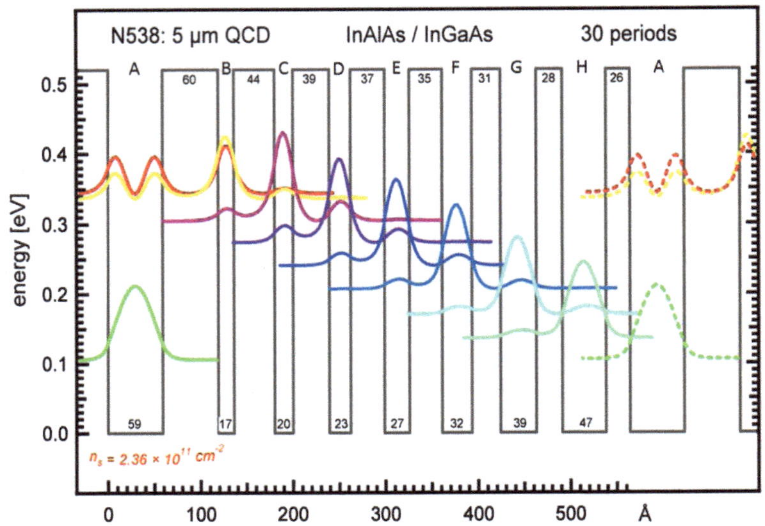

Fig. 3 Calculated conduction band profile of a QCD for 5.3 µm. QW A is the active QW, QWs B to H form the extractor cascade

goes up, the responsivity will start to decrease: in total, we will again get a maximum in the detectivity.

3 QCD design

The quantum-mechanical polarization selection rule [17] dictates that only the electric field component perpendicular to the QW layers interacts with ISB transitions. Therefore, vertical incidence of the incoming radiation has to be avoided. Possible device geometries taking into account the polarization selection rule include a 45° wedge multi-pass geometry, the Brewster geometry, or a surface grating. The particular choice of sample preparation is determined by factors such as the absorption strength of the sample or the specific application case of the detector.

The most crucial design aspect of ISB devices is the quantum-mechanical band structure. It can be calculated using numerical Schrödinger equation solvers. A typical band structure of a QCD is shown in Fig. 3. QW A is the active doped QW, while QWs B–H form the nominally undoped electron extraction cascade. The transport from the active QW into the extraction region is guaranteed via a resonant tunneling process between the excited state in the active QW, A, and the ground state of QW B; this allows for a thick barrier between the active QW and the extractor. To achieve an efficient electron extraction through phonon assisted scattering, the energy difference between the individual extractor states should be close to the longitudinal optical phonon energy E_{LO} (GaAs: $E_{LO} = 36$ meV, $In_{0.53}Ga_{0.47}As$: $E_{LO} = 32$ meV).

4 QCD devices

There are various photodetector applications in the mid-infrared spectral range. They include pyrometry, spectroscopy, night vision, and sensing of hot spots. In the first part of this section, experimental results obtained from QCDs sensitive at wavelengths between 4.7 µm and 9.8 µm (126 meV–268 meV) are presented. The investigated samples consist of $In_{0.53}Ga_{0.47}As$ QWs and $In_{0.52}Al_{0.48}As$ barriers lattice matched to InP substrates. All layer structures were grown by molecular beam epitaxy (MBE) in order to

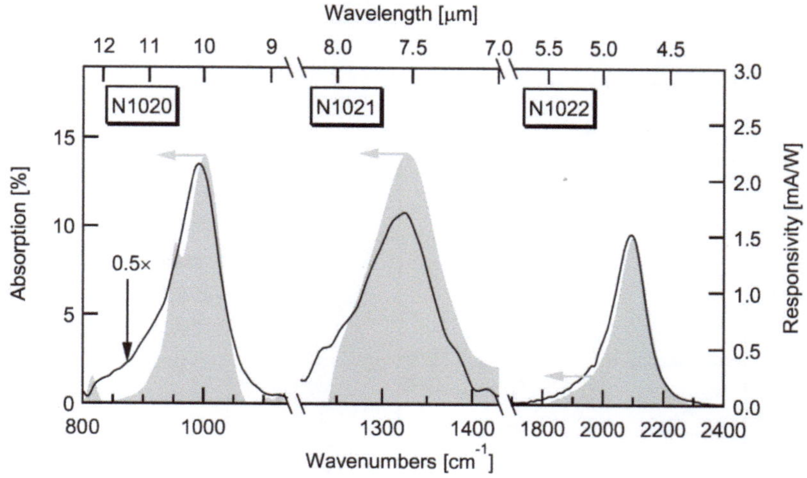

Fig. 4 Responsivity of the ISB main transition of a 9.8 µm device (*left*), a 7.5 µm device (*center*), and a 4.7 µm QCD (*right*) at 300 K along with the room temperature absorption spectra

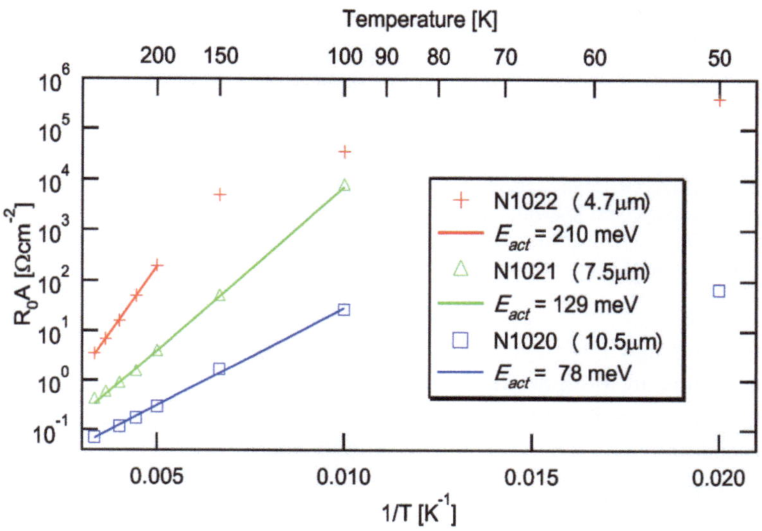

Fig. 5 Logarithmic representation of the R_0A-product of the three QCDs (4.7 µm: *cross symbols*, 7.5 µm: *upward triangles*, 10 µm: *squares*) shown in the previous figure as a function of inverse temperature. We observe activation energies of 210 meV for the 4.7 µm detector, 129 meV for the 7.5 µm device, and 78 meV for the 10.5 µm QCD

achieve a high interface quality between QWs and barriers. Growth started with a 6000 Å thick $In_{0.53}Ga_{0.47}As$ lower contact layer followed by 30 repetitions of the active region and a 2000 Å thick $In_{0.53}Ga_{0.47}As$ upper contact layer. Details on the layer thicknesses can be found in reference [18]. From a design point of view, the samples shown here are based on the QCDs presented in reference [19]. They have, however, thicker barriers to increase device resistance. After growth, the samples were polished into 45° multi-pass waveguides for absorption tests; for photocurrent measurements, mesas were processed using standard photolithography and wet etching. Contacting was obtained through evaporated metal contacts.

Figure 4 shows the measured responsivity spectra of the three samples for temperatures at 300 K. These responsivity spectra need to be compared with the grey-shaded absorption spectra. Excellent agreement is observed. For the two shorter wavelength samples, the peak detection energies of 268 meV/2144 cm^{-1} and 168 meV/1344 cm^{-1} correspond well to the simulated ones of 266 meV/2128 cm^{-1} and 165 meV/1320 cm^{-1}, respectively. For the longer wavelength sample, the observed value of 127 meV/1016 cm^{-1} is slightly too large with respect to the designed 118 meV/944 cm^{-1}. Figure 5 shows the R_0A-products for all three samples as a function of inverse temperature. Activation energies of 210 meV, 168 meV, and 78 meV were observed for the 4.7 µm, 7.6 µm, and 9.8 µm samples, respectively. Via (2), we calculated a background-limit temperature and detectivity of $T_{BLIP} = 45$ K and 2×10^{11} Jones, respectively, for the 4.7 µm QCD. Because of their lower device resistances, the two other QCDs show detectivities remains below D^*_{BLIP} for all investigated temperatures. They reach values of 5×10^9 Jones at 100 K for the 7.5 µm QCD and 9×10^8 Jones at 100 K for the 9.8 µm QCD.

The conduction band discontinuity of the lattice-matched $In_xGa_{1-x}As/In_yAl_{1-y}As$ heterostructures can be increased from the lattice-matched value of 520 meV at $x = 0.53$, $y = 0.52$ by raising the In content above 53% in the InGaAs QW and reducing it below 52% in the InAlAs barrier. However, the modified In contents introduce strain be-

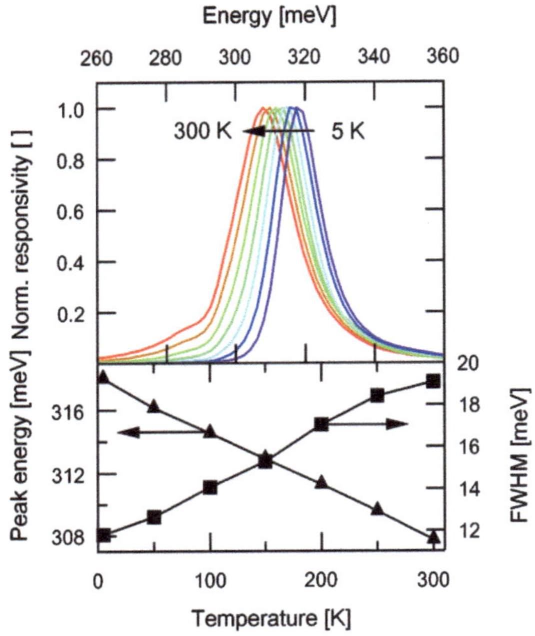

Fig. 6 *Top*: temperature evolution of the responsivity spectra of the strained InGaAs/InAlAs QCD between 5 and 300 K. *Bottom*: energy of peak responsivity and spectral linewidth as a function of temperature. The relative linewidth increases from 4.4% at 5 K to 6.6% at 300 K and the temperature tuning coefficient is $-0.275\ \mathrm{cm}^{-1}/\mathrm{K}$

tween the barrier and QW layers and the InP substrate. Since these strains are of opposite sign, namely tensile in the barrier and compressive in the QW, a nearly strain compensated pseudomorphic active region can be obtained by choosing appropriate layer thicknesses and material compositions. Using strained $In_{0.61}Ga_{0.39}As/In_{0.42}Al_{0.55}As$ with a band discontinuity of 610 meV [20], a 10 period QCD with a peak detection energy of 319 meV (3.88 μm) could be fabricated [21]. Its active region consisted of 60% barrier material; hence a small residual strain of 0.1% towards the InP substrate remained. Due to the relatively small total thickness of the detector, this small lattice mismatch did not lead to a relaxation of the crystal. The measured responsivity shown in Fig. 6 (top) has a maximum value of 10.7 mA/W at 318 meV (3.9 μm) and 150 K. The measured transition energy corresponds well to the simulated ISBT energy of 319 meV, demonstrating that both growth and simulation of 0.5% strained InGaAs/InAlAs heterostructures have a high maturity just as the lattice-matched InGaAs/InAlAs. Inserting the measured room temperature peak absorption per double pass of 6% and the corresponding peak responsivity of 5.8 mA/W at 300 K into (1) results in a 30% escape probability p_e of photoexcited electrons from the main QW into the extractor. The Johnson noise limited detectivity D_J^* of this short wavelength QCD was calculated with the measured R_0A and peak responsivities; it equals 4.9×10^7 Jones at 300 K and reaches the background limited detectivity $D_{\mathrm{BLIP}}^* = 1.2 \times 10^{11}$ Jones (for 300 K background temperature and a hemispherical FOV) at $T_{\mathrm{BLIP}} = 108$ K. The evolution of the linewidth and the responsivity peak position is shown at the bottom of Fig. 6.

A potential application for fast mid-infrared photodetectors is heterodyne spectroscopy, which allows distinguishing spectrally close absorption lines. This spectroscopic method requires mid-infrared detectors capable of multi-GHz operation. In order to obtain a preliminary estimation on the frequency response of a 5.46 μm QCD, such a device was tested using an optical heterodyne beat setup [22]. For this experiment, the detector was held at room temperature. The optical input signal was generated by two collinearly overlaid CW operated distributed feedback QCLs with identical emission wavelengths around 5.4 μm. By modifying the drive current of one of the QCLs, its emission frequency was slightly shifted resulting in a beat signal with a repetition rate equal to the difference frequency between the QCLs. The detector response showed a second order low-pass characteristic which could be described with an RLC circuit taking into account the input impedance of the amplifier R, the parasitic capacitance of the QCD, C, and the bond wire inductance, L. The corner frequency was 4 GHz and the cutoff frequency 23 GHz. Our simple theoretical model confirmed that the frequency response was limited by the sample mount. Due to the ultra-fast phonon-electron scattering, a cutoff frequency of roughly 65 GHz is expected for these detectors, similar as with the QWIPs presented in reference [2].

5 Outlook

Figure 7 shows an overview of the different QCDs fabricated so far in our labs. One can see that there exist three material families, which fall into the three interesting wavelength ranges. There are GaAs/AlGaAs devices for the far-infrared (84 μm, not shown in the figure), InGaAs/InAlAs detectors for the mid-infrared (3.9–17.5 μm), and InGaAs/AlAsSb QCDs for the near-infrared wavelength range (2.1–2.5 μm). It is quite obvious that the detectivities of the shorter mid-infrared devices are closest to the theoretical maximum. Therefore, any future effort will without any doubt concentrate on an improvement of the detectivity of QCDs in the technologically most important 5–10 μm range. Samples N973 and N975 show also a very high detectivity, but they had to be cooled to 5 K to achieve these good values. The reason for their high resistance and thus nearly ideal detectivity was the use of regular superlattice extraction regions with a relatively poor conductivity in growth direction. For each specific application of a QCD, a different set of parameters needs to be optimized. For a detector in pyrometry, the responsivity close to room temperature should be as large as possible, while spectrally narrow detection

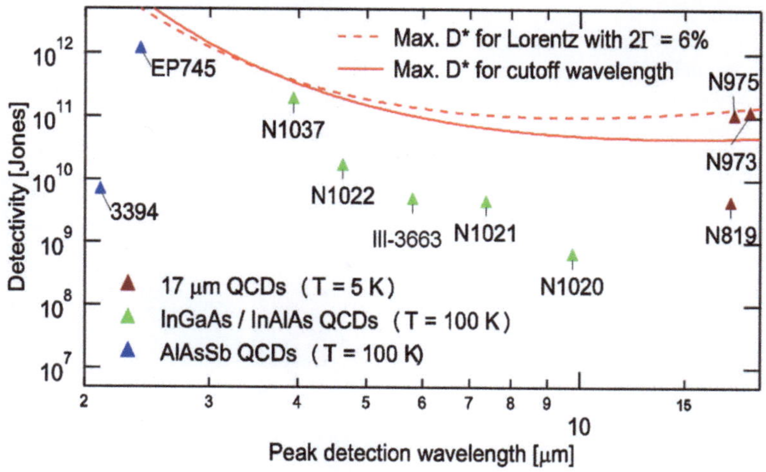

Fig. 7 Detectivity as a function of detection wavelength for all near- and mid-infrared QCDs presented in this article. It is obvious that the devices between 4.7 μm and 10 μm offer a considerable potential for improvement

is not necessarily an advantage. In highly sensitive spectroscopy applications, a narrow detection window together with a good detectivity would be highly desirable. For applications such as heterodyne spectroscopy, high frequency operation would be required. In a night vision system, finally, one would like to have a good detectivity, but not necessarily at room temperature. Such a system could very well function at cryogenic temperatures and at low frequencies. Several of the requirements of such a hypothetical wish list are contradictory; therefore, one has to make trade-offs. For instance, someone could be tempted to increase the room temperature resistance by making very thick barrier layers. This would result in a better Johnson noise behavior, but since thick barriers will decrease the tunneling probabilities, the detector would certainly end up slower than before. Or in an effort to increase the responsivity against spectrally broad blackbody radiation, one might try to make a broadband detector. But this will increase the background noise floor and therefore decrease the *BLIP* detectivity of the QCD. Here, we review briefly some possibilities for improved detectors:

Recently, a spectrally broad mid-infrared QCD based on lattice-matched InGaAs/InAlAs with a relative linewidth of 27% was designed and fabricated [23]. It made use of 26 carefully designed active region stages spanning a wavelength range between 4.7 and 7.4 μm. The device worked with a responsivity in excess of 10 mA/W at 150 K and 1.5 mA/W at 300 K. The detectivity was somewhat lower than in a standard narrowband QCD, but still acceptably high ($D^*_{\mathrm{BLIP}} = 2 \times 10^{10}$ Jones at $T_{\mathrm{BLIP}} = 40$ K).

As already stated above, pyrometers should possess a good detector response at temperatures close to 300 K. For this purpose, one has first to grow thick barriers to reduce the electron-phonon scattering rates between adjacent periods and to secondly push all extractor states towards higher energy using thinner QW layers. Due to a reduced ground state coupling and less thermal activation of electrons, these measures will result in a considerably higher device resistance, especially at room temperature. Since the resistance does not only improve the Johnson noise limited detectivity, but also prevent leakage currents in the device, this will definitely increase QCD performance in terms of response. On the other hand, thicker barriers will slightly reduce the detection speed. In Fig. 8, we show simulated resistance-area products for the 4 μm QCD shown previously, and for an improved version of it. It is obvious that there occurs a factor of 100 in the R_0A-product which results in a factor of 10 improvement in the Johnson noise limited detectivity.

For high-speed operation, it is clear that phonon resonance and efficient scattering processes in the extraction region are an absolute necessity. For this reason, barrier layer thicknesses should not be too large. As the 23 GHz cutoff frequency of a QCD in this article have shown, a more sophisticated sample mounting adapted to high frequency operation will have a positive effect on the detector speed. Together, these measures should enable detector operating frequencies closer to the theoretical maximum.

6 Conclusions

QCDs are a promising technology which has proven to work well at wavelengths from the near infrared to the THz region as presented in Fig. 7. The design process of QCDs is reliable and robust. Especially in the mid-infrared range around 4 to 17 μm, well established semiconductor material systems and processing methods are available. Accordingly, we demonstrated QCDs in the near-IR fabricated from InGaAs/AlAsSb, in the mid-IR using InGaAs/InAlAs, and in the THz-region using GaAs/AlGaAs materials. A chirped QCD design is a first step towards semiconductor-based ISB detectors for spectrally broad applications. Together with the high speed and design flexibility, this makes QCDs very interesting candidates for applications in this wavelength range. Considerable margin for improvement, especially in terms of room temperature detectivity and responsivity, has been identified.

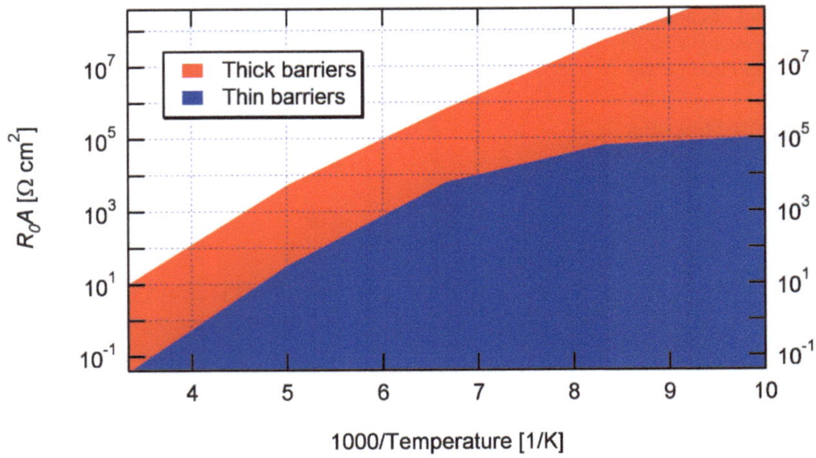

Fig. 8 Simulated resistance-area-products for the 4 μm QCD shown in Fig. 6 in its current configuration (thin barriers, low extractor states) and in the improved version (thick barriers, pushed up extractor states)

Acknowledgements The authors acknowledge the generous financial support of the Swiss National Science Foundation. This work was initiated by the Gebert-Rüf foundation in Basel, Switzerland. Technical assistance and valuable discussions with Marcel Graf, Milan Fischer, Marcella Giovannini, and Jérôme Faist are also gratefully acknowledged.

References

1. B.F. Levine, K.K. Choi, C.G. Bethea, J. Walker, R.J. Malik, Appl. Phys. Lett. **50**, 1092 (1987)
2. P. Bhattacharya, *Semiconductor Optoelectronic Devices*, 2nd edn. (Prentice-Hall, Englewood Cliffs, 1997), pp. 364–368
3. H.C. Liu, J.M. Li, E.R. Brown, K.A. McIntosh, K.B. Nichols, M.J. Manfra, Appl. Phys. Lett. **67**, 1594 (1995)
4. H. Schneider, H.C. Liu, *Quantum Well Infrared Photodetectors* (Springer, Berlin, 2006)
5. H.C. Liu, J. Appl. Phys. **73**, 3062 (1993)
6. K.M.S.V. Bandara, J.W. Choe, M.H. Francombe, A.G.U. Perera, Y.F. Lin, Appl. Phys. Lett. **60**, 3022 (1992)
7. S.D. Gunapala, S.V. Bandara, J.K. Liu, C.J. Hill, S.B. Rafol, J.M. Mumolo, J.T. Trinh, M.Z. Tidrow, P.D. LeVan, Infrared Phys. Technol. **47**, 67 (2005)
8. H. Schneider, P. Koidl, F. Fuchs, B. Dischler, K. Schwarz, J.D. Ralston, Semicond. Sci. Technol. **6**, C120 (1991)
9. H. Schneider, K. Kheng, M. Ramsteiner, J.D. Ralston, F. Fuchs, P. Koidl, Appl. Phys. Lett. **60**, 1471 (1992)
10. H. Schneider, J. Appl. Phys. **74**, 4789 (1993)
11. M. Graf, G. Scalari, D. Hofstetter, J. Faist, H. Beere, E. Linfield, D. Ritchie, G. Davies, Appl. Phys. Lett. **84**, 475 (2004)
12. H.C. Liu, H. Luo, C. Ying Song, Z. Wasilewski, A. SpringThorpe, J. Cao, IEEE J. Sel. Top. Quantum Electron. **14**, 374 (2008)
13. L. Gendron, M. Carras, A. Huynh, V. Ortiz, C. Koeniguer, V. Berger, Appl. Phys. Lett. **5**, 2824 (2004)
14. M.A. Reed, US patents #4'878'104 (1987) and #5'165'065 (1992)
15. L. Gendron, C. Koeniguer, V. Berger, X. Marcadet, Appl. Phys. Lett. **86**, 121116 (2005)
16. M. Helm, *The Basic Physics of Intersubband Transitions*. Semiconductors and Semimetals, vol. 62 (Academic Press, London, 2000)
17. H.C. Liu, M. Buchanan, Z. Wasilewski, Appl. Phys. Lett. **72**, 1682 (1998)
18. F.R. Giorgetta, E. Baumann, M. Graf, Q. Yang, C. Manz, K. Köhler, H.E. Beere, D.A. Ritchie, A.G. Davies E Linfield, Y. Fedoryshyn, H. Jäckel, M. Fischer, J. Faist, D. Hofstetter, IEEE J. Quantum Electron. **45**, 1039 (2009)
19. M. Graf, N. Hoyler, M. Giovannini, J. Faist, D. Hofstetter, Appl. Phys. Lett. **88**, 241118 (2006)
20. D. Hofstetter, M. Beck, T. Aellen, J. Faist, Appl. Phys. Lett. **78**, 396 (2001)
21. F.R. Giorgetta, E. Baumann, R. Theron, M.L. Pellaton, D. Hofstetter, M. Fischer, J. Faist, Appl. Phys. Lett. **92**, 121101 (2008)
22. D. Hofstetter, M. Graf, T. Aellen, J. Faist, L. Hvozdara, S. Blaser, Appl. Phys. Lett. **89**, 061119 (2006)
23. D. Hofstetter, F.R. Giorgetta, E. Baumann, Q. Yang, C. Manz, K. Köhler, Appl. Phys. Lett. **93**, 221106 (2008)

Spectral and modulation properties of a largely tunable MEMS-VCSEL in view of gas phase spectroscopy applications

S. Schilt · K. Zogal · B. Kögel · P. Meissner · M. Maute · R. Protasio · M.-C. Amann

Received: 15 September 2009 / Revised version: 16 December 2009 / Published online: 9 February 2010
© Springer-Verlag 2010

Abstract An extensive characterization of the spectral properties of a largely tunable laser in the 1.56-μm spectral range is reported. This device combines a vertical-cavity surface-emitting laser (VCSEL) with a micro-machined (MEMS) Bragg mirror in a very compact arrangement. The large tunability obtained by an electro-thermal actuation of the MEMS mirror makes this device very attractive for high-resolution spectroscopy. Relevant laser parameters for the implementation of wavelength modulation spectroscopy techniques in gas sensing, such as tuning and modulation properties, are presented. A preliminary gas spectroscopy experiment performed with this laser is also shown.

S. Schilt (✉) · R. Protasio
IR Microsystems SA, PSE-C, 1015 Lausanne, Switzerland
e-mail: stephane.schilt@unine.ch
Fax: +41-21-7182511

Present address:
S. Schilt
Laboratory for Time and Frequency, Institute of Physics, University of Neuchâtel, 2009 Neuchâtel, Switzerland

K. Zogal · B. Kögel · P. Meissner
Fachgebiet Optische Nachrichtentechnik, Technische Universität Darmstadt, Merckstrasse 25, 64283 Darmstadt, Germany

Present address:
B. Kögel
Department of Microtechnology and Nanoscience, Chalmers University of Technology, 412 96 Göteborg, Sweden

M. Maute · M.-C. Amann
Walter Schottky Institut, Technische Universität München, Am Coulombwall, 85748 Garching, Germany

1 Introduction

Widely- and fast-tunable lasers find applications in many different areas, such as optical telecommunications, atomic or molecular spectroscopy [1–3], or optical sensing [4]. In industrial gas sensing applications, for instance, a broad tuning range enables to address several molecules with a single laser source, facilitating selective multigas monitoring of heterogeneous gas samples alongside with the remote extraction of process parameters, such as temperature and pressure, in a simple and cost-effective configuration. Current technologies to achieve a large tunability include the use of an external cavity or the combination of a large number of diode lasers, each of them having a limited tuning range.

External cavity diode lasers (ECDL) are routinely used in spectroscopy [1, 5], in optical telecommunications [6, 7], or in instrumentation. This mature technology is commercially available for a long time and has been extended to mid-infrared quantum cascade lasers for a few years [8, 9]. In these systems, the broad gain curve of a Fabry–Perot diode laser is combined with a mechanically adjustable Bragg-grating cavity to provide a large tunability, which can amount up to 15% of the central wavelength [9]. In addition to the large tunability, the external grating cavity also provides single-mode emission with a narrow instantaneous linewidth (typ. < a few 100 kHz). Additionally, a large optical power (up to 100 mW continuous wave) is achievable in an ECDL [9]. However, these lasers need a very precise alignment of the external cavity to achieve mode-hop-free tuning, which makes external-cavity-based lasers expensive. The tuning is usually performed by the combined use of a stepper motor (for coarse wavelength adjustment) and a piezo-transducer (for fine adjustment). Other disadvantages of ECDLs are their large size compared to semiconductor diode lasers and the relative sensitivity of the external cavity to vibrations and acoustic noise, which increases the flicker ($1/f$) noise of the laser both in terms of intensity noise and frequency noise. As a consequence, an additional stabilization loop (e.g., using an atomic or molecular absorption line) may be necessary in some applications requiring very-high stability [10]. Finally, tuning of an ECDL over a large range is usually quite slow, as it is a mechanical process accomplished with the stepper motor.

Another type of commercially available system in the optical telecommunication spectral range (S-, C-, and L-band), which is worth to mention, is made up of an array of several (typ. 8–16) slightly detuned DFB laser chips. The lasers are coupled into an optical fiber using a MEMS mirror [11] or a passive optical combiner [12]. Owing to the individual 3–4-nm tuning range of a single DFB laser and to the number of different lasers integrated in these devices, a total tuning range of up to ~40 nm is achievable, with a typical output power of 10–50 mW. This tuning is achieved by activating a particular element of the array (coarse tuning) and by adjusting the laser temperature (medium tuning) and the injection current (fine tuning). Since all the chips are generally mounted on the same thermoelectric cooler, tuning over a large range is usually quite slow. Additionally, the number of DFB chips needed to reach the large tuning range makes these systems costly.

In this paper, we report on an alternative type of widely tunable laser obtained by combining two technologies in an integrated device, namely a vertical-cavity surface-emitting laser (VCSEL) and a micro-machined (MEMS) movable Bragg mirror membrane. This so-called MEMS-VCSEL technology combines the main advantages of VCSELs (inherent longitudinal single-mode behavior, compactness, low power consumption, etc.) with the large tunability provided by the thermal actuation of the membrane which acts as a movable external mirror. The potential advantages of the reported technology comprise a very compact size incorporable into a TO-package as commonly used for standard laser diodes, a tuning range of several tens of nanometers provided by an electro-thermal or electrostatic actuation of the membrane and a fast tuning over the entire range at millisecond timescales. A continuous static tuning range scaling up to 60 nm has been previously demonstrated for MEMS-VCSELs in the optical telecommunications C-band [2] with a 3-dB tuning bandwidth of 180 Hz [3, 13]. To achieve such a large tunability, the device is designed to operate in a so-called extended-cavity configuration, in which the top facet of the half-VCSEL is anti-reflection-coated to prevent a residual reflection at the semiconductor–air interface so as to deal with a single optical cavity between the bottom distributed Bragg reflector (DBR) and the MEMS mirror. The work reported here shows preliminary results obtained in the frame of the European project Subtune.[1] The main objectives of this development are: (1) to enlarge the spectral coverage of this type of MEMS-VCSEL to both shorter (down to 800 nm) and longer (up to 2.1 μm) wavelength using GaAs and InP materials, respectively; (2) to increase the tuning frequency (full spectral scan) to ~100 kHz; (3) to ensure well-defined polarization stability using a sub-wavelength grating. These developments are performed in close connection with different applications of interest for these lasers, namely optical telecommunications, fiber Bragg grating sensor interrogation, and gas sensing, in order to fulfil their respective requirements. In this paper, we focus on the gas sensing application and we present some properties of a 1.56-μm MEMS-VCSEL that are relevant in view of the future application of this device in wavelength modulation spectroscopy (WMS). In particular, the tuning and modulation characteristics of the device are shown, as well as preliminary results of gas spectroscopy obtained with this laser.

2 Laser design and fabrication

The tunable VCSEL is a hybrid two-chip assembly of a micro-electro-mechanical system with concave AlGaAs/GaAs mirror membrane and an InP-based semiconductor cavity with a tunnel junction aperture [13, 14]. The MEMS technology is very promising in respect to the achievement of very good spectral properties. By the use of a half-symmetric resonator, the fundamental Gaussian mode of the laser can be supported. Essential for this approach is the coincidence of the phase of the optical beam with the curvature of the membrane (see Fig. 1). The included air-gap can be expanded by using the electro-thermal MEMS actuation and the cavity resonance can be tuned to a longer wavelength.

Figure 2a presents the schematic of the tunable MEMS-VCSEL showing the upper movable Bragg mirror with the actuating contacts, an air-gap, the semiconductor cavity, and the bottom mirror terminated with a gold p-contact, which additionally provides efficient heat exchange. The top DBR consists of molecular beam epitaxy grown GaAs/AlGaAs effective quarter wavelength layers with a difference in refractive index of 0.39. The concave bending of the membrane is caused by targeted mechanical compressive strain embedded into the DBR layers during the epitaxial deposition of the material. The inclusion of indium up to 5% at the upper part of the mirror causes a slight lattice mismatch of GaAs/AlGaAs layers which then results in a membrane deflection within the range of 6–12 μm (air-gap) and a radius of curvature within the range of 2–3 mm. The MEMS-VCSEL used in this work has an air-gap of 6.1 μm and a

[1] www.subtune.eu.

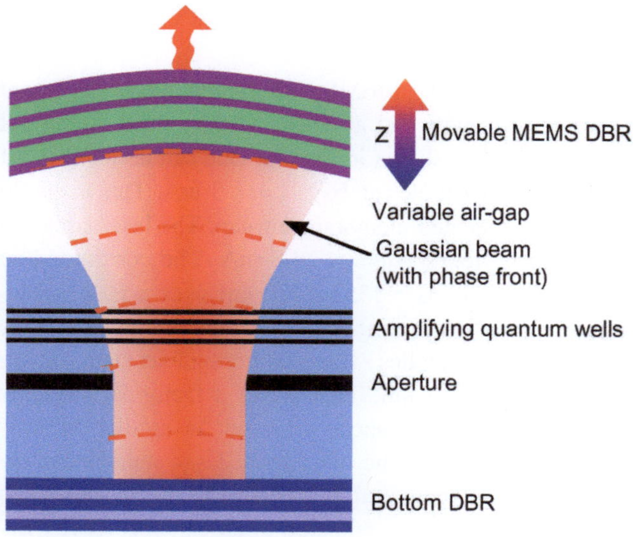

Fig. 1 The half-symmetric VCSEL resonator with concave bended top DBR (distributed Bragg reflector)

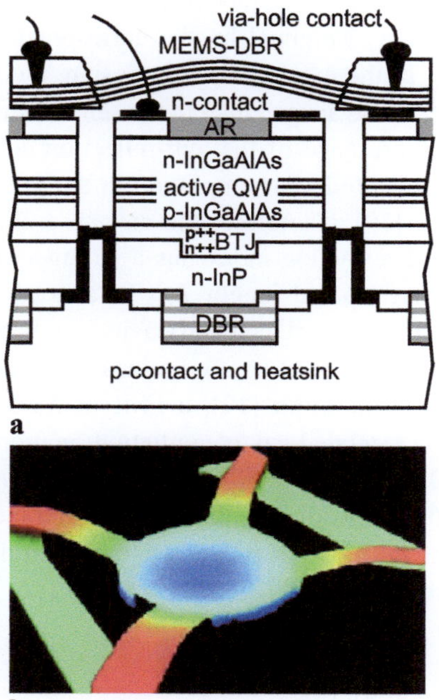

Fig. 2 (**a**) Cross-section view of the MEMS tunable VCSEL. (**b**) Confocal microscope 3D picture of the membrane

membrane diameter of 190 µm. Additionally, the semiconductor DBR layers are doped with silicon during the epitaxial process to enable an electrical conductivity and thus tuning of the membrane. The current flowing through via-holes contacts causes power dissipation in the membrane and thus temperature increase which is then responsible for the elongation of the four membrane support beams (89 µm length × 50 µm width) and mirror bending. Owing to the symmetric four-beam design, the deflection remains stable and there is no tilt of the membrane which would cause

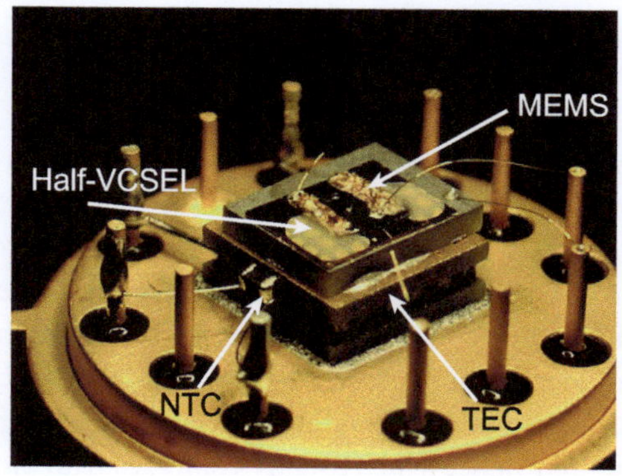

Fig. 3 Assembly of the MEMS-VCSEL into a T08-package with an active temperature control. TEC: thermoelectric cooler; NTC: negative temperature coefficient thermistor

losses in the resonator (see Fig. 2b). The membrane resistivity is 340 Ω. The active region consists of seven compressively strained InGaAlAs quantum wells embedded between an n-type and a thin p-type InGaAlAs confinement layer. A centrally limited buried tunnel junction aperture (Ø 10–20 µm) defines the active size of the laser. The dielectric bottom DBR consists of 3.5 pairs of evaporated calcium fluoride and zinc sulphide (CaF_2/ZnS) with a gold coating and an electroplated substrate. Such back mirror has limited size, slightly larger than the buried tunnel junction aperture.

In the reported experiments a tunable VCSEL in a semiconductor-coupled-cavity configuration is used, i.e., without anti-reflection coating on the top facet of the half-VCSEL. The optical thickness of the semiconductor cavity is $22\lambda/4$ and the buried tunnel junction diameter is 10 µm. The optical length of the resonator is $L_0 = 16.66$ µm ($22\lambda/4$ semiconductor cavity + $2.5\lambda/4$ back mirror + 1-µm penetration depth of the optical field into the MEMS-DBR + 6.1-µm air-gap), which corresponds to the calculated free spectral range (FSR) of the laser of 73 nm. However, the measured FSR in our tunable VCSELs without an anti-reflection coating is 39.8 nm. The deviation is caused by the coupled cavities (the uncoated interface forms a third mirror inside the cavity), which makes the previous simple calculation imprecise. A valid calculation of the FSR in a coupled-cavity VCSEL would require a 1D simulation. However, an "effective" cavity length of $L_{\mathrm{eff}} = 30$ µm can be derived from the measured value of the FSR in our VCSEL.

The laser emits up to 1 mW of optical power in continuous-wave operation at room temperature (20°C). Half VCSEL and MEMS are fixed after active alignment using UV-curable adhesive. The laser is mounted in a TO-8 package with an integrated thermoelectric cooler. In these preliminary experiments, the TO package is kept open, i.e., it is not sealed with a protective window. A fully assembled and packaged MEMS-tunable VCSEL is shown in Fig. 3.

3 Laser characterization

Spectral, tuning, and modulation characteristics are important properties of a laser in view of its application to gas phase high-resolution spectroscopy, such as trace gas sensing. Spectral purity and tunability are both needed for high selectivity sensing: spectral purity (i.e., single-mode emission) enables a gas absorption line to be probed with a narrowband, ideally monochromatic radiation; tunability is required either to precisely adjust the laser wavelength at a fixed required value, which usually coincides with the center of the analyzed absorption line in line-locked mode of operation, or to scan the laser through the absorption line in a scan mode. In the case of a largely tunable laser for multigas sensing, the tunability should enable to simultaneously scan several lines of different species or to quickly switch between lines of different species. The modulation capabilities of the laser are important to implement modulation-based sensitive detection techniques, such as wavelength modulation spectroscopy (WMS, [15, 16]) or frequency modulation spectroscopy (FMS, [17, 18]). A relevant parameter in these techniques is the frequency (or wavelength) modulation that is induced by a direct modulation of the laser. In the MEMS-VCSEL, high-frequency modulation may be applied to the laser injection current just as in the case of standard laser diodes. Additionally, modulation of the membrane current is also applicable but with a much lower bandwidth.

The tuning properties of the MEMS-VCSEL have been measured using a wavemeter (Burleigh WA-1000) with 1-pm resolution, while spectral and modulation properties have been measured using a high-resolution (0.01 nm) optical spectrum analyzer (OSA, Advantest Q8384). For all these measurements, the output power of the VCSEL is coupled into a single-mode optical fiber pigtail. The laser temperature remains constant at $T = 20°C$ and the laser and membrane current varies throughout the experiments.

3.1 Spectral and tuning properties

The spectrum of the MEMS-VCSEL obtained at different membrane currents and at constant laser injection current $I_{\text{laser}} = 18$ mA is shown in Fig. 4a. One can notice that the laser emission initiates at $I_{\text{MEMS}} = {\sim}5$ mA. This is due to slightly mismatched membrane air-gap. The laser emission is present up to $I_{\text{MEMS}} = {\sim}14.5$ mA. The corresponding spectral coverage is ~1552–1572 nm with >50 dB side-mode suppression ratio (SMSR) of higher order transverse and longitudinal modes. The total continuous static tuning range is thus ~20 nm. This is yet a factor 2–3 smaller than previous performances achieved for similar MEMS-VCSELs [2, 3]. For the considered laser semiconductor-coupled-cavity configuration, the free spectral range is typical 15–20 nm smaller compared to an extended-cavity design with the same optical length L_0. This results from the simple resonator theory: $FSR = \lambda^2/2L_{\text{eff}}$, where $L_{\text{eff}} = \sum n_i L_i$ is the effective optical length of the resonator that takes into account the refractive index n_i in each section L_i of the cavity, weighted by the corresponding field intensity [19]. In the extended-cavity design, the field intensity is higher in the air-gap than in the semiconductor cavity owing to the presence of the anti-reflection coating while these fields are of similar amplitude in the coupled-cavity configuration [19]. Therefore, the air-gap section that exhibits a small refraction index has a stronger contribution to the effective optical length L_{eff} in an extended-cavity configuration than in a coupled-cavity. In a comparative study of the two types of configurations for an almost identical physical dimension [19], a FSR of 58 nm was obtained for the extended-cavity configuration (corresponding to $L_{\text{eff}} = 20.7$ μm) compared to 42 nm in the coupled-cavity ($L_{\text{eff}} = 28.5$ μm). Additionally the gain spectral distribution together with the cavity length define the upper limitation of the tuning range.

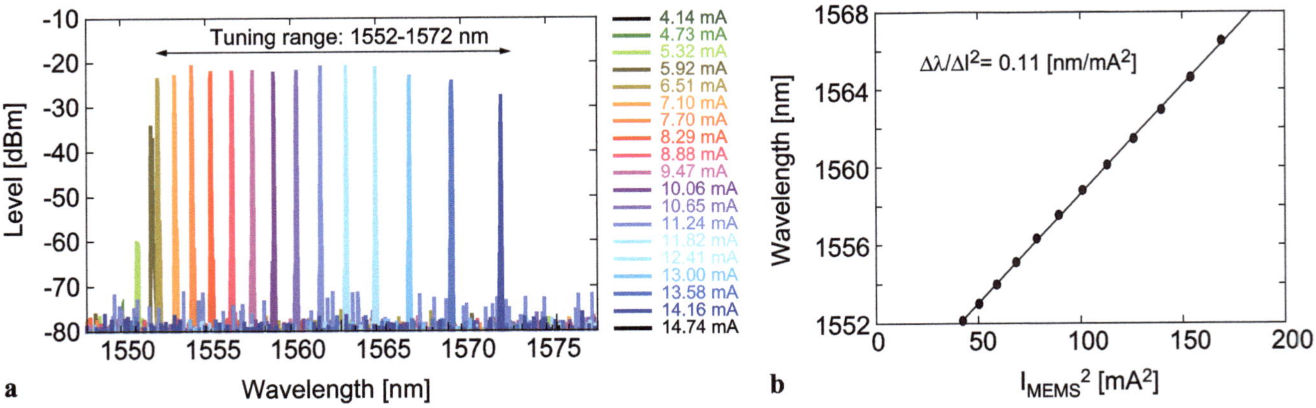

Fig. 4 (**a**) Superimposed spectra of the MEMS-VCSEL obtained at different membrane bias currents ($I_{\text{laser}} = 18$ mA). A total spectral range of 20 nm (1552–1572 nm) is covered by the laser with SMSR > 50 dB. (**b**) Dependency of the laser wavelength as a function of the squared membrane bias current. *Points* represent experimental data and *the line* is the result of a linear fit

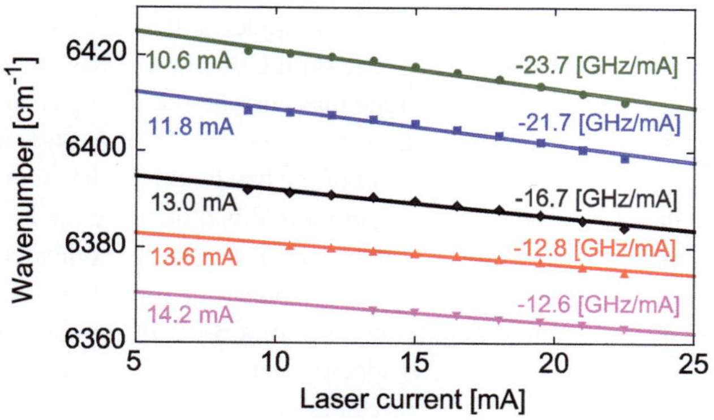

Fig. 5 Static tuning as a function of the laser injection current at different membrane bias currents. Measurement performed with an optical spectrum analyzer with the laser temperature stabilized at $T = 20°\mathrm{C}$

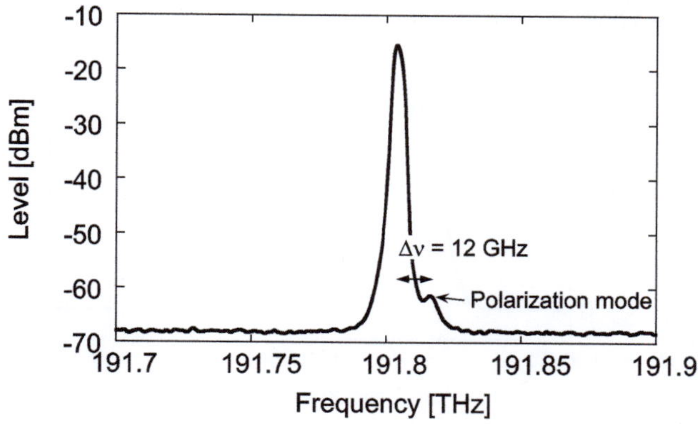

Fig. 6 The high-resolution view of the laser spectrum showing a small (PMSR > 40 dB) polarization mode shifted ~12 GHz from the main peak. Measurement obtained at $T_{\mathrm{laser}} = 20°\mathrm{C}$, $I_{\mathrm{laser}} = 21$ mA, $I_{\mathrm{MEMS}} = 11.8$ mA

The laser wavelength scales quite linearly with the square of the membrane current (with an average slope of 0.11 nm/mA^2), as expected for a thermal actuation (see Fig. 4b). But higher order nonlinearities are known to occur in MEMS-VCSELs operated in semiconductor-coupled-cavity configurations and a closer look at the tuning curve shows a weak inverted s-shaped dependence that is characteristic of these devices [20]. Coarse adjustment of the laser wavelength is thus performed by the electro-thermal actuation of the MEMS membrane. Finer adjustments may be achieved through the laser injection current as shown in Fig. 5. Tuning with injection current is also essentially a thermal effect characterized by a quadratic response curve. However, the nonlinearity is weak in that case so that the observed laser emission frequency scales almost linearly with the laser injection current, at a rate that depends on the MEMS bias current: the laser tuning coefficient $\Delta\nu/\Delta I_{\mathrm{laser}}$ tends to decrease with increasing MEMS bias current. This behavior results from the fact that static or low-frequency tuning with laser injection current is mainly a thermally-induced effect: the optical length of the semiconductor cavity expands with increasing injection current due to heating (resulting from the combination of a dilatation of the physical length of the cavity and an increase of the medium refractive index, which is the dominant process), which shifts the wavelength to longer values. The important parameter that governs the laser current tuning coefficient is the relative expansion of the entire cavity formed between the bottom DBR and the membrane mirror. This cavity is composed of two subsections, the semiconductor half-VCSEL and the air-gap. The absolute expansion of the first subsection (semiconductor cavity) may be considered as constant for a given injection current variation whatever is the MEMS current. But the second subpart (air-gap) extends with the MEMS current due to the membrane deformation. As a result, the relative expansion of the entire laser cavity due to a variation of the injection current drops with the MEMS current, and so does the laser injection current tuning coefficient. Values ranging from -24 to -12 GHz/mA have been obtained in static mode, which is typically 4–5 times smaller than the tuning rate of standard VCSELs in the same spectral range [21, 22].

A closer view of the laser peak obtained at $I_{\mathrm{laser}} = 21$ mA is shown in Fig. 6. A very small (~45 dB attenuation) lateral peak is observed ~12 GHz away from the main peak. This tiny mode corresponds to the orthogonal polarization mode, which, however, does not affect the general behavior of the laser since the polarization remains stable (no polarization mode switching or partition) and shows a stable polarization mode suppression ratio of PMSR > 40 dB. The resolution of the measured spectrum is insufficient to resolve the laser linewidth, but previous measurements obtained by the delayed self-heterodyne technique on a similar type of MEMS-VCSEL have shown a linewidth in the 30-MHz range [23].

3.2 Frequency modulation characteristics

The frequency modulation (FM) characteristics of the laser have been quantified for a modulation of either the laser injection current or the membrane current. In all cases, a modulation of a varying amplitude or frequency is applied to the device (through the injection current or membrane current). The amplitude $\Delta\nu$ of the induced FM is determined from the optical spectrum, measured with the high-resolution OSA. The determination of the FM amplitude is based on a similar principle commonly used to measure the laser FM response using a scanning Fabry–Perot interferometer [24]. In the case of a high FM index as encountered in our experiments (a low modulation frequency in conjunction with a large instantaneous frequency deviation $\Delta\nu$), the observed modulation spectrum shows a typical U-shape and the frequency interval between the two maxima of this spectrum is a direct measurement of the peak-to-peak instantaneous frequency deviation ($2\Delta\nu$) of the laser (see Fig. 7).

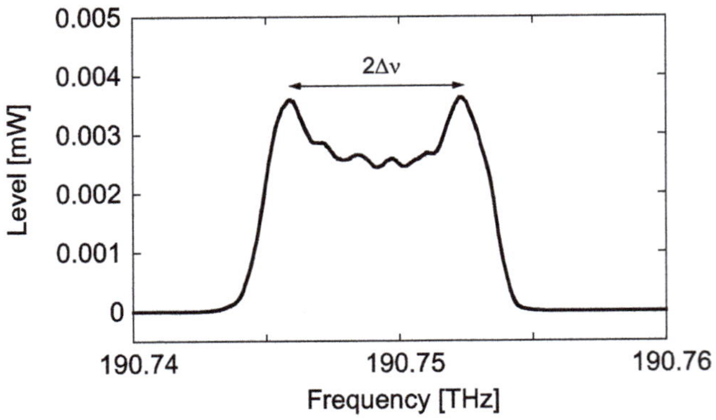

Fig. 7 Example of a laser FM spectrum measured with the optical spectrum analyzer ($I_{laser} = 21$ mA, $I_{MEMS} = 14.1$ mA, laser injection current modulation at $f = 25$ kHz, peak-to-peak amplitude 0.46 mA). The peak-to-peak variation of the laser instantaneous frequency ($2\Delta\nu = 6.6$ GHz) corresponds to the distance between the two maxima of the U-shape spectrum

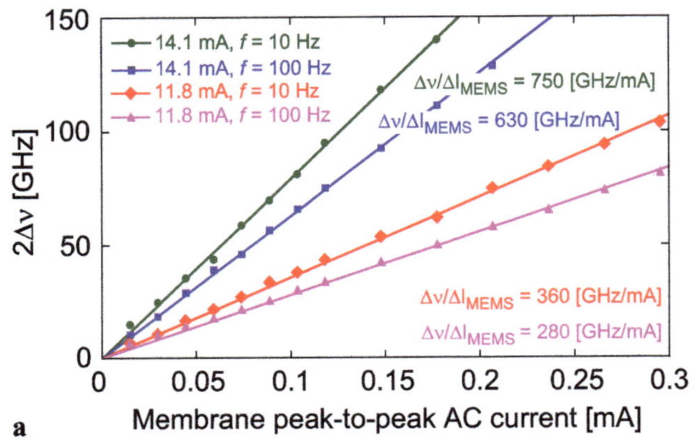

a

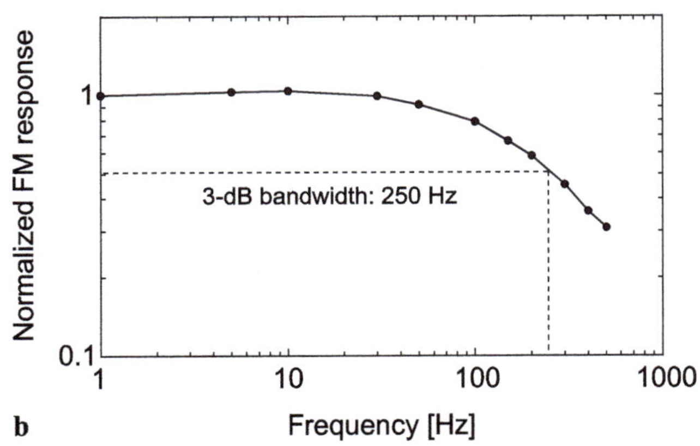

b

Fig. 8 (**a**) Amplitude of the laser FM as a function of the membrane AC current obtained at two different modulation frequencies (10 Hz and 100 Hz) and for two MEMS bias currents ($I_{MEMS} = 11.8$ mA and $I_{MEMS} = 14.1$ mA). (**b**) Frequency response of the MEMS modulation. The 3-dB bandwidth is 250 Hz

Modulation of the MEMS current was investigated at a constant laser injection current $I_{laser} = 18$ mA and at two different modulation frequencies, $f = 10$ Hz and $f = 100$ Hz. Figure 8a shows that the amplitude of the laser FM scales linearly with the amplitude of the AC current applied to the membrane. The tuning coefficient has some dependence on the membrane bias current: the tuning is approximately two times larger at $I_{MEMS} = 14.1$ mA than at $I_{MEMS} = 11.8$ mA. The effect of the bandwidth of the membrane modulation is also apparent in this plot, since the FM response is typically 20% lower at $f = 100$ Hz compared to $f = 10$ Hz. The frequency response of the membrane is shown in more details in Fig. 8b: a 3-dB bandwidth of 250 Hz is achieved. The electro-thermal actuation of the membrane thus enables to scan large spectral interval (up to the entire spectral coverage of the laser) in <10 ms timescale or to quickly switch between different distant wavelengths in the range of 10–20 nm for the present laser (or up to 50 nm for MEMS-VCSELs with a larger tuning range as reported in [2, 3]). These features are very attractive for multigas sensing.

Much faster modulation is achievable by modulating the laser injection current but with a strongly reduced spectral extend, which can nevertheless cover a few hundred gigahertz for a large variation of the laser current. This is usually by far sufficient to scan over one or a few neighboring absorption lines at atmospheric pressure, where lines have a typical width of a couple of GHz. For a (small) modulation of the laser injection current at ~25 kHz an almost linear FM response is obtained, as shown in Fig. 9. Here again, an influence of the membrane bias current is observed on the laser FM response. The tuning coefficient is more or less constant ($\Delta\nu/\Delta I = \sim 30$ GHz/mA in this case) up to $I_{MEMS} = \sim 12$ mA but it drops at higher MEMS bias current. The origin of this behavior is similar to the case of the static tuning with the laser injection current at different membrane bias currents discussed in Fig. 5. The highest value of the AC tuning coefficient observed in Fig. 9 compared to the static tuning previously shown in Fig. 5 is due to different laser injection currents (18 mA in Fig. 5 vs. 21 mA in Fig. 9). It is well known that the tuning coefficient of a VCSEL increases significantly with the laser DC current [21].

4 Preliminary gas spectroscopy results

Preliminary gas sensing experiments have been performed with our MEMS-VCSEL in order to evaluate the potential of such a light source for high-resolution spectroscopy. The experimental scheme used in this experiment is depicted in Fig. 10. Light emitted from the MEMS-VCSEL is coupled into a single-mode optical fiber, which facilitates the acquisition of additional spectral measurements using the OSA without modifying the setup. At the fiber exit, the light is collimated and transmitted through a single-pass 1-m long absorption cell sealed with Brewster-angled BK7 windows. The transmitted light is finally detected with a photodiode. Pure carbon dioxide (CO_2) is used as a test gas for preliminary WMS measurements. WMS is implemented using a custom electronics board that was originally developed for cost-effective gas sensors based on standard VCSELs. This home-designed electronics performs laser current and temperature control, injection current modulation at ~25 kHz, second-harmonic ($2f$) lock-in demodulation, and slow (step-by-step) scans of the laser DC current to measure the direct absorption signal or the $2f$ spectrum. The control of the membrane bias current is done externally using a laboratory current source.

Figure 11a shows an example of the observed $2f$ spectrum of a CO_2 absorption line. A characteristic second derivative signal is observed but the quality of the trace is low and the signal-to-noise ratio rather poor. Many oscillations and hops are observed in this figure resulting from instabilities in the laser spectrum. This is confirmed by additional measurements of the laser spectrum made with the OSA. Figure 11b displays several measurements of the laser line successively acquired in a total timescale of a few tens of seconds (without any modulation on the laser). It is evident from this figure that the laser wavelength is not stable and moves by several gigahertz. This jitter is large compared to the typical width of 3 GHz (half width at half maximum) of the analyzed CO_2 line at atmospheric pressure and is responsible for the noise observed in the $2f$ spectrum (this spectrum was acquired in a timescale of ~100 s, since the laser DC current is changed step-by-step in our driving electronics rather than ramped by a faster sawtooth waveform). Different factors may contribute to this spectral instability. Since the laser used in this experiment was packaged in an open TO can, i.e., not sealed with a protective window, it was first believed that mechanical noise on the membrane, due to vibrations, acoustic noise or air drafts, was the main origin of this problem. However, a simple model calculation of the "acoustic force" that would be required to cause the observed wavelength fluctuations shows that this assumption is not realistic: for a free spectral range of the cavity of $FSR \approx 5$ THz (corresponding to the effective cavity length $L_{eff} \approx 30$ μm), the membrane displacement needed to produce a typical frequency fluctuation $\Delta\nu = 10$ GHz (as observed in Fig. 11b) is $\Delta L = \Delta\nu \cdot \lambda/(2FSR) \cong 1.6$ nm. With a typical spring constant of the membrane of $k = 680$ N/m [23], the required force is $F = k\Delta L \approx 10^{-6}$ N. Using the area of the movable MEMS structure $A = 3.7 \times 10^{-8}$ m^2 (taking into account half of the beam length), a corresponding pressure $p = F/A = 28.7$ Pa is obtained. This value is by far too high to appear realistic so that acoustic noise cannot be the origin of the wavelength fluctuations observed in the laser spectrum.

Another possible reason for these instabilities is a fluctuation of the membrane bias current, since it was previously shown that the laser wavelength changes considerably with the membrane current (by many hundreds GHz/mA so that current noise of a few μA can induce frequency noise of >1 GHz). The use of a low-noise laser driver as a current source for the membrane bias current did not bring significant improvement on the laser wavelength stability. The noise spectral density of this current source, measured with

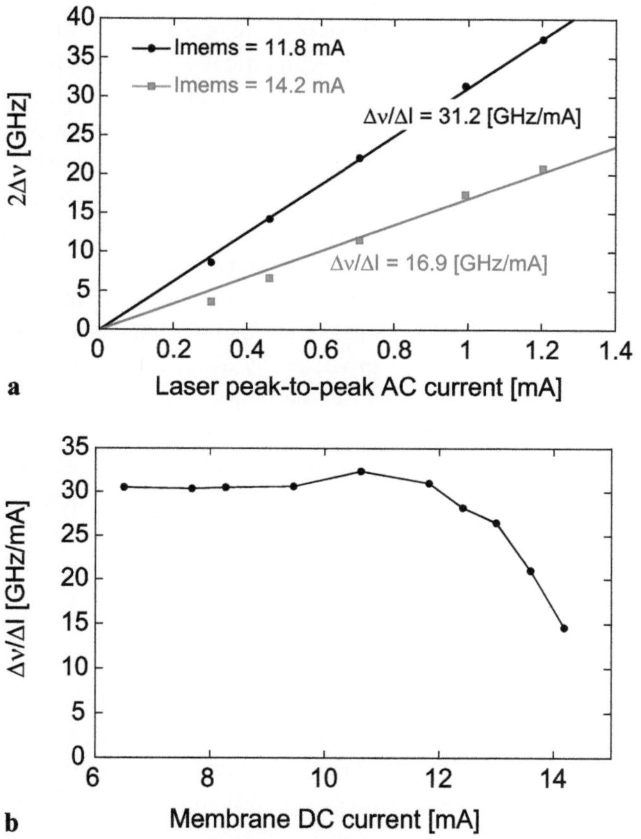

Fig. 9 (**a**) Amplitude of the laser FM as a function of the AC component of the injection current ($f = 25$ kHz) obtained for two different values of the MEMS bias current (at $I_{laser} = 21$ mA). (**b**) Influence of the membrane bias current on the laser FM response for a laser injection current modulation at $f = 25$ kHz ($I_{laser} = 21$ mA)

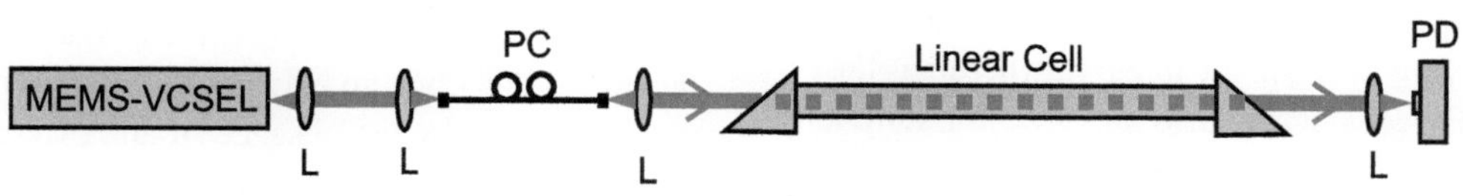

Fig. 10 Scheme of the optical setup used for preliminary gas sensing experiments. L: lens; PC: polarization controller; PD: photodiode. The polarization controller is used to have TM polarization on the Brewster-angled cell windows

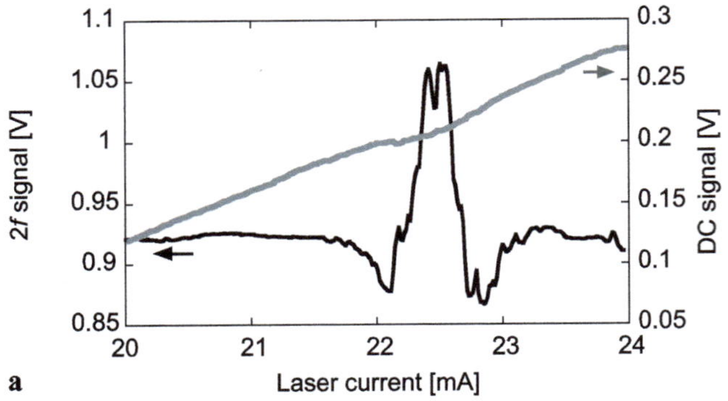

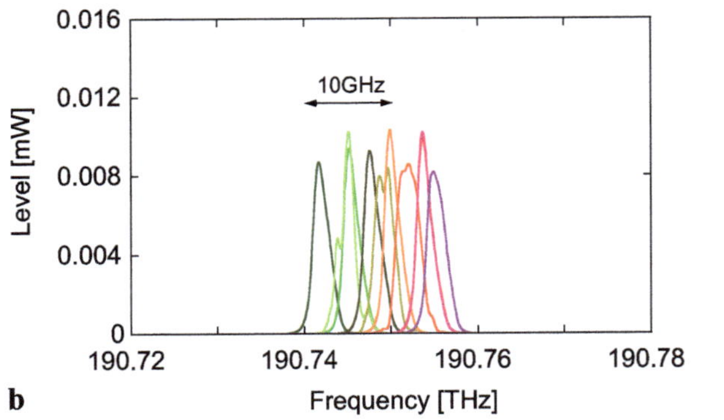

Fig. 11 (**a**) Direct absorption signal and $2f$ spectrum of a CO_2 line obtained when slowly scanning the laser via its injection current; $I_{MEMS} = 14$ mA. (**b**) Successive measurements of the (unmodulated) laser spectrum obtained over a total timescale of a few tens of seconds showing large spectral instabilities of the laser; $I_{laser} = 21$ mA; $I_{MEMS} = 14$ mA

a FFT dynamic analyzer, has shown a dominant $1/f^{\alpha}$-type of noise at frequencies smaller than a couple of tens Hz. The power spectral density (PSD) of this low-frequency noise is described by $S_I(f) = K_I/f^{\alpha}$, where K is the current noise PSD at $f = 1$ Hz and α generally lies in the range $0 < \alpha < 2$ for $1/f$-type noise. A fit of the experimental current PSD in the range 0.1–30 Hz has shown a more pronounced frequency dependence with $\alpha \approx 2.6$ and $K_I \approx 1.5 \times 10^{-16}$ A^2/Hz at $I_{dc} = 14$ mA. Accounting for a constant membrane tuning coefficient in this frequency range of approximately 750 GHz/mA as previously measured (see Fig. 8), these values translate into a frequency noise PSD $S_f(f) = K_f/f^{\alpha}$ with $K_f = 8.5 \times 10^{13}$ Hz^2/Hz. This low-frequency noise gives rise to a laser linewidth that depends on the observation time and which can be made arbitrarily large for long enough observation. This results from the divergence of the integral of the frequency noise PSD for $f \to 0$. A quantitative estimation of the influence of the observation time can be obtained by considering a finite low cut-off frequency $f_1 = 1/\tau$ in the integral, where τ is the observation time. By considering also $f_2 \to \infty$ for simplicity, the amplitude of the frequency fluctuations over an observation time τ is given by $\Delta f_{rms} = \sqrt{K_f \tau^{\alpha-1}/(\alpha - 1)}$. This leads to typical frequency fluctuations of 110 MHz over 10 s and 750 MHz over 100 s. This is clearly much smaller than the frequency fluctuations experimentally observed so that the current noise is not a significant contributor to the observed laser spectral instabilities.

Among the other possible sources of noise in our experiment, there are mechanical vibrations (transferred via the optical bench), which can have a severe effect on the membrane, as well as fluctuations of the membrane temperature due, for instance, to air drafts. A simple calculation has been performed to evaluate the membrane temperature fluctuation that is required to produce the 1.6-nm membrane displacement previously calculated. For this estimation, the geometry of the curved-mirror membrane with its supporting beams was approximated by a trapezoid (the mirror constitutes the short base of the trapezoid and the beams represent the legs). With a mean thermal expansion coefficient in the GaAs/AlGaAs DBR of $\alpha \approx 5.6 \times 10^{-6}$ K^{-1}, the expansion of one beam can be geometrically calculated in the right-angled triangle formed by the 6.1-µm air-gap altitude of the trapezoid and the $L_{beam} = 89$ µm beam length (hypotenuse). If the air-gap increases by 1.6 nm, the hypotenuse needs to expand by $\Delta L_{beam} = 110$ pm. This requires a temperature change $\Delta T = \Delta L_{beam}/(\alpha L_{beam}) \cong 0.2$°C. Such a temperature fluctuation of the membrane due to air drafts seems reasonably possible owing to the small thermal capacity of the membrane and to the fact that the laser package was open, i.e., not protected by a window. Therefore, we believe that this effect is the major contribution to the observed spectral instabilities of the laser and a better temperature stabilization of the membrane (e.g., by protecting it from air drafts using a package sealed with a protective window) should improve the laser stability.

5 Conclusion

A tunable MEMS-VCSEL made of a bottom InP-based semiconductor half-VCSEL and a top MEMS concave AlGaAs/GaAs mirror membrane has been extensively characterized with respect to its spectral, tuning, and modulation properties. A major objective of this work was to quantify important characteristics of this type of lasers for modulation-based spectroscopic detection techniques, such as WMS, in order to assess their suitability for gas sensing applications. The large tuning of MEMS-VCSELs offers new perspectives for compact and cost-effective multi-gas sensors and makes these lasers very attractive for such applications. The 20-nm continuous static tuning range obtained for the 1.56-µm MEMS-VCSEL used in this work is mainly limited by the absence of an AR coating on the top facet of the semiconductor half-VCSEL cavity, which makes the device operate in a semiconductor-coupled-cavity configuration, which is not favorable for large tunability. Tunability enlarged by a factor of 3 has been previously obtained at the same wavelength with the use of an AR coating, so that significant improvement of our devices are expected in a near future in terms of tunability by using such an approach. Dynamic MEMS modulation showed a 3-dB bandwidth of 250 Hz, which is $\sim$40% higher than previous results obtained in anextended-cavity configuration. Finally, faster modulation capabilities have been shown when modulating the laser injection current. The corresponding injection-current-tuning coefficient was shown to be smaller compared to standard VCSELs and to be dependent on the membrane bias current, which is understandable by the fact that the injection current affects part of the laser cavity only, i.e., the semiconductor half-VCSEL. Anyway, the FM characteristics of the MEMS-VCSEL are fully suitable for WMS experiments both in atmospheric conditions and at reduced pressure.

Unfortunately, preliminary experiments of gas spectroscopy performed with this MEMS-VCSEL have been misleading due to insufficient spectral stability of the laser wavelength. Among the various possible sources of instability that have been evaluated either theoretically using simple model calculations (acoustic and thermal influence of air fluctuations) or experimentally (noise on the current source used to bias the membrane), thermal fluctuations of the membrane due to air drafts have been determined as a major contribution partly due to the open package used in this experiment. Mechanical vibrations transferred via the optical bench have also shown to produce instabilities in the membrane. The use of a laser mounted in a sealed packaged including a protective window and a better isolation of the experimental setup from mechanical vibrations constitute essential steps towards the achievement of quantitative spectroscopic results necessary to demonstrate the applicability of this MEMS-VCSEL in gas sensing.

Acknowledgement This work is supported by the European Commission via the EU project "Subtune" (FP7-ICT-224259) within the seventh Framework Programme.

References

1. M. Scotoni, A. Rossi, D. Bassi, R. Buffa, S. Iannotta, A. Boschetti, Appl. Phys. B **82**(3), 495 (2006)
2. M. Lackner, M. Schwarzott, F. Winter, B. Kögel, S. Jatta, H. Halbritter, P. Meissner, Opt. Lett. **31**(21), 3170 (2006)
3. B. Kögel, H. Halbritter, S. Jatta, M. Maute, G. Böhm, M.-C. Amann, M. Lackner, M. Schwarzott, F. Winter, P. Meissner, IEEE Sens. J. **7**(11), 1483 (2007)
4. M. Wippish, K.D. Dessau, Ind. Phys. **9**(3), 24 (2003)
5. Q.-V. Nguyen, R.W. Dibble, T. Day, Opt. Lett. **19**(24), 2134 (1994)
6. T. Day, in *Optical Fiber Communication Conference and Exhibit OFC 2001* (2001), pp. TuJ4-1–TuJ4-3
7. T. Sato, J. Lightwave Technol. **22**(7), 1782 (2004)
8. M. Pushkarsky, M. Weida, T. Day, in *Proc. of the Society of Photo-optical Instrumentation Engineers*, vol. 6871 (SPIE, Bellingham, 2008), pp. X8711–X8711
9. G. Wysocki, T. Tsai, S. So, Y. Wang, R. Lewicki, F.K. Tittel, D. Weidmann, Broadly tunable external cavity quantum cascade lasers and applications, in *Tunable Diode Laser Spectroscopy Conference*, July 13–17 2009, Zermatt, Switzerland, paper L2
10. C. Affolderbach, G. Mileti, Opt. Lasers Eng. **43**(3–5), 291 (2005)
11. B. Pezeshki, E. Vail, J. Kubicky, G. Yoffe, S. Zou, J. Heanue, P. Epp, S. Rishton, D. Ton, B. Faraji, M. Emanuel, X. Hong, M. Sherback, V. Agrawal, C. Chipman, T. Razazzan, IEEE Photonics Technol. Lett **14**(10), 1457 (2002)
12. C.-E. Zah, M.R. Amersfoort, B.N. Pathak, F.J. Favire, P.S.D. Lin, N.C. Andreadakis, A.W. Rajhel, R. Bhat, C. Caneau, M.A. Koza, J. Gamelin, IEEE J. Sel. Top. Quantum Electron. **3**(12), 584 (1997)
13. B. Kögel, M. Maute, H. Halbritter, F. Riemenschneider, G. Böhm, M.-C. Amann, P. Meissner, J. Opt. A: Pure Appl. Opt. **8**, S370 (2006)
14. F. Riemenschneider, M. Maute, H. Halbritter, G. Boem, M.C. Amann, P. Meissner, IEEE Photonics Technol. Lett. **16**(10), 2212 (2004)
15. R. Arndt, J. Appl. Phys. **36**(8), 2522 (1965)
16. J. Reid, D. Labrie, Appl. Phys. B **26**(3), 203 (1981)
17. G.C. Bjorklund, Opt. Lett. **5**(1), 15 (1980)
18. G.C. Bjorklund, M.D. Levenson, W. Lenth, C. Ortiz, IEEE J. Quantum Electron. **20**(9), 1045 (1984)
19. P. Debernardi, B. Kögel, K. Zogal, P. Meissner, M. Maute, M. Ortsiefer, G. Böhm, M.-C. Amann, IEEE J. Quantum Electron. **44**(4), 391 (2008)
20. M. Maute, B. Kögel, G. Böhm, P. Meissner, M.-C. Amann, IEEE Photonics Technol. Lett. **18**(5), 688 (2006)
21. A. Lytkine, W. Jäger, J. Tulip, Spectrochim. Acta A **63**(5), 940 (2006)
22. A. Lytkine, W. Jäger, J. Tulip, Proc. SPIE **5737**, 157 (2005)
23. B. Kögel, K. Zogal, S. Jatta, C. Grasse, M.-C. Amann, G. Cole, M. Lackner, M. Schwarzott, F. Winter, P. Meissner, Proc. SPIE **7266**, 72660O (2008). doi:10.1117/12.816272
24. H. Olesen, G. Jacobsen, IEEE J. Quantum Electron. **18**(12), 2069 (1982)

Tunable diode laser spectroscopy with optimum wavelength scanning

J. Chen · A. Hangauer · R. Strzoda · M.-C. Amann

Received: 15 October 2009 / Revised version: 15 February 2010 / Published online: 25 March 2010
© Springer-Verlag 2010

Abstract In this paper, an improved signal processing method is given, aiming at a sensitivity improvement of the tunable diode laser spectroscopy based gas concentration measurement. The gas absorption spectrum is probed with an optimum wavelength scanning function which enables a more efficient curve fit than for traditional linear wavelength scanning. The wavelength scanning function is determined to be optimum, in the sense that the variance of the estimated concentration noise is minimized. This optimum scanning function depends on the signal model used in curve fitting. Several models including interfering gas spectra and etalon fringes are examined. Compared to the gas absorption spectrum recorded by ramping the wavelength linearly, the optimum spectrum waveforms have a cascade structure, which means that the optimum scanning preferably samples important points of the spectrum. The new method theoretically enables a factor of ~2 improvement on detection sensitivity of the estimated concentration. Furthermore, direct spectroscopy is superior to second harmonic detection, because the concentration noise can be an additional factor of ~2 lower.

1 Introduction

Tunable diode laser spectroscopy (TDLS) detects gas concentration by taking a spectral fingerprint of the gas absorption. The gas absorption spectrum $A(\lambda)$ is given by the quotient of ingoing (I_0) and outgoing (I) laser light intensities through a gas sample:

$$A(\lambda) = \frac{I_0(\lambda) - I(\lambda)}{I_0(\lambda)}. \tag{1}$$

Usually, in the sensor the gas absorption is measured directly, e.g. by implementing (1) (direct spectroscopy), or indirectly by using derivative spectroscopy, e.g. wavelength modulation spectroscopy (WMS) [1].

A common task of a TDLS-based gas sensor is to achieve high sensing sensitivity. To improve the sensing sensitivity of a sensor the following basic methods exist:

- Probe a stronger absorption line, e.g. using lines in the fundamental gas absorption band in the mid-infrared range.
- Increase the optical path length, e.g. by using multi-pass cells or cavity-enhanced methods, e.g. Integrated Cavity Output Spectroscopy (ICOS) [2, 3].
- Lower the noise induced by laser and detector, e.g. by applying a low-noise laser and/or using a higher laser power to improve the detector signal to noise ratio, in the case laser noise is not dominating. In the other case, laser noise compensation methods can also be employed, e.g. balanced detection [4, 5].
- Average the concentration values over time to reduce the overall sensor bandwidth, which lowers the sensor noise level. This method corresponds to application of a filter with rectangular impulse response. More sophisticated methods employ time-variant filters with an adaptive filtering bandwidth like the Kalman filter [6–8].
- Use optimized signal processing, e.g. linear least squares curve fitting. At a given sensor bandwidth linear least squares curve fitting is the best linear method to filter the measurement noise, because it gives the smallest variance

J. Chen (✉) · A. Hangauer · M.-C. Amann
Walter Schottky Institut, Technische Universität München,
Am Coulombwall 3, 85748 Garching, Germany
e-mail: jia.chen@wsi.tum.de

J. Chen · A. Hangauer · R. Strzoda
Corporate Research and Technologies, Siemens AG,
Otto-Hahn-Ring 6, 80200 Munich, Germany

on the fitted concentration values among all linear digital filtering methods, e.g. low pass/high pass or Wiener filtering [9, 10]. However, the performance of curve fitting depends on the signal model for the measured data. Therefore, the sensor performance can be improved if the measurement data is recorded in an adjusted way and the corresponding signal model is changed accordingly. This method is discussed in the following sections.

Usually, the laser wavelength is linearly ramped with respect to time to scan the absorption, so a linear wavelength scanning function[1] as shown in Fig. 1a is used. The recorded spectrum waveform[2] is disregarding the x-axis unit (t instead of λ) equal to the gas absorption spectrum. This is shown in Fig. 1b for direct spectroscopy [11] and Fig. 1c for WMS second harmonic [1], respectively. After each spectral scan, a curve fit is applied to determine the gas concentration, which then is an output value of the sensor (Fig. 2).

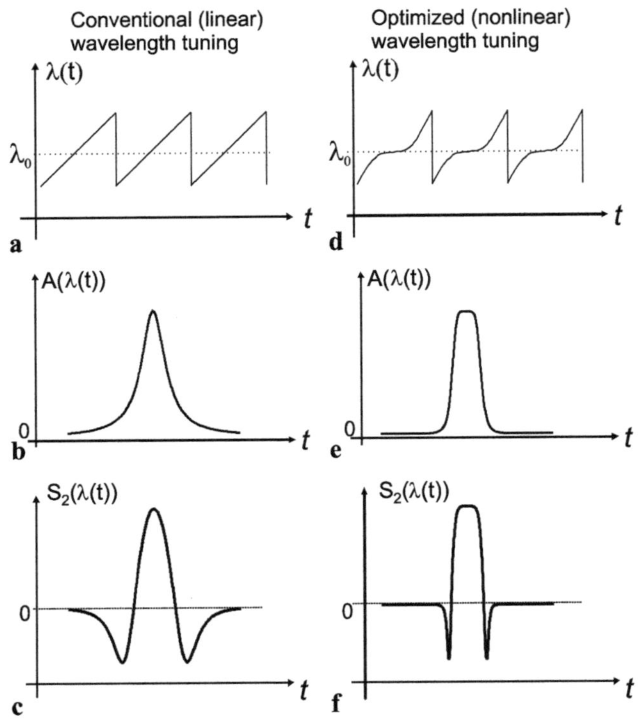

Fig. 1 Schematic illustration of the wavelength scanning function: conventional (linear) (**a**) and optimized (nonlinear) (**d**). The resulting spectrum waveform is shown for direct spectroscopy in (**b**) for linear scanning and (**e**) for nonlinear scanning as well as for WMS (second harmonic detection) in (**c**) and (**f**)

[1]The wavelength scanning function is the function describing the laser emission wavelength over time. The wavelength scanning function is applied to scan the gas absorption spectrum.

[2]The spectrum waveform is the signal obtained by scanning the gas absorption spectrum over time with a given wavelength scanning function.

However, with linear laser wavelength scanning the signal-to-noise ratio (SNR) on the estimated gas concentration may not be optimum. The SNR is defined as the averaged sensor output values versus their standard deviation at constant gas concentration. The variation of the sensor output originates from the measurement noise on the spectrum that 'passes through' the sensor signal processing, i.e. curve fitting. Basically, if a single line is probed, the gas absorption at center wavelength carries the highest information content of the gas concentration and thus should be sampled more often than the other spectral ranges. This is nonlinear wavelength sampling because the wavelength scanning function over time is nonlinear. An example is shown in Fig. 1d; the resulting spectrum waveform is shown in Fig. 1e and f for direct spectroscopy and WMS, which are more flat at the center wavelength, because more time is spent to sample the center wavelength. A special case of nonlinear scanning is applied in [12]. The wavelength scanning jumps over the spectral range where no absorption line is present and therefore optimizes scanning of two separate absorption lines. In the following, this approach is generalized and the optimum scanning method and corresponding SNR improvement is determined.

Whether a sensor is limited by white noise or slow drifts depends on the measurement rate of the sensor. At high measurement rate (low averaging time) the limitation is always given by white noise, as commonly seen on the characteristic slope of -1 in a typical Allan plot [13]. However, sensors for trace-gas analysis are often operated at their optimum averaging time (minimum of the Allan plot) where the sensor resolution is given by other noise sources than white noise, e.g. drifts or $1/f$ noise [14, 15]. White noise is always present (e.g. detector shot noise or amplifier thermal noise) and $1/f$ noise and drift have very different sources which strongly depend on the electrical and mechanical components. Furthermore, the contribution of $1/f$ noise generated in the detector or amplifier electronics is generally suppressed by technical measures, e.g. detection of a harmonic component in the kHz range or direct spectroscopy with scanning rates in the kHz range. In this paper only the white noise contribution of sensor noise is considered, which is the dominating noise source for sensors whose averaging time is well below the minimum of the Allan plot. The aim of this paper is to analyze the optimum case, i.e. to determine the upper limit of improvement that is possible to achieve by nonlinear scanning. This method may reduce the optimum averaging time (minimum of the Allan plot), because the white noise suppression is improved. For white noise limited systems (high measurement rate) the concentration resolution is enhanced. The nonlinear scanning method thus improves either sensor response time (for drift or $1/f$ noise limited systems) or concentration resolution (for white noise limited systems).

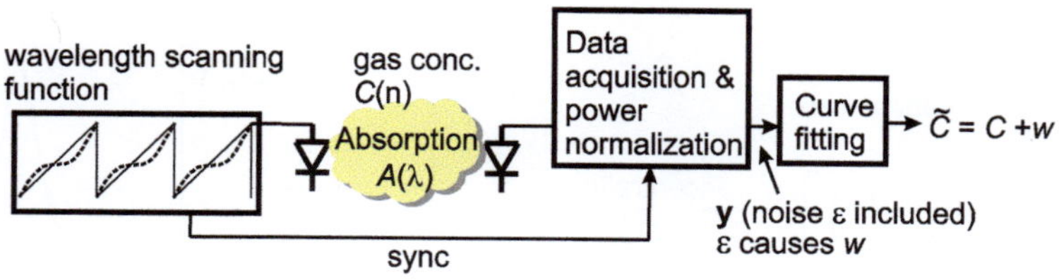

Fig. 2 Sensing block diagram: the laser is tuned with the wavelength scanning function: linear (*solid curve*) or nonlinear (*broken curve*). After power normalization the measured data set $\boldsymbol{y}$ is a pure measure of the gas absorption spectrum, and concentration estimation is done by a curve fit. ϵ represents the inevitable noise induced by the measurement apparatus, which is included in $\boldsymbol{y}$. The noise w on the estimated concentration values $\tilde{C}$ originates only from measurement noise ϵ

To maximize the SNR on the concentration values, the spectrum waveform has to be chosen properly. Basically, the spectrum waveform can be varied by:

- Scanning the laser current nonlinearly with respect to time so that certain wavelength ranges containing more information of the gas concentration are sampled more often than the less important wavelength ranges. The gas spectrum is not necessarily sampled at equally spaced wavelength points.
- Varying the modulation index $m = \lambda_a/a_L$ if WMS is applied. Here λ_a is the amplitude of the sinusoidal modulation in wavelength and a_L the half width at half maximum (HWHM) of the absorption line. m is typically set to a value aiming at the highest signal peak, which is around 2.2 for second harmonic detection for the Lorentzian gas profile [1]. But, it is not necessarily optimum in terms of minimal concentration noise because the scanned spectrum also contains other points, e.g. minima, which have then possibly lower signal amplitude and overcompensate the high center peak.
- Varying the wavelength range which is scanned. If the spectral baseline, i.e. slope, offset or parabolic terms, is unknown or the target gas absorption is interfered with other gas absorptions, a larger spectral range than only the absorption peak value should be scanned.

We present in this paper a universal criterion to determine the optimum spectrum waveform and thus also the corresponding wavelength scanning function. This includes how broad the scanned spectral range should be and—if WMS is applied—the optimum modulation index m. The optimum spectrum waveform depends on the signal model.

In most cases, the more components are included in the signal model and the more parameters are fitted, the less gain in SNR on the estimated target gas concentration value by the curve fit is achieved. No degradation in signal to noise ratio is observed if the additional components are orthogonal to the existing ones. Usually, the measured signal consists of several components, e.g. the spectral components belonging to the target gas absorption, the absorption spectrum of interfering gases, the spectral baseline consisting of offset, linear or parabolic terms (caused by the laser residual amplitude modulation (RAM) [16]) and etalon fringes caused by unwanted optical interference. These components for the measured data are used in the curve fit and form the 'signal model' which is needed for every curve fit. The components can be of both theoretical and experimental origin. Components which are quantitatively known from a reference measurement or a theoretical calculation are not included in the curve-fit model, because they are subtracted before curve fitting. One approach is described in [14], where repeated background measurements are performed and the determined background is subtracted from measurement data before the spectrum curve fitting. In this case the curve-fit model can contain additional components like offset or slope to compensate for incomplete background subtraction [15]. Another approach with an adaptive curve-fit model which includes several possible background patterns is described in [17]. Different signal models for the measurement data can be distinguished by the a priori information which is available in the specific situation. The 'value' of the a priori information can be determined, because different models give different SNR improvements in the curve fit. For example, it turns out that by knowing the spectral baseline the sensor resolution can theoretically be a factor of two better than without the knowledge of the spectral baseline (see Table 1).

In this paper, three signal models are discussed: (I) only the gas concentration is fitted (spectral baseline and spectral components and concentrations of interfering gases are known), (II) the target gas concentration and linear baseline are fitted and (III) the concentrations of target gas and interfering gas, spectral baseline and the amplitudes and phases of etalon fringes with known period lengths are fitted. We derive the optimum wavelength scanning function analytically for models I and II and numerically for model III. The solution for a linear model is also applicable for nonlinear models, by linearizing it with a Taylor series. This is discussed in Sect. 2.7.

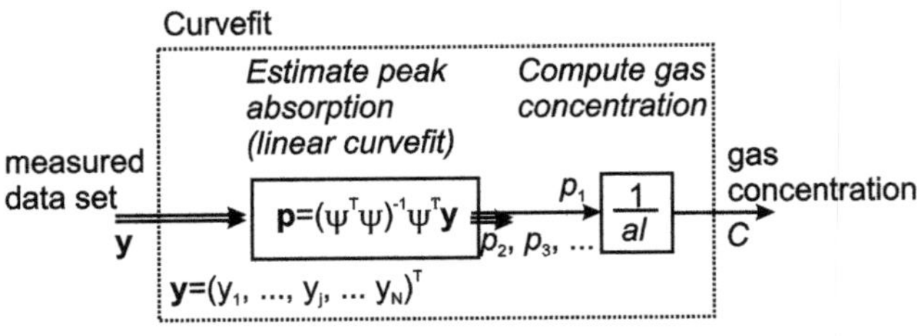

Fig. 3 The estimated gas concentration is the linear least squares fit solution of the measured data $\boldsymbol{y}$ divided by $a \cdot l$

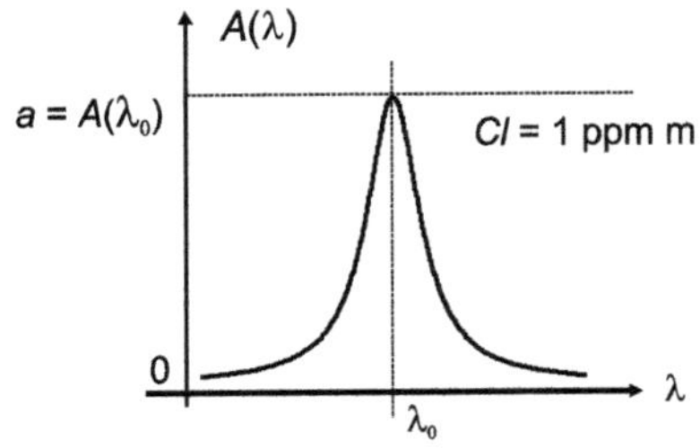

Fig. 4 Schematic explanation of the peak absorption coefficient a: it is the amplitude of a 1 ppm m gas absorption spectrum

2 Determination of the optimum spectrum waveform and scanning wavelength function

2.1 Optimization criterion

In Fig. 2, it is shown that the laser diode is tuned with a linear or nonlinear wavelength scanning function. The gas absorption spectrum recorded at the photodiode is power-normalized and is a pure measure of the gas absorption. A curve fit is carried out on the sampled measurement data $\boldsymbol{y}$ to estimate the gas concentration, whereas $\boldsymbol{\epsilon}$ represents the noise of $\boldsymbol{y}$ induced by the measurement system, i.e. laser noise, shot noise, thermal noise and electronics noise. The curve-fit result $\tilde{C}$ is the true gas concentration C plus noise w, which originates from $\boldsymbol{\epsilon}$. So, we have

$$\tilde{C} = C + w, \tag{2}$$

$$\sigma_w^2 = \langle w^2 \rangle, \tag{3}$$

with $\langle\rangle$ denoting the expectation value. To investigate the optimum spectrum waveform or the optimum wavelength scanning function the noise variance σ_w^2 is minimized. The sensor sensitivity (minimum detectable concentration) is equal to σ_w and is thus optimum. In the following, we derive an analytical expression for σ_w^2.

2.2 Signal model of curve fit

The vector $\boldsymbol{y}$ represents the measured data and is the sum of the sampled direct absorption spectrum or harmonic spectrum, spectral baseline, measurement noise, etc. It consists of N data points. The data sample rate is $R\,N$, with R (unit: Hz) being the measurement rate, i.e. the rate with which spectra are scanned and concentration values are output.

The vector $\boldsymbol{y}$ can be expressed by the following model:

$$\boldsymbol{y} = \underbrace{(\boldsymbol{\psi}_1 | \boldsymbol{\psi}_2 | \cdots)}_{\boldsymbol{\Psi}} \begin{pmatrix} p_1 \\ p_2 \\ \vdots \end{pmatrix} + \boldsymbol{\epsilon}. \tag{4}$$

$\boldsymbol{\psi}_1, \boldsymbol{\psi}_2, \ldots$ are the column vectors of the model matrix $\boldsymbol{\Psi}$, which represent the spectra components linearly weighted with the parameters $p_1, p_2, \ldots$ and superimposed with noise $\boldsymbol{\epsilon}$ to form the measured spectrum $\boldsymbol{y}$.

The parameters p_i are for example the peak absorption values of the target or interfering gases that are present in the optical cell; offset; slope and amplitude, phase of etalon fringes. p_1 is the peak gas absorption value of the desired gas, which is the product of gas concentration C, optical path length l and peak absorption coefficient a (see Fig. 4):

$$a = A(\lambda_0)|_{C \cdot l = 1\text{ppm m}}, \tag{5}$$

where λ_0 is the wavelength of the absorption line center or—in the case of a complex spectrum consisting of many overlapping lines of one gas—the wavelength of the peak value. The gas concentration C is estimated from $\boldsymbol{y}$ by a linear least squares curve fit [18] and has the following analytical expression:

$$C = \frac{p_1}{a\,l} = \frac{1}{a\,l}\left(\left(\boldsymbol{\Psi}^T \boldsymbol{\Psi}\right)^{-1} \boldsymbol{\Psi}^T \boldsymbol{y}\right)_1, \tag{6}$$

where the subscript notation $(\boldsymbol{X})_1$ denotes the first element of the vector $\boldsymbol{X}$. The curve-fit process is schematically explained in Fig. 3.

The $\boldsymbol{\psi}_i$ are the spectrum waveforms of the target gas and interfering gases, the spectral background (offset, slope) and etalon fringes. The first column $\boldsymbol{\psi}_1$ is the spectrum waveform of the target gas and has the following expression for direct spectroscopy:

$$\boldsymbol{\psi}_1 = \left(\frac{A(\lambda_1)}{a\,C\,l}, \frac{A(\lambda_2)}{a\,C\,l}, \ldots, \frac{A(\lambda_N)}{a\,C\,l}\right)^T \tag{7}$$

with the vector of sampled wavelength points:

$$\boldsymbol{\lambda} = (\lambda_1, \lambda_2, \ldots, \lambda_N)^T. \tag{8}$$

For WMS with second harmonic detection it is given by

$$\boldsymbol{\psi}_1 = \left(\frac{S_2(\lambda_1)}{a\,C\,l}, \frac{S_2(\lambda_2)}{a\,C\,l}, \ldots, \frac{S_2(\lambda_N)}{a\,C\,l}\right)^T. \tag{9}$$

$\boldsymbol{\psi}_1$ is independent of which gas is investigated, the gas concentration C as well as the optical path length l. In the case

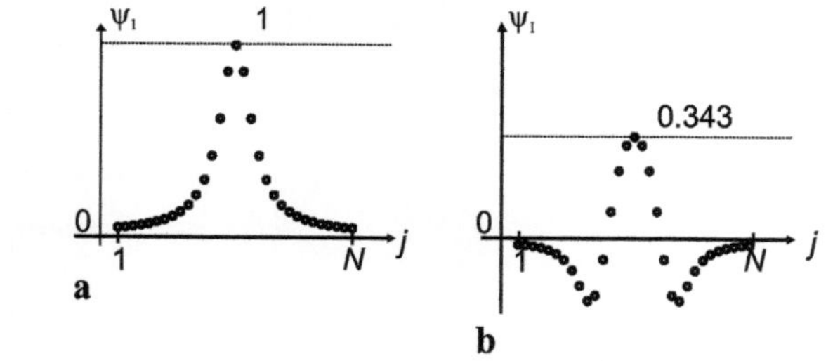

Fig. 5 Examples of sampled spectrum waveforms $\boldsymbol{\psi}_1$ for linear scanning in (**a**) direct spectroscopy and (**b**) WMS (second harmonic) are shown

of WMS the definition of the second harmonic spectrum $S_2(\lambda)$ is taken from [19]:

$$S_2(\lambda) = -\frac{1}{\pi}\int_{-\pi}^{\pi} A\big(\lambda + \lambda_a \cos(z)\big)\cos(2z)\,\mathrm{d}z, \tag{10}$$

with λ_a being the amplitude of wavelength modulation. $\boldsymbol{\psi}_1$ has a maximum value of 1 for direct spectroscopy and is less than 1 for WMS. Examples of linearly scanned $\boldsymbol{\psi}_1$ in direct spectroscopy and WMS (second harmonic) are shown in Fig. 5a and b, respectively.

2.3 Analytical expression for noise variance σ_w^2

As explained before, the measurement noise $\boldsymbol{\epsilon}$ is assumed to be white noise, i.e.

$$\langle \boldsymbol{\epsilon} \rangle = \mathbf{0}, \qquad \langle \boldsymbol{\epsilon} \cdot \boldsymbol{\epsilon}^T \rangle = \sigma^2 \boldsymbol{I}, \tag{11}$$

with $\boldsymbol{I}$ being the identity matrix. The noise variance σ^2 is expressed by the product of noise power density S_{yy} on measured data $\boldsymbol{y}$ and the noise bandwidth $R\,N$:

$$\sigma^2 = S_{yy}\,R\,N. \tag{12}$$

The relationship between σ^2 and σ_w^2 is given by taking the variance of (6) [18]:

$$\sigma_w^2 = \frac{1}{a^2 l^2}\left(\boldsymbol{\Psi}^T\boldsymbol{\Psi}\right)_{11}^{-1} \cdot \sigma^2 \tag{13}$$

$$= \frac{1}{a^2 l^2}\left(\boldsymbol{\Psi}^T\boldsymbol{\Psi}\right)_{11}^{-1} N \cdot S_{yy} \cdot R. \tag{14}$$

$(\boldsymbol{\Psi}^T\boldsymbol{\Psi})_{11}^{-1}$ is the left top element of the inverse matrix of $\boldsymbol{\Psi}^T\boldsymbol{\Psi}$. As clearly can be seen, σ_w^2 (unit: ppm^2) is a function of S_{yy} (unit: 1/Hz), N (unit: 1), R (unit: Hz), a (unit: 1/ppm m), optical path length l (unit: m) and the vectors $\boldsymbol{\psi}_i$. By scanning the wavelength nonlinearly or modifying the WMS modulation index m or varying the scanned wavelength range, the spectral components $\boldsymbol{\psi}_i$ change, which results in the variation of $(\boldsymbol{\Psi}^T\boldsymbol{\Psi})_{11}^{-1} N$ and possibly lower values of σ_w^2. Note that $(\boldsymbol{\Psi}^T\boldsymbol{\Psi})_{11}^{-1} N$ is constant for $N \to \infty$, because $(\boldsymbol{\Psi}^T\boldsymbol{\Psi})_{11}^{-1}$ is asymptotically inversely proportional to N, whereas the limiting case also represents the continuous situation without sampling.

In the following sections, the optimum wavelength scanning functions for three different models are discussed.

2.4 Signal model I: only target gas concentration is unknown

If the spectral baseline is known, and there is no interfering gas or the concentration and spectrum of interfering gas are known, the model is written as

$$\boldsymbol{y} = (\boldsymbol{\psi}_1)(C\,l\,a) + \boldsymbol{\epsilon}, \tag{15}$$

where $\boldsymbol{y}$ is the measured data with baseline and interfering gas spectrum removed.

Substituting the model in (14) gives

$$\sigma_w^2 = \frac{1}{a^2 l^2 |\boldsymbol{\psi}_1|^2} \cdot S_{yy} \cdot R \cdot N. \tag{16}$$

It is evident that (16) is minimized if $|\boldsymbol{\psi}_1|$ is maximized. This is achieved by sampling the absorption spectrum only at its absorption maximum (peak) value. This gives

$$\boldsymbol{\psi}_{1,\mathrm{opt}} = \left(\psi_{1,\max}, \ldots, \psi_{1,\max}\right)^T, \tag{17}$$

which results in

$$\sigma_{w,\mathrm{opt}}^2 = \frac{1}{a^2 l^2 \psi_{1,\max}^2} \cdot S_{yy} \cdot R. \tag{18}$$

The corresponding spectrum waveform carries the constant value $\psi_{1,\max}$, which is 1 for direct spectroscopy and 0.343 for second harmonic detection, with $m = 2.2$. The modulation index m is chosen for a maximum signal peak height. For this model with known baseline and known interfering gas spectrum if there is any, it is thus optimum to sample only the maximum value of the gas absorption spectrum or the WMS harmonic spectrum.

2.5 Signal model II: target gas concentration and baseline (slope, offset) are unknown

In this case the signal model is consisting of spectral components belonging to the target gas and a spectral baseline consisting of a slope and an offset. There is no interfering gas in the spectral range and no other spectral baseline components, e.g. cubic baseline or interference fringes, or they are known and subtracted from the measurement data as in [14, 15].

$$\boldsymbol{y} = (\boldsymbol{\psi}_1|\boldsymbol{\psi}_2|\boldsymbol{\psi}_3)\begin{pmatrix} C\,l\,a \\ \text{slope} \\ \text{offset} \end{pmatrix} + \boldsymbol{\epsilon}, \tag{19}$$

with the new baseline components for offset and slope:

$$\boldsymbol{\psi}_2 = (1 \ldots 1)^T \quad \text{(offset)}, \tag{20}$$

$$\boldsymbol{\psi}_3 = (-1 \ldots 1)^T \quad \text{(slope)}. \tag{21}$$

Note that the slope component of the spectral baseline also changes in a nonlinear fashion when the wavelength is scanned nonlinearly. Then σ_w^2 is given by (under the assumption that $\boldsymbol{\psi}_1$ is symmetric)

$$\sigma_w^2 = \frac{1}{a^2 l^2 \left(|\boldsymbol{\psi}_1|^2 - N\overline{\boldsymbol{\psi}}_1^2\right)} \cdot S_{yy} \cdot R \cdot N. \tag{22}$$

$\overline{\boldsymbol{\psi}}_1$ denotes the average value of the elements of vector $\boldsymbol{\psi}_1$. Equation (22) is minimized when $(|\boldsymbol{\psi}_1|^2 - N\overline{\boldsymbol{\psi}}_1^2)$ is maximized. From optimization theory it follows that this requires sampling the minimum and maximum points of the spectrum only. Knowing this, it follows by a simple calculation that the optimum is achieved by sampling the maximum and minimum of the spectrum equally often and in a symmetrical manner as required by the assumption

$$\boldsymbol{\psi}_{1,\text{opt}} = (\underbrace{\boldsymbol{\psi}_{1,\min}\cdots}_{N/4}\underbrace{\boldsymbol{\psi}_{1,\max}\cdots}_{N/2}\underbrace{\boldsymbol{\psi}_{1,\min}\cdots}_{N/4})^T. \tag{23}$$

The minimized noise variance then has the expression

$$\sigma_{w,\text{opt}}^2 = \frac{4}{a^2 l^2 (\psi_{1,\max} - \psi_{1,\min})^2} \cdot S_{yy} \cdot R. \tag{24}$$

For direct spectroscopy we have $\psi_{1,\max} = 1$ and $\psi_{1,\min} \approx 0$. For second harmonic detection, the peak to peak amplitude of the second harmonic spectrum is maximum when m is set to 3 [20], which gives $\psi_{1,\max} = 0.33$ and $\psi_{1,\min} \approx -0.22$. The optimum spectrum waveforms for target gas $\boldsymbol{\psi}_1$ and the wavelength scanning function are shown in Fig. 6.

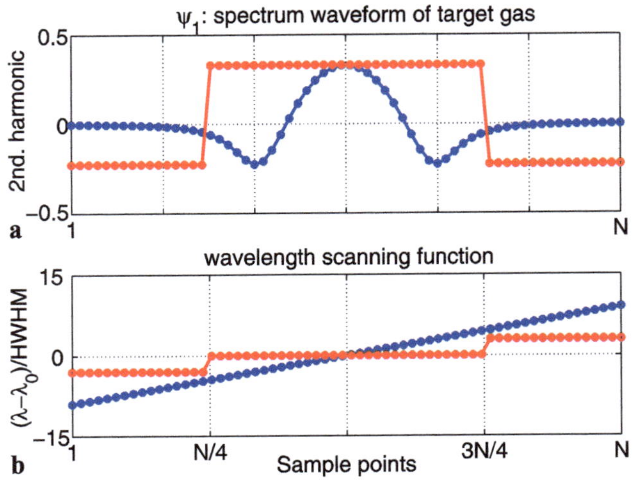

Fig. 6 The scanning waveform (second harmonic spectrum) for target gas (**a**) and the wavelength scanning function (**b**) for model II (including baseline, but no interfering gas). Traditional linear scanning (*blue*) and proposed optimum scanning (*red*) are both presented. The y-axis of (**b**) is the wavelength, which is relative to the line center λ_0 normalized with the HWHM of the gas absorption line

2.6 Signal model III: target gas, baseline, interfering gases and etalon fringes are unknown

In this section, an interfering gas spectrum and etalon fringes are also included in the model. A parabolic/cubic spectral baseline or polynomial arbitrary order can also be added to the model. Etalon fringes caused by interfering of light consist of a periodic sinusoidal structure in the measured spectrum; this structure can limit the sensitivity of a diode laser measurement for trace gas concentration measurements. Most of the etalon fringes are caused by parallel reflecting surfaces, between which the light is reflected back and forth several times. The reflected beams interfere at the photodetector constructively or destructively, causing the resulting intensity to be wavelength dependent. The period length of the sinusoidal structure depends on the distance between the two optical surfaces. If the distance is known, the etalon fringes can be included in the linear model, with amplitude and phase of the sinusoidal structure being fitted. This modeled spectral background consists of a linear baseline and interfering fringes with certain known period length. The amplitude of the linear baseline and amplitude and phase of the fringes are fitted continuously, so a time-dependent baseline change is handled. As an extension, the fringes could be modeled not only as a single sinusoid, but as many shifted and partially overlapping finite length sinusoids. This would even allow for fitting a fringe with varying phase and amplitude during a single spectrum scan. However, simulations have shown that the optimum scanning waveform for this extended model is very similar to the simplified case used here. The model is the following:

$$\boldsymbol{y} = (\boldsymbol{\psi}_1|\boldsymbol{\psi}_2|\boldsymbol{\psi}_3|\boldsymbol{\psi}_4|\boldsymbol{\psi}_5|\boldsymbol{\psi}_6)\begin{pmatrix} Cla \\ \text{slope} \\ \text{offset} \\ C_2 l \\ A \\ B \end{pmatrix} + \epsilon, \tag{25}$$

with $\boldsymbol{\psi}_4$ being the interfering gas spectrum and C_2 the concentration of the interfering gas. $\boldsymbol{\psi}_5$ and $\boldsymbol{\psi}_6$ are the components for the etalon fringes with unknown phase:

$$\boldsymbol{\psi}_5(\boldsymbol{\lambda}) = \sin\left(2\pi\,(\boldsymbol{\lambda} - \lambda_0)/\Delta\lambda\right)^T, \tag{26}$$

$$\boldsymbol{\psi}_6(\boldsymbol{\lambda}) = \cos\left(2\pi\,(\boldsymbol{\lambda} - \lambda_0)/\Delta\lambda\right)^T, \tag{27}$$

and $\Delta\lambda$ is the period length of the etalon fringes in wavelength, which is derived from the length of the etalon L. λ_0

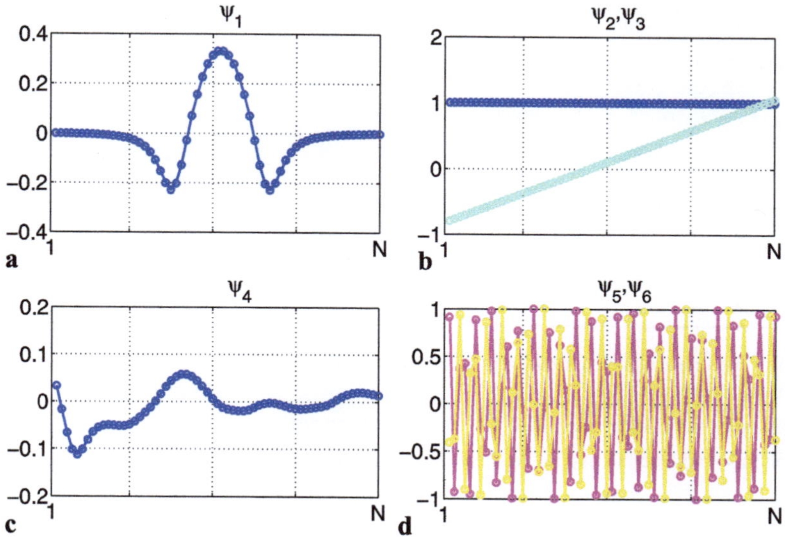

Fig. 7 Example of simulated spectral compositions ψ_i in model III: ψ_1 is the second harmonic spectrum of CO at 2.365 μm (**a**) with interfering gas methane ψ_4 (second harmonic spectrum of 20 methane lines) (**c**); ψ_2 and ψ_3 (**b**) are the spectral baseline: offset and slope; ψ_5 and ψ_6 (**d**) are the sampled sine and cosine components forming the etalon fringes

is the central wavelength of the laser light. We have

$$\Delta\lambda = \frac{\lambda_0^2}{2nL}. \tag{28}$$

By fitting the parameters A and B, the amplitude and phase of the sinusoidal structure are adjusted.

σ_w^2 can be only minimized numerically. The optimization problem is

$$\text{minimize} \quad \left(\boldsymbol{\Psi}^T\boldsymbol{\Psi}\right)_{11}^{-1} \tag{29}$$

$$\text{subject to} \quad \lambda_{\min} \leq \lambda_1 \leq \lambda_2 \leq \cdots \leq \lambda_N \leq \lambda_{\max}, \tag{30}$$

where the columns of the $\boldsymbol{\Psi}$ matrix ψ_i are dependent on the sampled wavelength λ_j. This algorithm for the constraint optimization problem [21] is described in [22, 23], and implemented in the MATLAB function 'fmincon'. We demonstrate the numerical optimization for CO detection at 2.365 μm with the interfering gas methane. For this case, the spectral compositions ψ_i are shown in Fig. 7. The results obtained are shown in Figs. 8 and 9 for direct spectroscopy and second harmonic detection, respectively. It is remarkable that also in the numerically optimized case the optimum wavelength scanning function has a cascade structure. The maximum is more often sampled than in the linear scanning case. The algorithm determined the modulation index m to be 3, so the peak to peak amplitude of the second harmonic spectrum is maximized as in model II. On the other hand, the transitions from minimum to maximum are also sampled to obtain the information of the methane spectrum and to average over one period of the etalon fringes.

2.6.1 Comparison between direct spectroscopy and WMS

The sensor sensitivity σ_w is multiplied with $a\,l/\sqrt{S_{yy}\,R}$ and the values are listed for comparison of linear scanning and

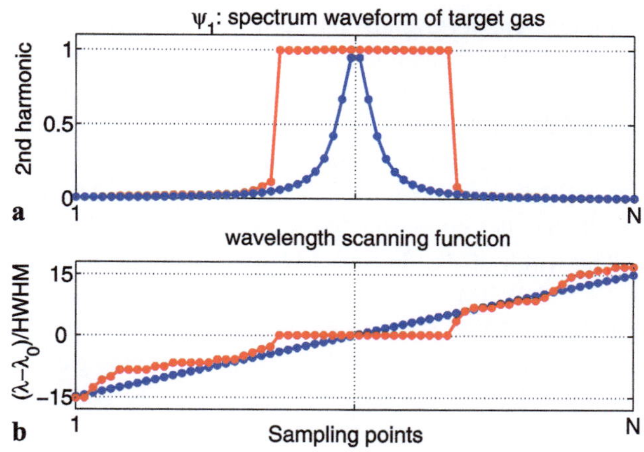

Fig. 8 (**a**) The spectrum waveform (direct absorption spectrum) for target gas CO and (**b**) the wavelength scanning function for model III. Traditional linear scanning (*blue*) and proposed optimum scanning (*red*) are both presented

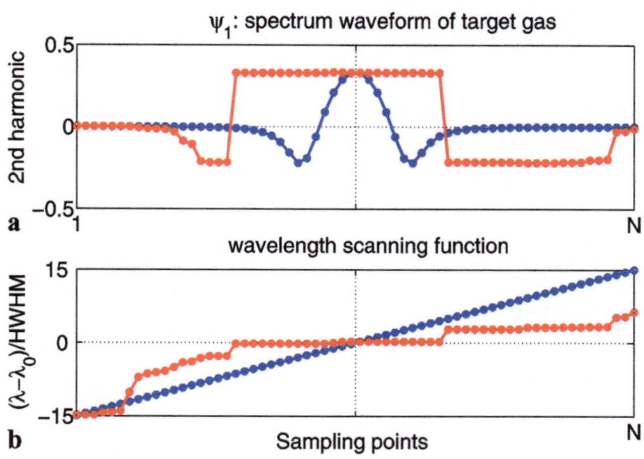

Fig. 9 (**a**) The spectrum waveform (second harmonic spectrum) for target gas CO and (**b**) the wavelength scanning function for model III. Traditional linear scanning (*blue*) and proposed optimum scanning (*red*) are both presented

Table 1 List of $\frac{\sigma_w a l}{\sqrt{S_{yy}} R}$ values for the comparison of linear scanning and nonlinear wavelength scanning applying direct spectroscopy or WMS. Lower values are better

$\frac{\sigma_w a l}{\sqrt{S_{yy}} R} = \sqrt{(\boldsymbol{\Psi}^T \boldsymbol{\Psi})_{11}^{-1} N}$	Direct spectroscopy		WMS (second harmonic)	
	Optimum nonlinear scanning	Linear scanning (typical width)[a]	Optimum nonlinear scanning	Linear scanning (typical width)[a]
I	1	2.8	2.9[b]	5.9
II	2	3.6	3.6[c]	5.9
III	2.2	4.2	4.2[c]	7.6

[a]The width of the linear wavelength scanning is chosen as 12 × HWHM of the gas absorption line for models I and II, and 20 × HWHM of the gas absorption line for model III

[b]Modulation index adjusted to maximize spectrum waveform peak ($m_{\text{opt}} \approx 2.2$)

[c]Modulation index adjusted to maximize spectrum waveform amplitude (peak to peak) ($m_{\text{opt}} \approx 3$)

nonlinear wavelength scanning applying direct spectroscopy or WMS (Table 1). This value is independent of the absorption line strength (given by a), optical path length l, sensor time resolution R and noise power density on measurement data S_{yy}. The lower values are better. Different models represent different levels of available a priori information. For model I (Sect. 2.4) the spectral baseline is known, for model II (Sect. 2.5) it is unknown and model III (Sect. 2.6) additionally handles interfering gases and interference fringes.

The width of the linear wavelength scanning is chosen as 12 × HWHM of the gas absorption line for models I and II, and 20 × HWHM of the gas absorption line for model III because of the involvement of interfering gas. According to Table 1, the standard deviation of the concentration noise/sensor detection sensitivity can be improved by a factor ~2 by applying nonlinear wavelength scanning instead of linear scanning for direct spectroscopy and WMS. Furthermore, an additional factor of ~2 lower standard deviation can be achieved theoretically if direct spectroscopy instead of second harmonic detection is applied.

With the values given in Table 1, we can also calculate the expected detection sensitivity of a sensor system. For example, assume a sensor with a measurement rate of 10 Hz, and a high enough light intensity at the detector so that the laser noise is dominating. Additionally, we assume a laser relative intensity noise (RIN) of –130 dB/Hz and therefore obtain $S_{yy} = \text{RIN} = 10^{-13}$. The CO line at 2.3655 μm has an absorption coefficient $a = 3.5714 \times 10^{-5}\,\text{ppm}\,\text{m}^{-1}$ at ambient conditions. With an optical path length of about 10 cm the standard deviation of the estimated concentration value is according to the table (model III; WMS) expected to be 2.1 ppm for linear scanning, which is in good agreement with the experimental data (standard deviation: 2–3 ppm) [24, 26]. This performance can theoretically be improved by a factor of 1.8, if nonlinear wavelength scanning with second harmonic detection is applied, or by a factor of 3.3 using direct spectroscopy with optimum nonlinear wavelength scanning.

2.7 Nonlinear model

The theory for the linear model is applicable for the general nonlinear curve-fit model by linearizing it at first. The nonlinear model contains nonlinear parameters, e.g. gas pressure, gas temperature or wavelength offset. To explain this procedure, we analyze the simplest case that the model contains two parameters: one is the gas concentration C; the other is a nonlinear parameter p (e.g. gas pressure or a wavelength shift). By applying Taylor series, the model is linearized around (C_0, p_0) to

$$\boldsymbol{y} = \boldsymbol{f}(C, p) + \boldsymbol{\epsilon} \tag{31}$$

$$\approx \boldsymbol{f}(C_0, p_0) + \frac{\partial \boldsymbol{f}(C_0, p_0)}{\partial C}(C - C_0) + \frac{\partial \boldsymbol{f}(C_0, p_0)}{\partial p}(p - p_0) + \boldsymbol{\epsilon} \tag{32}$$

$$= \frac{\partial \boldsymbol{f}(C_0, p_0)}{\partial C} C + \frac{\partial \boldsymbol{f}(C_0, p_0)}{\partial p}(p - p_0) + \boldsymbol{\epsilon} \tag{33}$$

$$= \underbrace{\left(\frac{\partial \boldsymbol{f}(C_0, p_0)}{\partial C} \,\middle|\, \frac{\partial \boldsymbol{f}(C_0, p_0)}{\partial p} \right)}_{\boldsymbol{\Psi}} \begin{pmatrix} C \\ p - p_0 \end{pmatrix} + \boldsymbol{\epsilon}, \tag{34}$$

where C_0 and p_0 should be as close as possible to the real values of C and p.

3 Conclusion and discussion

In summary, the theoretical investigation gives the optimum method to scan the laser wavelength to minimize the influence of white noise on the determined concentration values. For this, the gas absorption spectrum is probed with an optimized wavelength scanning function which enables a more efficient curve fit than for traditional linear wavelength scanning. The wavelength scanning function of the laser current/wavelength and the modulation index—if WMS is applied—are determined to be optimum, in the sense that the variance of the estimated concentration noise is minimized. This optimum scanning function depends on the signal model used in curve fitting. Several models including interfering gas spectra and etalon fringes are examined. While for the simple models, analytical solutions were found, for the more comprehensive cases like models including interfering gas components and etalon fringes the equation (14) were handled numerically. Compared to the gas absorption spectrum recorded by ramping the wavelength linearly, the optimum spectrum waveforms have a cascade structure, which means that the optimum scanning only samples important points of the spectrum. When the spectral baseline is unknown, the modulation index m for WMS is optimally chosen to maximize the peak to peak value of the signal ($m = 3$), not the maximum value ($m = 2.2$). The new method theoretically enables a factor of $\sim$2 improvement on detection sensitivity of the estimated concentration compared to conventional linear wavelength scanning with typical scanning width. Furthermore, direct spectroscopy is superior to second harmonic detection, because the concentration noise can be an additional factor of $\sim$2 lower.

To use this method, knowledge of the sampled wavelength and the laser current-to-wavelength relationship is essential. For WMS it is not obvious which wavelengths correspond to the minima of the curve, because their positions are also dependent on the modulation index m. To obtain the laser current-to-wavelength relationship, a wavelength calibration with a reference measurement can be carried out. This can be done with a separate reference cell or integrated reference cell filled with reference gas [24, 26] or a reference etalon [25]. A remaining open question is whether the nonlinear scanning may increase the influence of drifts and time-varying fringes compared to linear scanning. Unfortunately, this cannot be answered by this theory. It has to be examined by experiments. The result is partly confirmed by the observed SNR improvement by using jump scanning in [12] and the used formula (14) for the SNR after the curve fit is verified in [26]. There the computed maximized SNR by optimizing the scanning width of a spectrum is in good agreement with the experimental results. In addition to the 'jump scanning' experimental paper [12], this paper tells the optimum positions where the jumps have to occur and the obtained gain in signal-to-noise ratio.

References

1. J. Reid, D. Labrie, Appl. Phys. B, Lasers Opt. **26**, 203 (1981)
2. A. O'Keefe, Chem. Phys. Lett. **293**, 331 (1998)
3. A. O'Keefe, J.J. Scherer, J.B. Paul, Chem. Phys. Lett. **307**, 343 (1999)
4. C.B. Carlisle, D.E. Cooper, Opt. Lett. **14**, 1306 (1989)
5. D.M. Sonnenfroh, W.T. Rawlins, M.G. Allen, C. Gmachl, F. Capasso, A.L. Hutchinson, D.L. Sivco, J.N. Baillargeon, A.Y. Cho, Appl. Opt. **40**, 812 (2001)
6. R.E. Kalman, Trans. ASME, J. Basic Eng. **82**, 35 (1960)
7. H. Riris, C.B. Carlisle, R.E. Warren, Appl. Opt. **33**, 5506 (1994)
8. D. Leleux, R. Claps, W. Chen, F. Tittel, T. Harman, Appl. Phys. B, Lasers Opt. **74**, 85 (2002)
9. N. Wiener, *Extrapolation, Interpolation, and Smoothing of Stationary Time Series* (Wiley, New York, 1949)
10. H. Riris, C.B. Carlisle, R.E. Warren, D.E. Cooper, R.U. Martinelli, R.J. Menna, P.K. York, D.Z. Garbuzov, H. Lee, J.H. Abeles, N. Morris, J.C. Connolly, S.Y. Narayan, Spectrochim. Acta Part A, Mol. Biomol. Spectrosc. **52**, 843 (1996)
11. M. Sigrist, R. Bartlome, D. Marinov, J. Rey, D. Vogler, H. Wächter, Appl. Phys. B **90**, 289 (2008)
12. A. Fried, B. Henry, J.R. Drummond, Appl. Opt. **32**, 821 (1993)
13. P. Werle, R. Muecke, F. Slemr, Appl. Phys. B **57**, 131 (1993)
14. A. Fried, B. Henry, B. Wert, S. Sewell, J. Drummond, Appl. Phys. B, Lasers Opt. **67**, 317 (1998)
15. P. Werle, B. Scheumann, J. Schandl, Opt. Eng. **33**, 3093 (1994)
16. P. Kluczynski, O. Axner, Appl. Opt. **38**, 5803 (1999)
17. D. Bomse, D. Kane, Appl. Phys. B, Lasers Opt. **85**, 461 (2006)
18. C. Rao, H. Toutenburg, *Linear Models: Least Squares and Alternatives* (Springer, Berlin, 1995)
19. A. Hangauer, J. Chen, M.-C. Amann, Appl. Phys. B **90**, 249 (2008)
20. S. Schilt, L. Thévenaz, P. Robert, Appl. Opt. **42**, 6728 (2003)
21. R. Dechter, *Constraint Processing* (Morgan Kaufmann, San Mateo, 2003)
22. S. Han, J. Optim. Theory Appl. **22**, 297 (1977)
23. R. Fletcher, M. Powell, Comput. J. **6**, 163 (1963)
24. J. Chen, A. Hangauer, R. Strzoda, M. Ortsiefer, M. Fleischer, M.C. Amann, Compact carbon monoxide sensor using a continuously tunable 2.3 μm single-mode VCSEL, in *Proc. 21st Annu. Meet. IEEE Lasers and Electro-Optics Society*, vol. 9 (2008) pp. 721–722
25. D.D. Nelson, J.B. McManus, S.C. Herndon, M.S. Zahniser, B. Tuzson, L. Emmenegger, Appl. Phys. B, Lasers Opt. **90**, 301 (2008)
26. J. Chen, A. Hangauer, R. Strzoda, M.C. Amann, VCSEL-based calibration-free carbon monoxide sensor at 2.3 μm with in-line reference cell. Appl. Phys. B, Lasers Opt. (2010, to be published). doi:10.1007/s00340-010-4011-0

Laser-based measurements of line strength, self- and pressure-broadening coefficients of the $H^{35}Cl$ R(3) absorption line in the first overtone region for pressures up to 1 MPa

P. Ortwein · W. Woiwode · S. Wagner · M. Gisi · V. Ebert

Received: 19 October 2009 / Revised version: 30 November 2009 / Published online: 23 December 2009
© Springer-Verlag 2009

Abstract A high-resolution spectrometer based on a vertical-cavity surface-emitting laser (VCSEL) was developed and used to determine the line strength $S(T_0) = 12.53(11) \times 10^{-21}$ $cm^{-1}/(molec\,cm^{-2})$ and the self-broadening coefficient $\gamma^0_{HCl} = 0.021787(61)$ cm^{-1}/atm of the R(3) absorption line in the first rovibrational overtone ($2 \leftarrow 0$) band of $H^{35}Cl$. Furthermore, the first laser-based high-pressure study on the pressure broadening of HCl by He, N_2 and O_2($\gamma^0_{N_2} = 0.07292(5)$ cm^{-1}/atm, $\gamma^0_{He} = 0.02113(1)$ cm^{-1}/atm, $\gamma^0_{O_2} = 0.03978(6)$ cm^{-1}/atm) is presented covering pressures of up to 1 MPa. The results are compared to previously available low-pressure data.

1 Introduction

The detection of HCl plays an important role in atmospheric chemistry as well as in many technical processes. Especially in combustion and gasification plants high-temperature chlorine corrosion via HCl often shortens the plant lifetime therefore leading to considerable additional costs [1]. A fast detection of HCl by extractive gas analysis for process control purposes is difficult due to strong HCl adsorption, as well as large amounts of condensates or particles in the process gases. Sampling-free, optical analysis techniques like tunable diode laser absorption spectroscopy (TDLAS) are promising diagnostic tools, which take advantage of the very narrow laser line width, thus permitting excellent selectivity and sensitivity [2]. Other properties like high temporal resolution in combination with a relatively simple technical setup effort make TDLAS well suited for technical field applications. This could be successfully shown over the last years even under complex industrial conditions [3–6]. However, a sensitive HCl measurement in pressurized high-temperature gasification processes requires precise knowledge of the HCl-spectroscopic line parameters, like line strength or self- and foreign-broadening coefficients. Numerous studies published such parameters for the mid infrared i.e. the fundamental vibrational band or for the low-pressure range up to 0.1 MPa only [7–10]. But little data are available for the Near Infrared (NIR) overtone transitions or for high-pressure conditions. Besson et al. measured foreign broadening coefficients for the HCl R(4) absorption line in the first overtone ($2 \leftarrow 0$) using photo-acoustic methods at atmospheric pressure [11]. The only determination of broadening coefficients for higher pressures (up to 4 MPa) was performed by Boulet et al. using an FTIR spectrometer [12]. But he studied only the technically not really relevant Ar atom as collision partner. This situation can also be seen in the HITRAN 2008 line data base, where, due to the lack of suitable overtone line data, the corresponding parameters of the fundamental band ($1 \leftarrow 0$) were simply copied to the first overtone ($2 \leftarrow 0$) transitions [13]. In addition—though not unexpected for an atmospheric line data base—only self- and air-broadening coefficients are given.

To fill this gap, at least for technically interesting lines, i.e. those spectrally well isolated from water and CO_2, and to resolve the discrepancies between the literature values and the HITRAN 2008 data base for the broadening coef-

P. Ortwein · W. Woiwode · S. Wagner · M. Gisi · V. Ebert (✉)
Institute of Physical Chemistry, University of Heidelberg,
INF 253, 69120 Heidelberg, Germany
e-mail: volker.ebert@pci.uni-heidelberg.de
url: www.gasanalysis.org

V. Ebert
Physikalisch-Technische Bundesanstalt, Bundesallee 100,
38116 Braunschweig, Germany
e-mail: volker.ebert@ptb.de

ficients and line strength for the $H^{35}Cl$ ($2 \leftarrow 0$) R(3) absorption line [14, 15], we performed a high-resolution line parameter study for the R(3) line for numerous technically important perturbers. Furthermore we wanted to determine line parameters closer to the actual high-pressure conditions in technical gasifiers. E.g. biomass gasification for synthetic fuel production is done at up to 8 MPa [16, 17]. Hence we used a recently developed vertical-cavity surface-emitting laser (VCSEL) at 1.74 μm with a very wide tuning range of 10 cm^{-1} and developed a highly sensitive, high-pressure suitable spectrometer to realize the first, quantitative, laser-based high-pressure line shape analysis of the $H^{35}Cl$ R(3) absorption line in the first overtone ($2 \leftarrow 0$) rovibrational band at up to 1 MPa.

2 Experimental details

For our measurements a widely tunable laser was necessary to realize measurements with pressures of up to 1 MPa. DFB lasers, which are very frequently used for TDLAS, have a tuning range of about 1 cm^{-1} only. Thus they cannot cover strongly pressure broadened absorption lines. Recently developed, long-wavelength vertical-cavity surface-emitting lasers (VCSEL) in the range of 1.5 to 2 μm provide a considerably wider tuning range, combined with good spectroscopic characteristics such as a narrow line width of only a few 10 MHz [18–20].

Demanding technical applications like a sampling-free gas analysis in high-temperature gas flows require an application specific line selection in order to determine a target line with minimal spectral interference by other gas species in the matrix like H_2O, CO_2 or CH_4. Using our new software-based line selection program [21] we found the $H^{35}Cl$ R(3) line at 5739 cm^{-1} most suitable. This line provides good spectroscopic properties such as sufficient line strength, also at high temperatures, and minimized cross-sensitivity to CO_2 and H_2O. Therefore it is well applicable for the HCl detection in industrial processes.

The 1.74 μm VCSEL we used as light source offers a maximum optical output power of 0.7 mW. A highly precise measurement of the line strength as well as the broadening parameters requires a comprehensive knowledge of the optical, spectral and electronic characteristics of the VCSEL. The wavelength coverage of the respective laser is limited by the maximum range of operating temperatures (15–35°C) and operating currents (1–6 mA). Within this range we determined quasi-static tuning coefficients of $d\nu/dT = 0.45$ cm^{-1}/K ($I = 5$ mA) and $d\nu/dI = 1.75$ cm^{-1}/mA ($T = 20$°C and $T = 30$°C). Furthermore, we measured the modulation frequency dependent, so called "dynamic" tuning characteristics of the laser for modulation frequencies

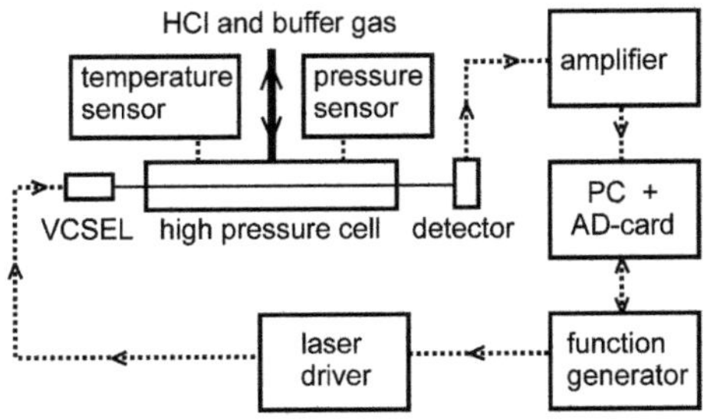

Fig. 1 Experimental setup of the spectrometer for the determination of the line strength, self- and pressure-broadening coefficients of the $H^{35}Cl$ R(3) absorption line based on a vertical surface-emitting laser at 1.74 μm

from 0.5 to 6000 Hz. Ideally, under linear current modulation a constant laser tuning coefficient $d\nu/dt$ would be desirable. However, most lasers deviate strongly from linearity-particularly for higher modulation frequencies-making an efficient tuning linearization algorithm indispensable [22]. In the present work the characterization was done using a planar, air-spaced ($L = 10$ cm) precision etalon and semi-automated fringe fitting software.

The spectrometer for the line strength and broadening measurements is shown in Fig. 1. It consisted of a function generator (Agilent 33220A) supplying the laser driver with a periodic, triangular signal. The laser is modulated with 10.4 Hz for the measurement of the line strength $S(T_0)$ and the self-broadening coefficient γ^0_{HCl} because this was done at low pressures and therefore at very narrow molecular line widths. For the measurement of the foreign broadening coefficients with He, N_2 and O_2 for pressures up to 1 MPa the line widths are much wider and therefore a modulation frequency of 139.8 Hz could be used to improve the signal to noise level. The laser is temperature-stabilized at 305 K using a built-in Peltier element that is mounted in the laser can. The laser light cone is collimated using an AR-coated aspheric lens ($f = 4.5$ mm, diameter 6.35 mm) and directed through a 24.5 cm long high-pressure cell. To avoid optical fringes the cell is equipped with wedged (1°) 1″ sapphire windows on both sides. The laser beam is detected by an extended InGaAs photodiode (diameter 1 mm) and the diode current is amplified by a low-noise transimpedance amplifier (Femto LCA-S, 10^3 V/A, 2 MHz bandwidth). The resulting signal is digitized with a 12 bit A/D-converter (60 MS/s) and evaluated using our LABVIEW-based line fitting code.

The HCl gas was generated immediately before the measurement by a reaction of sulfuric acid (H_2SO_4) and sodium chloride (NaCl) in an evacuated glass apparatus and directed via stainless steel pipes to the measurement cell. For the line strength and the self-broadening measurements the evacuated cell was filled successively with HCl up to a pressure of 1000 hPa. Foreign broadening coefficients were measured by filling the evacuated cell with 150 hPa HCl, followed by the addition of the respective perturber up to total pressures of 1 MPa.

The TDLAS principle is based on the measurement of the wavelength dependent light attenuation while illuminating a gas sample. In our case TDLAS can be described by an extended Lambert-Beer's law [23, 24]:

$$I(\nu,t) = I_0(\nu)\cdot Tr(t)\cdot \exp\big(-S(T)\cdot g(\nu-\nu_0)\cdot n\cdot L\big) + E(t) \qquad (1)$$

with n being the number density of the molecular absorbers, $I_0(\nu)$ the initial laser intensity and $I(\nu)$ the intensity detected after the absorbing medium of thickness L.

$Tr(t)$ describes the effect of variation in cell transmission and $E(t)$ that of additive detection signals, like background radiation. The absorption line is characterized by the temperature dependent, spectrally integrated line strength $S(T)$, and the normalized (area = 1) shape function $g(\nu,\nu_0)$, which is centered at the wavelength ν_0. Resolving this equation with respect to the absorber density n, while spectrally integrating the absorption line shape $g(\nu,\nu_0)$ over the wavelength ν [cm^{-1}] leads to [23, 24]

$$n = -\frac{1}{S(T)\cdot L}\int \ln\left(\frac{I(\nu,t)-E(t)}{I_0(\nu)\cdot Tr(t)}\right)\frac{\partial\nu}{\partial t}\,dt. \qquad (2)$$

Thus the number density can be determined without the need of concentration calibration, solely with knowledge of the line strength $S(T)$, the experimental boundary conditions $(p, T, Tr(t), E(t))$ and the laser tuning coefficient $d\nu/dt$.

The shape of the line form function $g(\nu,\nu_0)$ is dependent on the line width which is influenced by gas temperature, partial pressures and the collision partner. The form function can be described by a Voigt line shape model which is a convolution of the temperature dependent Gaussian contribution (Doppler effect) and the pressure dependent Lorentzian i.e. collision broadening. The collisional line width $\Delta\nu$ (in cm^{-1}) in a multi-component gas matrix is usually described by the following equation [24]:

$$\Delta\nu_c = \gamma^0_{\mathrm{self}} p_{\mathrm{self}}\left(\frac{T_0}{T}\right)^s + \sum_x \gamma^0_x p_x\left(\frac{T_0}{T}\right)^s. \qquad (3)$$

Here γ^0_{self} is the self-broadening coefficient, which is dependent only on temperature T and HCl partial pressure p_{self}. The second term describes the foreign pressure broadening for the foreign gas x. The temperature exponent s is 0.5 in all our calculations as derived from kinetic gas theory.

3 Results and discussion

Line strength and self-broadening coefficient of the $H^{35}Cl$ R(3) absorption line were determined at room temperature by measurements in pure HCl in a pressure range of 10 to 1000 hPa. The laser wavelength was modulated by a triangular shaped current with a frequency of 10.4 Hz to avoid an additional electrical line broadening at small pressures (<100 hPa) due to the limited bandwidth (2 MHz) of the transimpedance amplifier. All measured line profiles were averaged over 10 scans, corresponding to a time resolution of 1 second, and fitted with the Voigt model using a non-linear Levenberg–Marquart algorithm and a third order background polynomial. In order to clearly isolate the pressure-broadening coefficient we calculated the Gaussian width in all our experiments using the measured temperature in the high-pressure cell (typical Gaussian HWHM at 293 K was 200 MHz). This procedure resulted in excellent agreement between measured and fitted data with signal to noise levels often exceeding a few hundred.

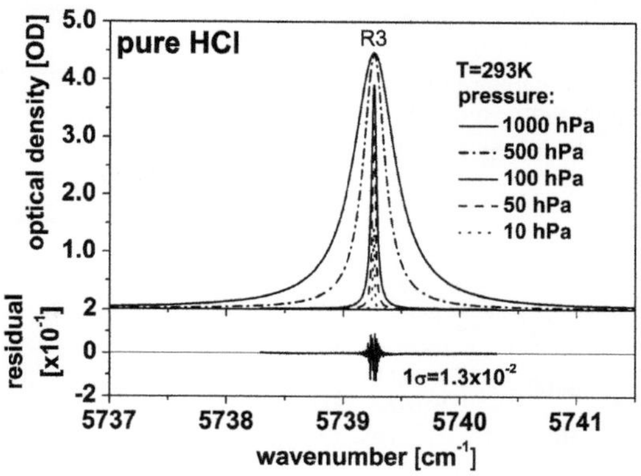

Fig. 2 Measured and fitted $H^{35}Cl$ R(3) line profiles for pure HCl at pressures from 10 to 1000 hPa. For 100 hPa the residual between measured and fitted data is depicted yielding a S/N of 300

Figure 2 shows a subset of the measured line profiles for different pressures for the case of pure HCl gas and nicely demonstrates the advantageous, wide tuning range of the VCSEL. From the low-pressure profiles we extracted the line strength $S(T)$. The line width progression was used to extract the self-broadening coefficient γ^0_{HCl}.

$S(T)$ can be calculated using (2), the measured parameters temperature T, pressure p, absorption path length L and the known concentration of HCl (99.0(1.0) vol.%) in the measurement cell.

The HCl purity was verified with a high-resolution (0.07 cm^{-1}) FTIR spectrometer and found to have a water content of less than 0.2 vol.%. The conversion of the measured value for $S(T)$ at the measurement temperature T to the standard temperature T_0 (= 296 K) results in $S(T_0) = 12.53(11)\times 10^{-21}$ cm^{-1}/(molec cm^{-2}) [25]. This result clearly deviates from the HITRAN 2008 data base (see Table 1) but is in very good agreement with DeRosa et al. [15], who measured at 1.3 hPa. Our more than 3 times better signal to noise level (relative error 0.9%) compared to the measurement of De Rosa et al. (relative error 3.2%) clearly shows the high accuracy of the developed VCSEL spectrometer.

Table 1 Comparison of our results with HITRAN 2008 and data from the literature for the line strength $S(T_0)$ of the $H^{35}Cl$ R(3) (2 ← 0) absorption line

	This paper	De Rosa et al.	HITRAN 2008
$S(T_0)$ [cm^{-1}/(molec cm^{-2})]	12.53 (11)	12.60 (40)	11.60 (58)
Relative error [%]	0.9	3.2	5.0
Deviation to our measurement [%]	–	+0.6	−8.0

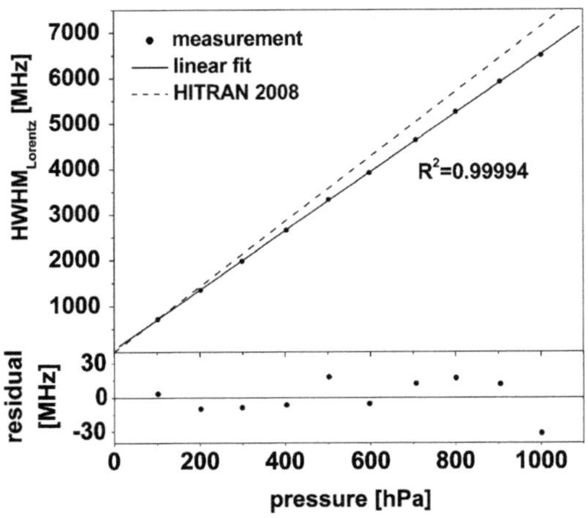

Fig. 3 Measured Lorentzian half width half maximum for the pressure dependent self-broadening of the $H^{35}Cl$ R(3) (2 ← 0) absorption line with residual between measured values and the linear fit compared with HITRAN 2008

The self-broadening coefficient γ^0_{HCl} was determined by evaluating the pressure dependence of the Lorentzian width by fitting Voigt line shapes with calculated Gaussian width using the measured gas temperature (Figs. 2 and 3). In the pressure range below 100 hPa the changes of the line width are small due to the dominating Doppler effect (at 10 hPa the Lorentzian HWHM is 120 MHz compared to 200 MHz for the Gaussian HWHM). For pressures above 100 hPa the line profiles widen significantly. The pressure dependence of the Lorentzian width is described by the self-broadening coefficient γ^0_{HCl} given by the first term in (3) (in cm^{-1}):

$$\Delta\nu_L = \gamma^0_{HCl} p_{HCl} \left(\frac{T_0}{T}\right)^{0.5}. \tag{4}$$

The respective Lorentzian widths for the pressures from 10 to 1000 hPa are plotted in Fig. 3.

The measured widths show very good linearity with pressure leading to a value of $R^2 = 0.9994$ for the linear fit that is described by

$$\Delta\nu_L = \gamma^0_{HCl} p_{HCl} \left(\frac{T_0}{T}\right)^{0.5} + \gamma^0_{laser} \tag{5}$$

with γ^0_{laser} being the offset of the fit indicating the laser line width of the used VCSEL. Thus we determined a self-broadening coefficient of $\gamma^0_{HCl} = 6.446(18) \times 10^{-2}$ MHz/Pa corresponding to $\gamma^0_{HCl} = 0.021787(61)$ cm^{-1}/atm. The offset of the linear fit was 56(11) MHz, which correlates well with typical VCSEL line width data in the literature [26]. Our self-broadening coefficient (see Table 2), on the other hand agrees well with the measurement by DeRosa et al. [15]. Again we can realize a three times higher accuracy, owing to our over six times higher pressure range. This result also verifies the nearly 10% discrepancy of DeRosa et al. to the HITRAN 2008 data, which is an expected (1 ← 0) result, as the broadening data of the fundamental band were simply copied to the first overtone (2 ← 0) band.

Finally the foreign broadening data for the $H^{35}Cl$ R(3) absorption line (2 ← 0) with Helium (He), Nitrogen (N_2) and Oxygen (O_2) as collision partners was, for the first time, determined over a pressure range from 100 hPa to 1 MPa. During these measurements the laser wavelength was modulated at 139.8 Hz to enhance the optical resolution and thus the signal to noise level. The line profiles were averaged for 25 scans yielding again a time resolution of 1 second. The measured data were fitted using the Voigt line shape model with the Gaussian width being calculated from the measured temperature. Due to the significant line widths at higher pressures above 0.1 MPa we used a multi-line Voigt fit to account for spectral overlap by the four neighboring HCl lines ($H^{37}Cl$ R(3), $H^{35}Cl$ R(2), $H^{37}Cl$ R(4), $H^{35}Cl$ R(4)). As these lines were located outside the absorption scan they had to be calculated using HITRAN 2008. This resulted in a significantly better fit yielding a 50 % better S/N and an improved accuracy of the measured Lorentzian widths as well as the HCl concentration. Figure 4 shows measured and fitted HCl line profiles for N_2 broadening over a pressure range from 100 hPa to 1 MPa.

The measured absorption profiles agree very nicely with the Voigt model using the multi-line fitting procedure. In the foreign broadening evaluation the self-broadening was taken into account using our measured coefficient γ^0_{HCl} and the absolute concentration derived from the line area measurement using our line strength $S(T)$.Typical HCl concentrations were 4937 ppm for 1 MPa. Typical average residuals were around 10^{-4} OD. Even for the N_2 case, which showed the strongest broadening (Fig. 4), we reached an average residual (1σ) of 3.85×10^{-4}, which corresponds to an excellent S/N of 400.

Figures 5 and 6 show typical line profiles and fits at 1 MPa for He and O_2 as collision partners. For O_2 we yield a S/N of 325 and for He even a S/N of 1000. Looking at

Table 2 Comparison of our results with HITRAN 2008 and data from the literature for the self-broadening coefficient γ^0_{HCl}, of the $H^{35}Cl$ R(3) (2 ← 0) absorption line

	This paper	De Rosa et al.	HITRAN 2008
Pressure range [hPa]	10–1000	60–150	–
γ^0_{HCl} [cm^{-1}/atm]	0.21787(61)	0.21710(220)	0.2410(482)
Relative error [%]	0.3	1.0	20.0
Relative deviation to this paper [%]	–	−0.4	+9.6

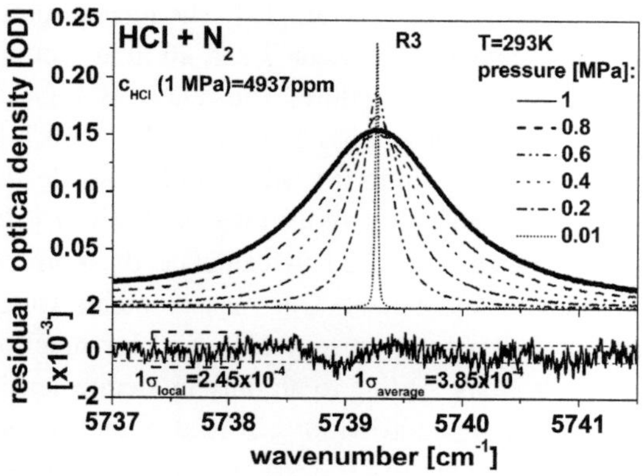

Fig. 4 Measured $H^{35}Cl$ R(3) line profiles in N_2 at pressures from 100 hPa to 1 MPa. The residual between measured and fitted data is depicted for 1 MPa total pressure yielding an average S/N of 400 and a local S/N of 637

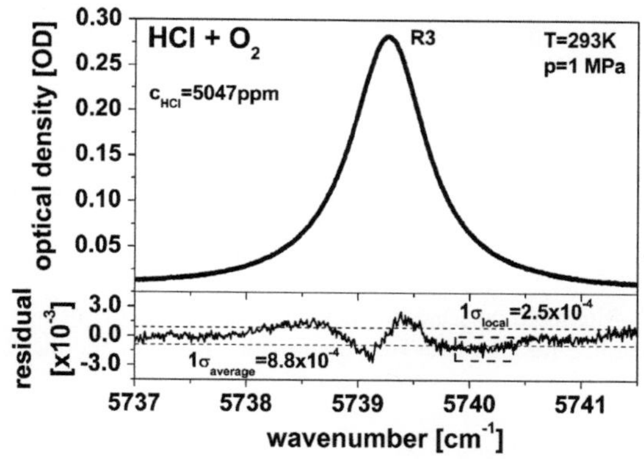

Fig. 6 Measured $H^{35}Cl$ R(3) (2 ← 0) absorption profile for HCl in a O_2 matrix at 1 MPa and room temperature yielding an average S/N of 325 and a local S/N of 1144

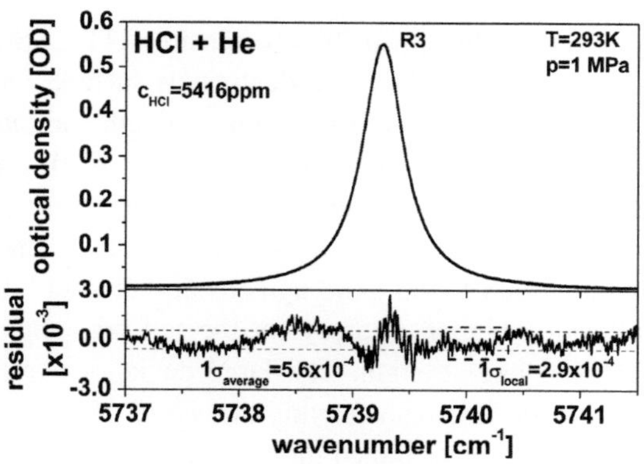

Fig. 5 Measured line profile of the $H^{35}Cl$ R(3) absorption line for HCl in He at a pressure of 1 MPa. The residual between measured and fitted data yields an average S/N of 1000 and a local S/N of 1931

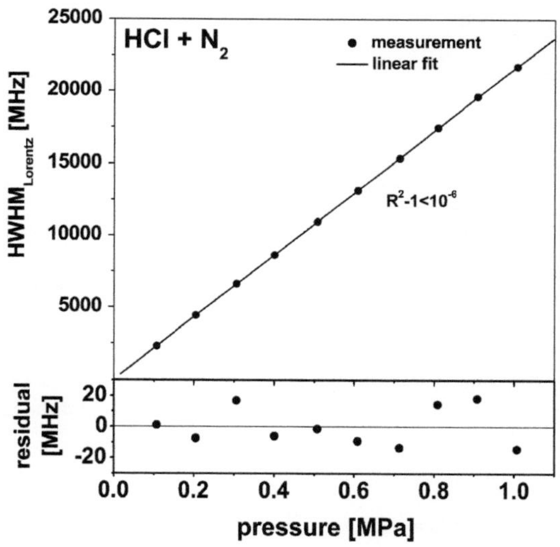

Fig. 7 Measured pressure broadening of the $H^{35}Cl$ R(3) (2 ← 0) absorption line for HCl with N_2 fitted with a linear model. The residual between fit and data is shown below

the noise-governed residual outside the line peak we derive even better "local" S/N ratios. The optimum S/N ratios can be estimated neglecting the residual structure at the line center that occurs from line shape effects which cannot be described by the Voigt line shape model yielding a S/N increase to 637 for N_2, 1931 for He and 1144 for O_2.

The respective foreign broadening coefficients are calculated using the pressure dependent Lorentzian width formula in (3). The self-broadening was taken into account using our measured self-broadening coefficient γ^0_{HCl}, the respective HCl partial pressure using the measured values for the line area and the line strength $S(T)$ for each pressure. The respective broadening data are corrected to HITRAN standard temperature (296 K). After subtracting the self-broadening the foreign broadening coefficient can be extracted by plotting the respective foreign broadened line widths against the pressure (Figs. 7, 8 and 9).

Very good agreement was found between the measurements and the linear model fits ($R^2 - 1$ better than 10^{-6} for all three species). Our offset terms (−4(12) MHz for He, 9(9) MHz for N_2 and 4(7) MHz for O_2) are found to

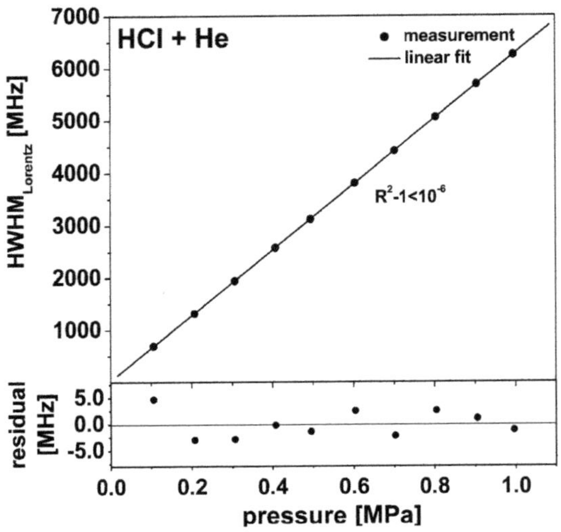

Fig. 8 Measured pressure broadening of the $H^{35}Cl$ R(3) ($2 \leftarrow 0$) absorption line for HCl with He. Again with a linear fit and the residual below

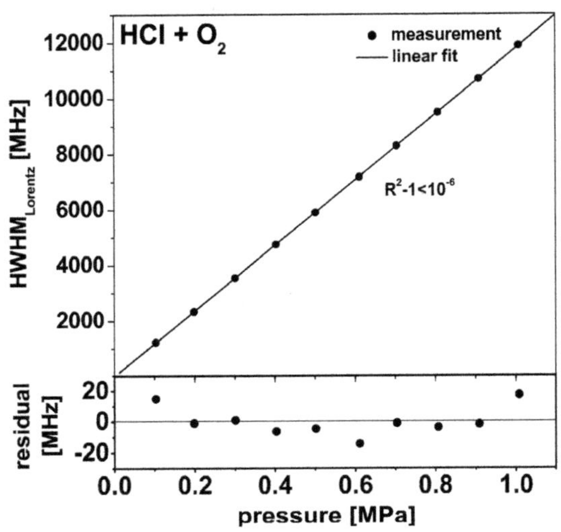

Fig. 9 Measured pressure broadening of the $H^{35}Cl$ R(3) ($2 \leftarrow 0$) absorption line for HCl with O_2. Again with a linear fit and the residual below

be consistent with a zero offset after correcting for laser line broadening. This was done by subtracting the laser line width (56(11) MHz), which was determined during the low-pressure self-broadening measurements. The foreign broadening coefficients as well as the comparison with data from the literature is shown in Table 3.

We find a significant deviation of 3.2% for N_2, 4.4% for He and 0.8% for O_2 between our measurements and the values measured by De Rosa et al. [15]. The relative precision of our measurements is between 7 and 40 times better than for the data from the literature. The relatively high deviations to De Rosa et al. might be explained by a much lower pressure range (only up to 0.1 MPa), which also leads to a higher relative contribution of the measured self-broadening coefficient, which is needed to calculate the correct foreign broadening coefficient. Our measurements were performed for pressures at up to 1 MPa and for much lower HCl concentrations leading to an only moderate sensitivity for self-broadening contributions. It is also unclear and not explained in the paper of De Rosa et al. if their values were corrected for the temperature dependence.

4 Conclusion

Using a newly developed, highly sensitive TDLAS spectrometer based on a 1.74 μm vertical-cavity surface-emitting laser we precisely determined the line strength and the self-broadening coefficient of the $H^{35}Cl$ R(3) ($2 \leftarrow 0$) absorption line at room temperature over a range from 10 to 1000 hPa. We compared our results with existing data from the literature and found good agreement, but could deliver a significantly higher accuracy for the self-broadening coefficient and line strength. The large deviations from the HITRAN 2008 line data for this line could be confirmed. Furthermore, we determined the foreign broadening coeffi-

Table 3 Comparison of our results with data from the literature for the foreign broadening coefficients of the $H^{35}Cl$ R(3) ($2 \leftarrow 0$) absorption line with He, N_2 and O_2

	This paper	De Rosa et al.
$\gamma^0_{N_2}$ [cm^{-1}/atm]	0.07292(5)	0.07530(60)
Pressure range [MPa]	0.01–1	0.01–0.047
Relative error [%]	0.07	0.8
Relative deviation to this paper [%]	–	+3.2
γ^0_{He} [cm^{-1}/atm]	0.02113(1)	0.02210(40)
Pressure range [MPa]	0.01–1	0.026–0.1
Relative error [%]	0.05	1.8
Relative deviation to this paper [%]	–	+4.4
$\gamma^0_{O_2}$ [cm^{-1}/atm]	0.03978(6)	0.04010(40)
Pressure range [MPa]	0.01–1	0.023–0.084
Relative error [%]	0.15	1.0
Relative deviation to this paper [%]	–	+0.8

cients for the R(3) line with N_2, He and O_2 as collision partners for pressures of up to 1 MPa. These results present the first high-pressure laser data measured for the $H^{35}Cl$ R(3) ($2 \leftarrow 0$) absorption line with these collision partners. The advantage can be seen clearly in our up to 40 times higher accuracy for the foreign broadening coefficients compared to data from the literature. The comparison showed significant deviations to the data from the literature of up to 4.4%. Our results will be useful for absorption measurements in all environmental and industrial applications particularly in gasification processes where high pressures can occur. Further measurements for other collision partners and for higher temperatures are in progress.

Acknowledgement This work was funded by the German Ministry of Education and Research (BMBF) under the project number 03SF0320D.

References

1. O. Bjoroey, K.H. Haugholt, T. Jaeger, Quantum Electron. **26**, 1090 (1996)
2. C. Schulz, A. Dreizler, V. Ebert, J. Wolfrum, in *Springer Handbook of Experimental Fluid Dynamics*, ed. by C. Tropea, J. Foss, A. Yarin (Springer, Berlin, 2007), pp. 1241–1316
3. V. Ebert, H. Teichert, P. Strauch, T. Kolb, H. Seifert, J. Wolfrum, Proc. Comb. Inst. **30**, 1611 (2005)
4. P. Ortwein, M. Gisi, M. Eberhard, T. Kolb, S. Wagner, V. Ebert, 2.3 µm Diode Laser Absorption Spectrometer for Absolute CO Measurements in Gasification Processes, VDI Berichte 2047, ISBN 978-3-18-092047-4, pp. 195–200 (2008)
5. P. Ortwein, W. Woiwode, M. Gisi, M. Eberhard, T. Kolb, S. Wagner, V. Ebert, Absolute laserbasierte in situ Bestimmung von HCl im Reaktionsraum von Verbrennungs- und Vergasungsprozessen, VDI Berichte 2056, ISBN 978-3-48-092056-6, pp. 263–269 (2009)
6. S. Wagner, B.T. Fisher, J.W. Fleming, V. Ebert, Proc. Comb. Inst. **32**, 839 (2009)
7. D. Hurtmans, A. Henry, A. Valentin, C. Boulet, J. Mol. Spectrosc. **254**, 126 (2009)
8. I. Morino, K.M.T. Yamada, J. Mol. Spectrosc. **233**, 77 (2005)
9. A.S. Pine, J.P. Looney, J. Mol. Spectrosc. **122**, 41 (1987)
10. D. Ramachandra Rao, T. Oka, J. Mol. Spectrosc. **122**, 16 (1987)
11. J.P. Besson, S. Schilt, L. Thevenaz, Appl. Phys. B **90**, 191 (2008)
12. C. Boulet, P.M. Flaud, J.M. Hartmann, J. Chem. Phys. **120**, 11053 (2004)
13. L.S. Rothman, I.E. Gordon, A. Barbe, D.C. Benner, P.F. Bernath, M. Birk, V. Boudon, L.R. Brown, A. Campargue, J.P. Champion, K. Chance, L.H. Coudert, V. Dana, V.M. Devi, S. Fally, J.M. Flaud, R.R. Gamache, A. Goldman, D. Jacquemart, I. Kleiner, N. Lacome, W.J. Lafferty, J.Y. Mandin, S.T. Massie, S.N. Mikhailenko, C.E. Miller, N. Moazzen-Ahmadi, O.V. Naumenko, A.V. Nikitin, J. Orphal, V.I. Perevalov, A. Perrin, A. Predoi-Cross, C.P. Rinsland, M. Rotger, M. Simecková, M.A.H. Smith, K. Sung, S.A. Tashkun, J. Tennyson, R.A. Toth, A.C. Vandaele, J. Vander Auwera, J. Quant. Spectrosc. Radiat. Transfer **110**, 533 (2009)
14. C. Corsi, M. Inguscio, S. Chudzynski, K. Ernst, F. D'Amato, M. De Rosa, Appl. Phys. B **68**, 267 (1999)
15. M. De Rosa, C. Nardini, C. Piccolo, C. Corsi, F. D'Amato, Appl. Phys. B **72**, 245 (2001)
16. M. Balat, M. Balat, E. Kirtay, H. Balat, Energy Convers. Manag. **50**, 3158 (2009)
17. L.G. Dodge, J.A. Biaglow, J. Eng. Gas Turbine Power **108**, 209 (1986)
18. M. Lackner, F. Winter, G. Totschnig, M. Ortsiefer, J. Rosskopf, M.C. Amann, R. Shau, Technisches Mess. **70**, 294 (2003)
19. G. Totschnig, M. Lackner, R. Shau, M. Ortsiefer, J. Rosskopf, M.C. Amann, F. Winter, Meas. Sci. Technol. **14**, 472 (2003)
20. M. Lackner, G. Totschnig, F. Winter, M. Ortsiefer, M.C. Amann, R. Shau, J. Rosskopf, Meas. Sci. Technol. **14**, 101 (2003)
21. K. Wunderle, T. Fernholz, V. Ebert, Selektion optimaler Absorptionslinien für abstimmbare Laserabsorptionsspektrometer, VDI Berichte 1959, ISBN 3-18-091959-0, pp. 137–148 (2006)
22. A. Lytkine, W. Jäger, J. Tulip, Spectrochim. Acta, Part A **63**, 940 (2006)
23. H. Teichert, T. Fernholz, V. Ebert, Appl. Opt. **42**, 2043 (2003)
24. V. Ebert, J. Wolfrum, in *Optical Measurements—Techniques and Applications* (Springer, Heidelberg, 2001), pp. 227–265
25. R.R. Gamache, S. Kennedy, R. Hawkins, L.S. Rothman, J. Mol. Struct. **517–518**, 407 (2000)
26. C. Lauer, M. Ortsiefer, R. Shau, J. Rosskopf, G. Böhm, R. Meyer, M.-C. Amann, Phys. Status Solidi C **1**, 2183 (2004)

Impact of water vapor on 1.51 µm ammonia absorption features used in trace gas sensing applications

S. Schilt

Received: 15 September 2009 / Revised version: 15 February 2010 / Published online: 16 March 2010
© Springer-Verlag 2010

Abstract Water-vapor-induced pressure broadening is reported for two NH_3 absorption features at 6612.7 and 6596.4 cm^{-1} that are exploitable for gas sensing applications at atmospheric pressure. Absorption spectra of different NH_3–H_2O–N_2 mixtures were measured at an elevated temperature of 70°C to enable high H_2O concentrations to be reached. Line parameters were determined from a fitting procedure. The significantly greater values obtained for the H_2O-broadening coefficients of the two lines compared to N_2-broadening leads to cross-sensitivity effects in NH_3 trace gas sensors based on spectroscopic techniques that are sensitive to the width of the analyzed absorption line, as is the case in a simple implementation of wavelength modulation spectroscopy or in photoacoustic spectroscopy. In such a case, cross-sensitivity results in inaccurate gas concentration retrieval when the composition of the diluting gas changes. H_2O represents a potentially significant cross-sensitivity source as its concentration may be subject to large variations, especially in high-temperature applications where concentrations up to a couple of tens of percent may be encountered. In contrast to interference which can be minimized by an appropriate choice of the analyzed transitions, cross-sensitivity affects the entire spectrum of the analyte and is thus unavoidable in the mentioned type of gas sensors.

S. Schilt (✉)
Laboratory for Time and Frequency, Institute of Physics, University of Neuchâtel, Av. de Bellevaux 51, 2009 Neuchâtel, Switzerland
e-mail: stephane.schilt@unine.ch
Fax: +41-21-7182511

1 Introduction

Sensitive and continuous ammonia sensing at trace level is needed in a variety of applications, ranging from environmental monitoring, industrial process control [1], and medical breath analysis [2], requiring ppb sensitivities, to DeNO*x* processes [3], leak detection in industrial refrigerator installations, or air monitoring in animal production facilities [4, 5], where detection at ppm level is sufficient. Laser spectroscopic techniques have demonstrated their relevance for such applications and have been implemented in various forms, e.g., rapid-scan direct absorption [6, 7] with standard or balanced detection, wavelength-modulation spectroscopy (WMS) [8, 9], frequency-modulation spectroscopy (FMS) [10], cavity ring-down spectroscopy [11, 12], photoacoustic spectroscopy (PAS) [1, 4, 5, 13] and using different types of near-infrared or mid-infrared distributed-feedback (DFB) lasers [5, 6, 8–10, 13], vertical-cavity surface-emitting lasers (VCSEL) [7], or quantum cascade lasers (QCL) [11, 12] depending on the sensitivity requirements. All these techniques benefit from the recognized properties of laser spectroscopic techniques, such as sensitive and selective contactless measurements. Interferences, i.e., spurious signals induced at the sensor operating wavelength by nearby absorption lines from other species, can most often be minimized or even suppressed by a proper choice of the operation wavelength [10], sometimes also combined with reduced pressure operation. Since absorption lines become narrower at low pressure, possible overlaps are reduced accordingly. However, even in the absence of interference, a spectroscopic gas sensor may be affected by the presence of other species as a result of a modified line broadening that gives rise to *cross-sensitivity* effects. A variation of the width of an absorption line affects the peak absorption accordingly but does not change the integrated area of the

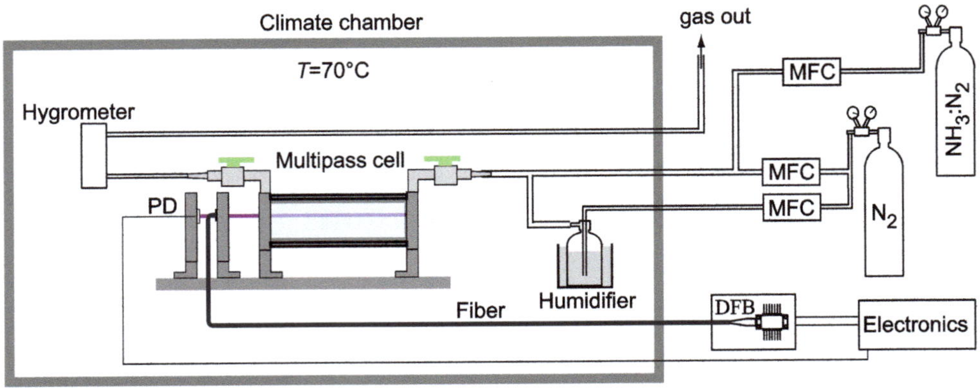

Fig. 1 Schematic of the experimental setup. *PD*: photodiode; *MFC*: mass-flow controller

line since the linestrength is pressure independent. If such an effect does not affect gas concentration monitoring when the entire absorption profile is measured and analyzed (which is the case for the rapid-scan method for instance), it can nevertheless have a significant influence when only the signal at line center is considered, as is often the case in the simplest implementation of WMS with second-harmonic ($2f$) detection (see for instance Ref. [9]) or in PAS [1, 4, 23]. In this case, an inaccurate gas concentration is retrieved from the measured on-line absorption signal; the induced error varies when the composition of the diluting gas changes.

In the case of ammonia, N_2-broadening of several NH_3 lines in the 1.5-μm spectral range has been shown to be ~10% higher in average than air-broadening [14, 15], illustrating the importance of a sensor calibration made in proper conditions when using a spectroscopic method that is sensitive to the width of the analyzed absorption line, such as encountered with on-peak WMS or PAS techniques. The influence of water vapor on NH_3 lines can also be important in applications where water vapor may significantly vary, especially for high-temperature applications where high H_2O concentrations up to 25–30% may be encountered (for instance, 100% relative humidity (RH) at 60°C corresponds to ~20% absolute H_2O concentration in air).

Unfortunately, poor data are available about the broadening parameters of NH_3 lines in the near-infrared. For instance, the 1.5 μm $\nu_1 + \nu_3$ band of NH_3 is absent from Hitran spectroscopic database [16], so that FTIR data from Lundsberg-Nielsen et al. [17] were the only available data during a long time. Webber et al. [18], in one of the earliest analysis of ammonia near-infrared absorption lines at 1.5 μm, generally considered for each NH_3 line the average published broadening parameters $g_{NH_3-N_2} = 0.10\ cm^{-1}\ atm^{-1}$ and $g_{NH_3-NH_3} = 0.45\ cm^{-1}\ atm^{-1}$. More recently, assignments [19, 20], linestrength [21], and broadening coefficients [14, 15, 22, 23] have been reported for many NH_3 lines in the 1.5-μm range. However, no data about H_2O-broadening coefficients of NH_3 lines has ever been published to the best knowledge of the author, even in other spectral absorption bands. Therefore, the expected influence of H_2O on a NH_3 sensor is difficult to quantify. In a first approximation, one could expect that the H_2O-broadening coefficient of NH_3 lines would be close to the NH_3 self-broadening factor, since H_2O and NH_3 molecules have similar molecular weights and dipole moments (to which collision-broadening effects are related). However, it is desirable to experimentally quantify the effect of H_2O on NH_3 lines, which is the purpose of the reported work. Two different NH_3 lines in the 1512–1516 nm range that are exploitable for gas monitoring applications at atmospheric pressure are analyzed with respect to their sensitivity to water vapor. In order to obtain precise quantitative data, a measurement of the direct absorption signal was preferred to the WMS-$2f$ signal, since the width of this later signal is not univocally related to the width of the absorption lines but still involves the normalized frequency modulation index m that varies itself with linewidth [24]. To obtain a measurable direct absorption signal with a sufficient signal-to-noise ratio, a long enough pathlength was necessary, which was achieved using a 4-m-pathlength multipass cell. The experiment has been performed at elevated temperature (70°C) in a climate chamber in order to enable high H_2O partial pressures to be reached without condensation issues. For practical reasons (possibility to keep the laser out of the climate chamber at room temperature, easy coupling of the laser beam into the cell, etc.) a fiber-coupled DFB laser was used in this experiment.

2 Experimental setup

A schematic of the experimental setup is shown in Fig. 1. The multipass cell (Infrared Analysis, model 2.4-PA) equipped with a wedged CaF_2 window to prevent etalon fringes is placed in a climate chamber stabilized at 70°C for stable high-temperature operation. Light from the DFB laser (NEL NLK1556STG) located out of the climate chamber (at ambient temperature) is coupled into the cell using a fiber collimator. The optical power exiting from the cell is detected using a large surface (3 mm diameter) Ge photodiode. The detector is tilted to reduce etalons fringes induced by the detector window. The laser beam is carefully aligned into the cell to minimize scattered light and to maximize the detector signal. A coarse optical alignment was made using a visible laser coupled into the same optical fiber and a finer adjustment was made by optimizing the detector signal. In this arrangement, a small residual free-space optical path arises out of the cell. This path amounts to ∼13 cm (5 cm between the collimator and the cell window and 8 cm between the window and the photodiode) but it has a negligible influence on the measured spectra compared to the much longer (4 m) internal contribution of the multipass cell. However, it might contribute to a small observable H_2O signal when low H_2O concentrations are considered in the cell.

A gas mixing installation made of three mass-flow controllers (MFC, model MKS 1179, flow range 100 and 500 sccm [standard cubic centimeter per minute]) driven by a multigas controlling unit (MKS 647C) was used to prepare the different mixtures flown into the cell. A mixture of 204 ppm NH_3 was prepared by mixing a flow of 70 sccm from a 496 ppm NH_3:N_2 cylinder with 100 sccm of pure N_2. The N_2 flow was split into two parts using two MFCs. The first part was directly mixed with the NH_3 flow and was carried to the climate chamber in a first PTFE tube. PTFE was used to reduce adsorption effects in the tube due to the sticky nature of NH_3 molecules [25] and to minimize the response time when changing the NH_3 concentration. The remaining N_2 flow passed in a different tube and was used to add humidity in the gas mixture. This flow passed through a humidifier, i.e., a cuvette located in the chamber and filled with liquid water at thermal equilibrium ($T = 70$°C). The flow exiting the humidifier, saturated in water vapor ($P_{sat} = 311$ mbar at 70°C), was mixed to the dry NH_3:N_2 flow and the resulting mixture was injected into the multipass cell for analysis. The amount of water vapor in the mixture was varied by adjusting the two N_2 flows, keeping their sum constant to 100 sccm. A hygrometer (Rotronic Hygroclip) placed at the cell exit was used to measure the relative humidity and the temperature of the gas. The absolute water-vapor concentration was then determined from these two values.

Multipass cells are known to be an important source of etalons fringes in laser spectroscopy systems [26]. Due to the multiple reflections on the mirrors surface, interferences usually arise between adjacent spots on the mirrors, thus corresponding to a cavity length that is twice the physical distance between the mirrors. Such fringes, corresponding to a 20-cm cavity (twice the 10 cm cell length), have been experimentally observed as shown in Fig. 2. The use of a small modulation (at a frequency of about 20 kHz) applied to the laser current enables to average out these fringes as demonstrated by the strong reduction of the fringes contrast in Fig. 2. It is also observed that the considered modulation does not modify the absorption line measured in the

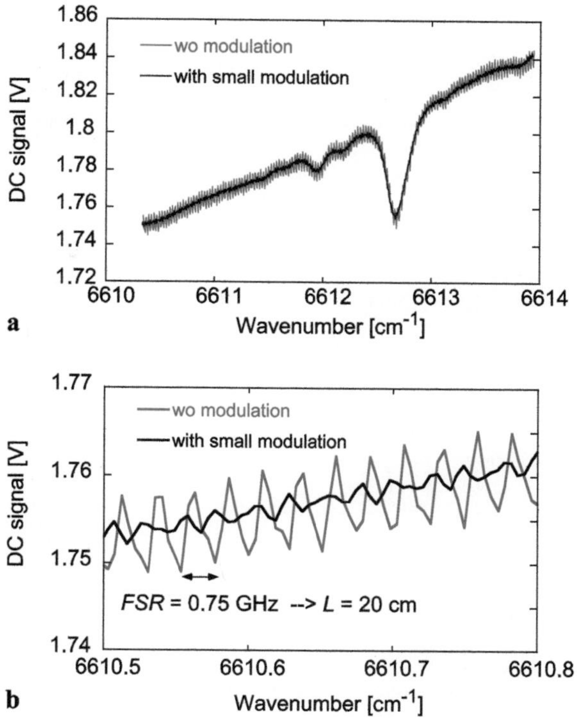

Fig. 2 Effect of a small modulation applied to the laser current in order to reduce etalon fringes created in the multipass cell. (**a**) Scan of a NH_3 line showing that the modulation does not modify the width and depth of the absorption line. (**b**) Narrower scan showing the benefit of the modulation to reduce the etalon fringes

DC signal, i.e., the modulation is small enough so that the induced broadening of the laser linewidth remains negligible compared to the width of the absorption lines at atmospheric pressure. Such a modulation was thus used in all measurements in order to reduce etalons fringes and to improve the signal-to-noise ratio of the recorded spectra.

3 Ammonia and water-vapor spectra

The laser temperature-tuning characteristic was measured using a wavemeter (Burleigh WA-1000) with 0.01 cm^{-1} resolution and the measured data were approximated by a third-order polynomial curve. The laser average linear tuning coefficient is −14.3 GHz/K.

Broad absorption spectra of NH_3 and H_2O have been measured to locate the different NH_3 absorption features to be studied in detail. In these measurements, the optical power transmitted through the cell was recorded when the laser temperature was slowly scanned from 5 to 60°C with the cell filled with a mixture of 204 ppm NH_3 in dry N_2 in a first step (at $T = 70$°C). In a second step, the mixture was moisturized to 59% relative humidity (at $T = 70$°C, thus corresponding to ∼184,000 ppm H_2O). The measured transmission curves were then processed to determine the

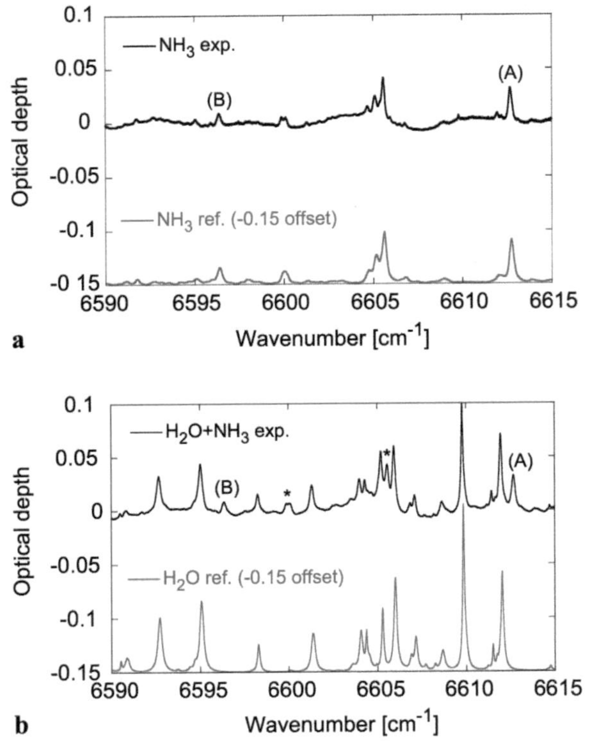

Fig. 3 Experimental spectra obtained at $T = 70°C$ for 204 ppm NH_3 in dry N_2 (**a**) and for the same mixture moisturized at 59% relative humidity (**b**). Reference absorption spectra are shown for comparison and are vertically offset for clarity of the figure. The studied NH_3 absorption features are labelled (*A*) and (*B*) in the graphics. In plot (**b**), *stars* (*) indicate other NH_3 lines discernible in the experimental spectrum

absorption spectra. For this purpose, a third-order polynomial curve was fitted to the non-absorbing parts of the experimental curves in order to determine the baseline and the transmission curves were normalized to this baseline. The previously measured laser tuning curve was used to represent the spectra in terms of wavenumber. The use of a nonlinear baseline was necessary to account for the following effects: (1) some nonlinearities in the laser emission power in a temperature scan, (2) higher order nonlinearities in the laser tuning response, and (3) the presence of residual long-period etalon fringes in the spectra. In this context, a third-order polynomial baseline proved to better approximate the background signal than a lower-order polynomial curve and was thus chosen to fit the baseline of the different spectra.

Figure 3 displays the NH_3 and H_2O spectra measured in the range 6590–6615 cm^{-1} and compares them with reference spectra from Hitran database [16] for H_2O and PNNL database [27] for NH_3. A good qualitative agreement is observed between experimental and reference spectra. However, quantitative discrepancies occur for several reasons. First, experimental spectra have a nonzero baseline due to the presence of residual long-period etalon fringes. The amplitude of the lines in the reference spectra is also stronger than in the experimental spectra. Even if experimental curves were obtained at higher temperature (70°C) than the reference spectra (23°C), this disparity is believed to mainly result from the different carrier gas considered in the spectra: reference spectra are given in air, whereas experimental spectra were measured in nitrogen. Nitrogen-broadening coefficients are generally larger than air-broadening factors for both NH_3 [14, 15, 22, 23] and H_2O lines [28–30] resulting in broader lines with a smaller peak value. Finally, the small horizontal shift observed between the experimental and the reference spectra is due to a tiny inaccuracy in the laser wavenumber measured by the wavemeter. The two NH_3 absorption features that are analyzed in detail here after are labeled (A) and (B) in Fig. 3. These features are centered at 6612.7 (1512.2 nm) and 6596.4 cm^{-1} (1516 nm), respectively.

4 Quantitative analysis of the influence of humidity on the NH_3 absorption features

Narrower temperature scans were performed at a higher spectral resolution for each of the selected NH_3 absorption features and for different humidity rates ranging between 1 and 59% RH (at 70°C). The data were processed as previously by fitting a third-order polynomial baseline in the region out of the absorption lines in order to determine the absorption spectrum (see Fig. 4). Here again, the use of a non-linear baseline was necessary to account for nonlinearities in the laser emission power in the temperature scan, as well as for the presence of long-period etalon fringes. The choice of a third-order polynomial curve proved to better approximate the nonabsorbing background signal than a second-order curve. The obtained absorption spectra were then fitted by multi-peak Lorentzian functions using the Igor Pro software to extract the parameters (position, area, width) of the individual lines contributing to the spectra. The most relevant absorption lines in each spectral region according to the data of spectroscopic databases were considered in these fits as labeled in Tables 1 and 2. The other lines were not taken into account in these fits due to their much weaker linestrength and due to the fact that they are completely masked by stronger overlapping lines of the other species at the atmospheric pressure considered in this study (for instance, some weak NH_3 lines are masked by stronger H_2O lines). The peaks considered in the fitting routine were taken according to Hitran08 database [16] for H_2O and to Lundsberg-Nielsen's data [17] for NH_3.

Since the individual peaks composing an absorption feature may be very close together and fully overlap in atmospheric pressure conditions as considered here, the parameters retrieved from the multipeak fitting procedure are poorly accurate as the fit function has too many independent parameters. It was thus decided to fit each absorption feature

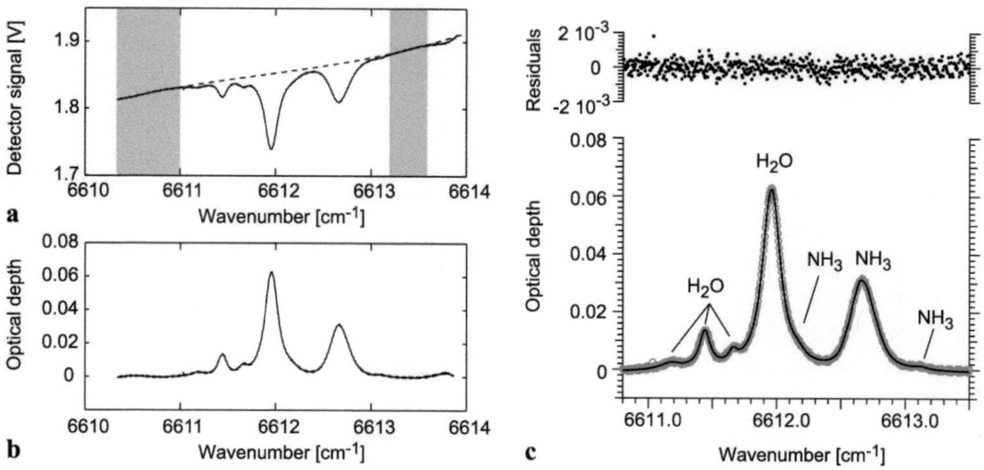

Fig. 4 Example of data processing in the 6612 cm^{-1} spectral range: (**a**) a third-order polynomial baseline (*dashed line*) is fitted on the measured transmission curve in the regions out of the absorption lines (*grey areas*); (**b**) calculated absorption spectrum; (**c**) multipeak Lorentzian fit of the data using Igor Pro; *grey circles* are experimental points and the line is the fitted curve. This measurement corresponds to 204 ppm NH_3 and 51% RH at 70°C and atmospheric pressure

Table 1 Parameters of all individual NH_3 lines tabulated by Lundsberg-Nielsen [17] in the spectral windows considered in this study. The lines labelled in *bold* are the strongest which were accounted in the multipeak fitting of the spectra

Spectral window A		Spectral window B	
wn [cm^{-1}]	*S* [cm^{-1}/(mol cm^{-2})]	*wn* [cm^{-1}]	*S* [cm^{-1}/(mol cm^{-2})]
6611.032	1.30E−22	6594.508	7.61E−23
6611.477	9.84E−23	6594.602	8.49E−23
6611.705	2.45E−22	**6594.643**	**3.84E−22**
6611.862	6.09E−23	**6594.754**	**1.03E−22**
6611.956	8.85E−22	6594.915	1.95E−22
6612.019	8.00E−23	6594.945	9.99E−23
6612.243	**7.98E−22**	6595.063	9.13E−22
6612.285	9.15E−23	6595.241	5.09E−22
6612.392	6.29E−23	**6595.616**	**3.13E−22**
6612.615	**6.98E−22**	6595.705	5.91E−23
6612.665	**1.48E−21**	**6595.923**	**8.04E−22**
6612.672	**1.45E−21**	6596.206	8.42E−23
6612.717	**2.69E−21**	**6596.296**	**8.68E−22**
6612.827	**2.12E−21**	**6596.343**	**6.85E−22**
6612.934	**2.06E−22**	**6596.382**	**9.36E−22**
6613.163	**2.82E−22**	**6596.419**	**9.57E−22**
6613.421	5.96E−23	6596.534	2.91E−22
		6596.905	1.76E−22

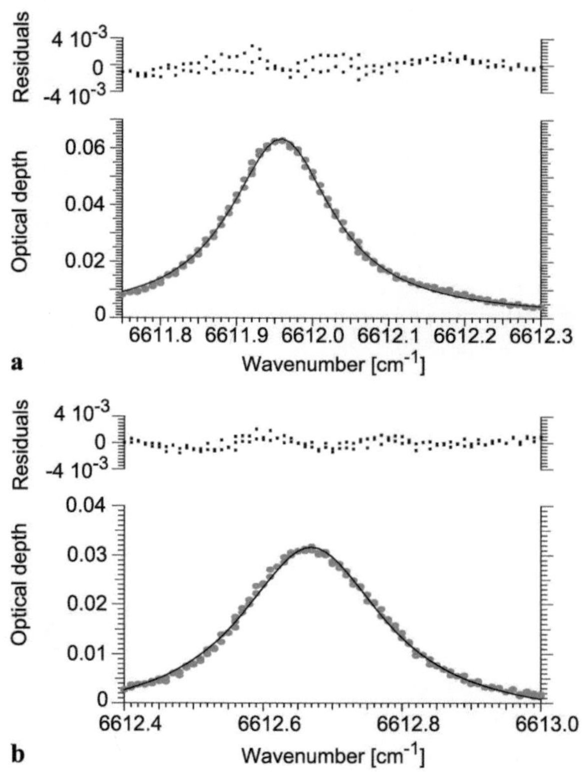

Fig. 5 Example of fits of an absorption feature by a single Lorentzian profile: (**a**) H_2O line at 6612.0 cm^{-1}; (**b**) NH_3 line at 6612.7 cm^{-1}. This measurement corresponds to 204 ppm NH_3 and 51% RH at 70°C and atmospheric pressure

by a single Lorentzian profile when studying the influence of humidity. This is a good approximation at atmospheric pressure as proved by an example of such fits obtained in spectral window (A) for the 6612.7 cm^{-1} NH_3 line and the 6612.0 cm^{-1} H_2O line shown in Fig. 5. Similar fits were performed for the other NH_3 and H_2O lines in spectral window (B).

4.1 Influence of humidity on NH_3 line A (6612.7 cm^{-1})

The effect of humidity on the 6612.7 cm^{-1} NH_3 absorption feature is shown in Fig. 6. Moistures ranging from 1 to 59% RH (at 70°C) were measured with the hygrometer when the ratio between the two N_2 flows was varied from 1:0 to 0:1. This corresponds to H_2O concentrations up to 18.4%. The fact that a zero humidity was not measured when purging the cell with dry NH_3:N_2 may be due to the limited precision of the hygrometer at low relative humidity. The residual H_2O line that is nevertheless observed in the spectrum in this case mainly arises from the small external pathlength which is not purged with N_2. The effect of humidity on the H_2O and

Table 2 Parameters of all individual H_2O lines tabulated in Hitran08 database in the different spectral windows considered in this study (at $T = 296$ K). The lines labelled in *bold* are the strongest which were accounted for in the multipeak fitting of the spectra

Spectral window A				Spectral window B			
wn [cm^{-1}]	*S* [$cm^{-1}/(mol\ cm^{-2})$]	g_{air} [cm^{-1}/atm]	g_{self} [cm^{-1}/atm]	*wn* [cm^{-1}]	*S* [$cm^{-1}/(mol\ cm^{-2})$]	g_{air} [cm^{-1}/atm]	g_{self} [cm^{-1}/atm]
6611.232	**1.03E−25**	**0.050**	**0.29**	6594.4242	8.07E-26	0.0396	0.3009
6611.5016	**5.45E−25**	**0.014**	**0.205**	**6594.6982**	**4.32E-25**	**0.0479**	**0.267**
6611.721	**2.02E−25**	**0.017**	**0.2143**	6594.7413	3.23E-26	0.0813	0.4228
6612.0186	**7.95E−24**	**0.055**	**0.25**	6594.8402	1.21E-26	0.0377	0.3009
6612.0341	**2.65E−24**	**0.052**	**0.4291**	6594.9009	2.16E-26	0.082	0.43
				6595.0784	**1.19E-23**	**0.0803**	**0.407**
				6595.2849	3.01E-26	0.0485	0.2373
				6595.7422	2.22E-26	0.0808	0.393
				6595.9144	9.32E-27	0.0774	0.31

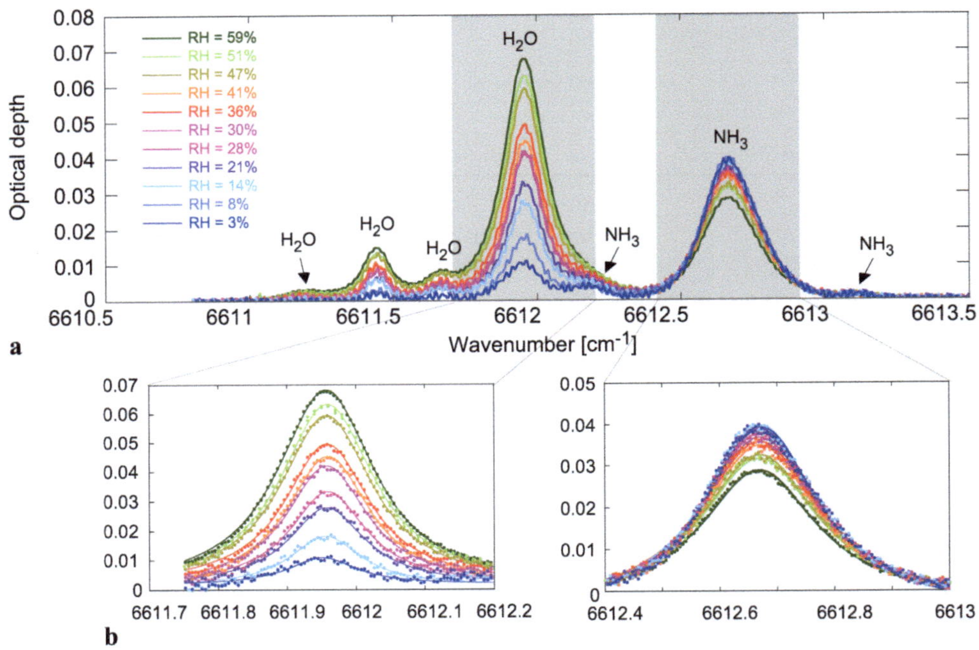

Fig. 6 (**a**) Absorption spectra measured in the neighbourhood of NH_3 line (A) at different humidity levels and for a nominal NH_3 concentration of 204 ppm. (**b**) Detailed view of the effect of humidity on the H_2O and NH_3 absorption features. *Points* represent experimental data and *lines* are the result of single Lorentzian curve fits

NH_3 absorption features is shown in more details in Fig. 6b with the corresponding Lorentzian fits.

The dependence of the width and area of the 6612.7 cm^{-1} NH_3 absorption feature extracted from the fits with respect to the H_2O partial pressure P_{H_2O} (calculated from the hygrometer measurement) is shown in Fig. 7. The same parameters are also shown for the neighbouring 6612.0 cm^{-1} H_2O absorption feature.

From the results shown in Fig. 7, several observations can be made:

- The integrated area of the H_2O absorption feature (which depends only on the H_2O concentration at fixed temperature) scales linearly with H_2O partial pressure, which confirms the correctness of the humidity measurement made with the hygrometer.
- The width of the H_2O absorption feature extrapolated to $P_{H_2O} = 0$ is $\Delta\nu = 0.058 \pm 0.001$ cm^{-1} (the error correspond to the uncertainty obtained in the linear regression of Fig. 7). This is in a good qualitative agreement with the air-broadening coefficients of two overlapping H_2O lines composing this absorption feature that are tabulated in Hitran as listed in Table 3 ($g_{air} = 0.055$ cm^{-1}/atm for the strongest $(\nu_1, \nu_2, \nu_3) J K_a, K_c = (1,2,0)4 4 1 \leftarrow (0,0,0)5 5 0$ transition at 6612.019 cm^{-1} and $g_{air} = 0.0515$ cm^{-1}/atm for the smaller neighboring (1,2,0) $4\ 4\ 0 \leftarrow (0,0,0)5 5 1$ transition at 6612.034 cm^{-1}). However, a quantitative discrepancy occurs for the following reasons:

Table 3 Comparison of the broadening factors of the 6612.0 cm^{-1} H_2O absorption feature experimentally measured and tabulated in Hitran. Errors on the experimental values correspond to the uncertainties retrieved from the linear regression in Fig. 7

Transition	Reference						Experimental	
	wn [cm^{-1}]	g_{air} @ 23°C [cm^{-1}/atm]	g_{air} @ 70°C [cm^{-1}/atm]	g_{self} @ 23°C [cm^{-1}/atm]	g_{self} @ 70°C [cm^{-1}/atm]	n	g_{N_2} @ 70°C [cm^{-1}/atm]	g_{self} @ 70°C [cm^{-1}/atm]
(1,2,0) 4 4 1 ← (0,0,0) 5 5 0	6612.01857	0.055	0.0497	0.25	0.2258	0.69	0.058 ±0.001	0.205 ±0.008
(1,2,0) 4 4 0 ← (0,0,0) 5 5 1	6612.03414	0.0515	0.0465	0.4291	0.3876	0.69		

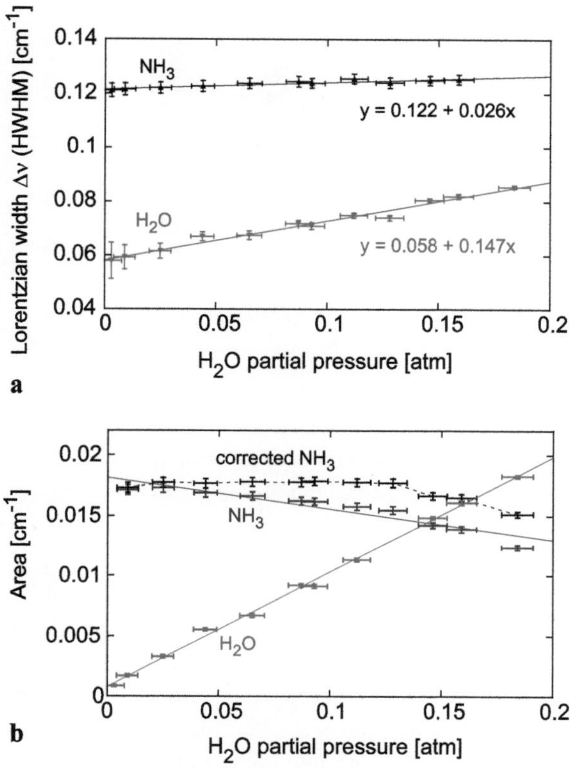

Fig. 7 Lorentzian width (**a**) and area (**b**) of the 6612.7 cm^{-1} NH_3 and 6612.0 cm^{-1} H_2O absorption features retrieved from the fits as a function of the H_2O partial pressure. *Horizontal error bars* represent the imprecision on the calculated H_2O partial pressure due to the incertitude in the relative humidity (±1.5% RH) and temperature (±0.3°C) measured with the hygrometer; *vertical error bars* represent the imprecision (standard deviation) on the spectral parameters retrieved from the line fitting process. *Straight lines* correspond to a linear regression of the data, which provides the self- and H_2O-broadening coefficients

1. The width of the H_2O absorption feature is dominated by transition (1,2,0) 4 4 1 ← (0, 0, 0)5 5 0, which is more than three times stronger than the overlapping (1,2,0) 4 4 0 ← (0, 0, 0)5 5 1 transition. However, the presence of this second transition, which is slightly shifted from the centre of the first one, slightly broadens the overall absorption feature compared to the width of the individual lines;
2. The experimental width was obtained at $T = 70°C$, whereas Hitran parameters are tabulated at 23°C; widths calculated at $T = 70°C$ using the temperature coefficient n tabulated in Hitran are ∼10% smaller compared to the 23°C reference values as indicated in Table 3.
3. The experimental width was obtained with nitrogen as a diluting gas, whereas Hitran data are referenced to air. N_2-broadening coefficients are typically 10% higher than air-broadening coefficients according to values reported for other near-infrared H_2O transitions (see for instance [28–30]).

All these reasons explain the small difference observed between experimental and tabulated values for the width of the H_2O absorption feature extrapolated to $P_{H_2O} = 0$.

- The self-broadening coefficient (g_{self}) of the H_2O absorption feature can be obtained from the slope of the linear regression of the H_2O linewidth $\Delta\nu_{H_2O}$ as a function of the H_2O partial pressure: $\Delta\nu_{H_2O} = P_{N_2}g_{N_2} + P_{H_2O}g_{self} = P_{tot}g_{N_2} + P_{H_2O}(g_{self} - g_{N_2})$, where the contribution of the NH_3 foreign-broadening of the H_2O line has been neglected due to the small NH_3 concentration that is considered. A value $g_{self} = 0.205 \pm 0.008$ cm^{-1}/atm is obtained experimentally, which is close to, but slightly smaller than the self-broadening coefficient of the strongest (1,2,0) 4 4 1 ← (0, 0, 0)5 5 0 H_2O transition tabulated in Hitran ($g_{self} = 0.25$ cm^{-1}/atm at 23°C and 10% smaller at 70°C if the same temperature coefficient n is considered as for the dependence of the air-broadening coefficient). The smaller (1,2,0) 4 4 0 ← (0, 0, 0)5 5 1 overlapping transition has a larger self-broadening coefficient ($g_{self} = 0.43$ cm^{-1}/atm at 23°C).
- The N_2-broadening coefficient of the 6612.7 cm^{-1} NH_3 absorption feature, obtained by extrapolation to $P_{H_2O} = 0$, is $g_{N_2} = 0.122 \pm 0.001$ cm^{-1}/atm. This is slightly larger than the average value of the air-broadening coefficient of NH_3 transitions considered by Webber et al. (0.1 cm^{-1}/atm) [18], but this difference is partly due to the fact that the considered absorption feature is made of the overlap of several lines, the two strongest of which being separated by ∼0.11 cm^{-1} from Table 1. Furthermore, N_2-broadening coefficients are typically 10% higher than air-broadening coefficients according to values reported for other NH_3 transitions at close wavelengths (see for instance [14, 15, 22, 23]).

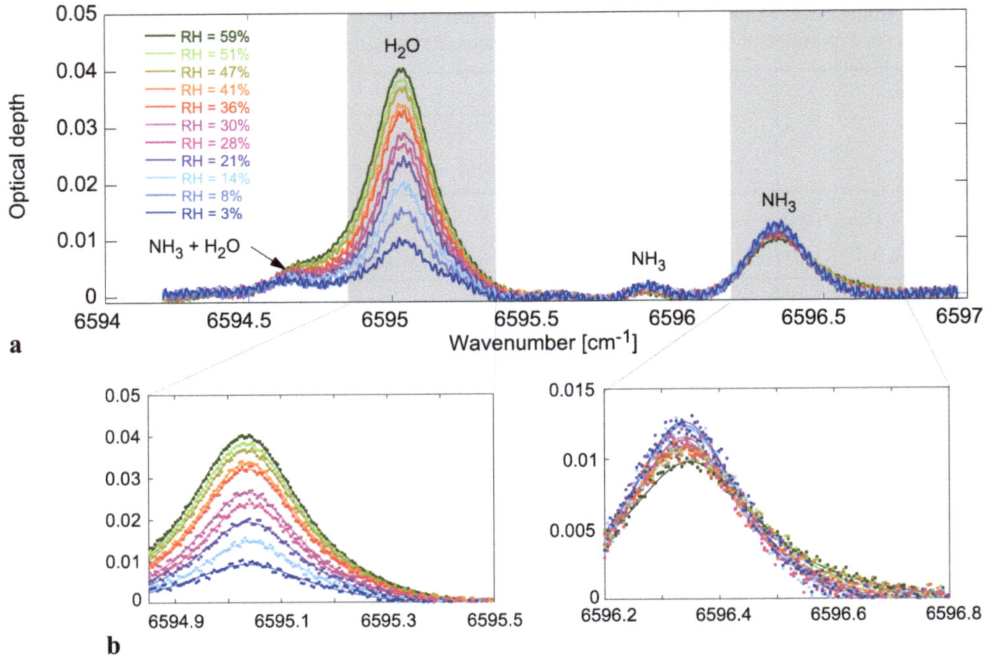

Fig. 8 (**a**) Absorption spectra measured in the neighborhood of NH_3 line (B) at different humidity levels and for a nominal NH_3 concentration of 204 ppm. (**b**) Detailed view of the effect of humidity on the H_2O and NH_3 absorption features. *Points* represent experimental data and *lines* are the result of single Lorentzian curve fits

- The influence of water vapor on the 6612.7 cm^{-1} NH_3 absorption feature can be obtained from the slope of the linear regression of the width $\Delta\nu_{NH_3}$ of the NH_3 feature retrieved from the line fitting process at different humidity levels: $\Delta\nu_{NH_3} = P_{N_2}g_{N_2} + P_{H_2O}g_{H_2O} = P_{tot}g_{N_2} + P_{H_2O}(g_{H_2O} - g_{N_2})$. A H_2O-broadening parameter $g_{H_2O} = 0.148 \pm 0.003$ cm^{-1}/atm is obtained.
- The nominal 204 ppm NH_3 concentration in the gas mixture used in the experiment decreases when humidity increases since H_2O molecules are added to the gas mixture. As a result, the absorption curves measured for the different moistures do not correspond to the same real NH_3 concentration. This is confirmed in Fig. 7b by the decrease of the integrated area of the NH_3 absorption profile (which is independent of broadening effects) with respect to the H_2O concentration. When the absorption line area is normalized by a corrected NH_3 concentration ($C_{NH_3corr} = C_{NH_3}/(1 - C_{H_2O})$), a flatter curve is obtained, except in the upper range of humidity where the area of the NH_3 line still decreases for an unknown reason. The fact that the NH_3 concentration is not kept constant during this experiment is responsible for a large part of the variation of the NH_3 peak absorption observed in Fig. 6 when humidity increases. A small contribution to this effect is due to the H_2O-induced broadening of the NH_3 line, but this contribution amounts to only ~4% in the range $P_{H_2O} = 0$–0.2 atm according to the H_2O-broadening coefficient of the NH_3 line retrieved from the experimental data. This effect is not too significant. It should be stated that the variation of the NH_3 concentration observed in this experiment with increasing humidity does not affect the measurement of the width of the NH_3 absorption line, which was the main goal of this study. The NH_3 concentration is weak enough so that the contribution of the self-broadening coefficient to the width is negligible compared to the N_2- and H_2O-broadening coefficients.

4.2 Influence of humidity on NH_3 line B (6596.4 cm^{-1})

The effect of humidity on the 6596.4 cm^{-1} NH_3 absorption feature is shown in Fig. 8. As previously discussed, the observed variation of the peak amplitude of the NH_3 absorption feature at increasing humidity is mainly due to the reduction of the NH_3 concentration in the gas mixture that results from the experimental conditions. The effect of humidity on the H_2O and NH_3 absorption features is shown in more details in Fig. 8b with corresponding fits.

The same analysis was performed as previously to extract the dependence of the width and area of the NH_3 and H_2O absorption features for the different H_2O mixing ratios using a single Lorentzian curve in each case. Results are shown in Fig. 9. The following observations are made from these data:

- The measured N_2-broadening coefficient of the 6595.1 cm^{-1} H_2O line is $g_{N_2} = 0.116 \pm 0.001$ cm^{-1}/atm compared to a value of 0.08 cm^{-1}/atm at 23°C reported in Hitran for the air-broadening coefficient of this isolated line. The discrepancy between these values is believed to result in a large part from the difference in the diluting gas.
- The measured self-broadening coefficient of the H_2O absorption line ($g_{self} = 0.23 \pm 0.01$ cm^{-1}/atm) is smaller by ~40% than the value at 70°C calculated from Hitran.

Table 4 Comparison of the broadening factors of the 6595.1 cm^{-1} H_2O absorption feature experimentally measured and tabulated in Hitran. Errors on the experimental values correspond to the uncertainties retrieved from the linear regression in Fig. 9

Transition	Reference						Experimental	
	wn [cm^{-1}]	g_{air} @ 23°C [cm^{-1}/atm]	g_{air} @ 70°C [cm^{-1}/atm]	g_{self} @ 23°C [cm^{-1}/atm]	g_{self} @ 70°C [cm^{-1}/atm]	*n*	g_{N_2} @ 70°C [cm^{-1}/atm]	g_{self} @ 70°C [cm^{-1}/atm]
(2,1,0) 4 3 1 ← (0,0,0) 5 4 2	6595.0784	0.0803	0.0725	0.407	0.3676	0.69	0.116 ±0.001	0.23 ±0.01

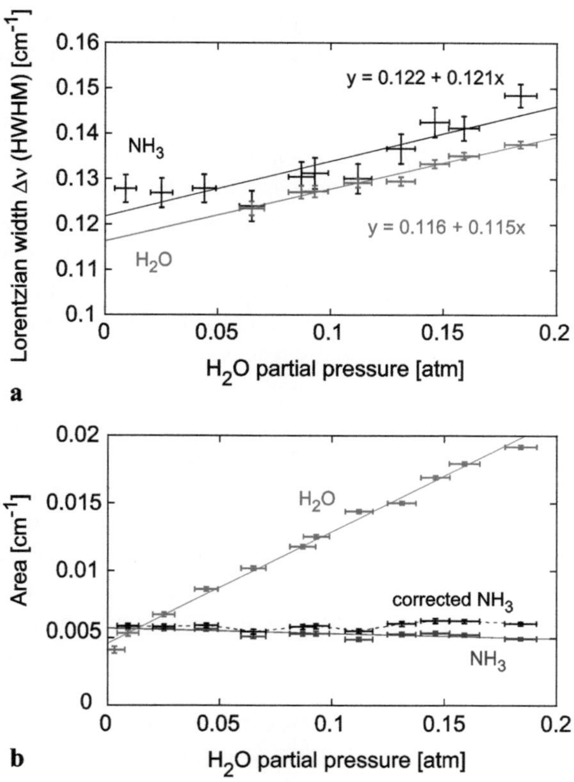

Fig. 9 Lorentzian width (**a**) and area (**b**) of the 6596.4 cm^{-1} NH_3 and 6595.1 cm^{-1} H_2O absorption features retrieved from the fits as a function of the H_2O partial pressure. Error bars represent the imprecision on the H_2O partial pressure and on the spectral parameters retrieved from the fitting process (see Fig. 7)

- The N_2-broadening coefficient of the 6595.4 cm^{-1} NH_3 absorption feature obtained at $P_{H_2O} = 0$ is $g_{N_2} = 0.122 \pm 0.002$ cm^{-1}, similar to the first analyzed NH_3 absorption feature at 6612.7 cm^{-1}. Here again, this is a consequence of several overlapping transitions that constitute the absorption feature, among which the four strongest cover a spectral interval of 0.12 cm^{-1}.
- The influence of water vapor on the 6595.4 cm^{-1} NH_3 absorption feature is characterized by a H_2O-broadening coefficient of $g_{H_2O} = 0.24 \pm 0.02$ cm^{-1}/atm. This value is larger than for the first NH_3 absorption feature at 6612.7 cm^{-1}.
- Here again, the integrated area of the NH_3 line decreases by ~15% when the gas mixture is moisturized from 0 to 0.18 atm H_2O partial pressure. This results from the 18% reduction of the NH_3 concentration arising when H_2O is added to the mixture. When the absorption line area is corrected for the change of the NH_3 concentration, a more or less constant area is obtained.

Table 5 Comparison of the broadening coefficients obtained for the two investigated NH_3 absorption features. Values obtained at $T =$ 70°C and atmospheric pressure

Transition	*wn* [cm^{-1}]	g_{N_2} [cm^{-1}/atm]	g_{H_2O} [cm^{-1}/atm]	g_{H_2O}/g_{N_2}
A	6612.7	0.122 ±0.001	0.148 ±0.003	1.21
B	6596.4	0.122 ±0.002	0.24 ±0.02	1.97

5 Discussion

The broadening coefficients of the two NH_3 absorption features studied in this work at $T = 70$°C are summarized in Table 5, together with the uncertainties retrieved from the linear regression of the line width as a function of the H_2O partial pressure. These values have been obtained from a fit of the experimental absorption features by a single Lorentzian function, which stands for broadened features obtained at atmospheric pressure as considered here. The present evaluation does not take into account the numerous individual lines of different rotational quantum number that overlap to form the broad features observed at atmospheric pressure. Therefore, the values reported here are valid around atmospheric pressure only and not for low pressure operation where some of the individual lines start to be resolved. However, this restriction is not too severe, as many NH_3 sensing applications concern the monitoring of ambient air at atmospheric pressure (see for instance Refs. [1, 4, 5]). Furthermore, these applications are the most susceptible to be affected by H_2O cross-sensitivity since atmospheric H_2O concentration may significantly vary in such conditions, especially when temperature increases.

Experimental results of Table 5 demonstrate the occurrence of H_2O cross-sensitivity at atmospheric pressure for the two investigated NH_3 absorption features. For NH_3 line (A) at 6612.7 cm^{-1}, the measured H_2O-broadening coefficient is ~21% higher than the N_2-broadening factor, which leads to a $\sim$4% additional broadening of the NH_3 line for a 0–20% change of H_2O concentration (in N_2). For NH_3 line (B) at 6596.4 cm^{-1}, the H_2O-broadening coefficient is almost a factor two higher than the N_2-broadening factor, leading to a $\sim$19% additional line broadening for a humidity variation of 20% (0 to 20% H_2O mixing ratio). The impact of H_2O cross-sensitivity is expected to be even slightly higher when air is considered as the diluting gas rather than N_2 as used in the present experiment, since air-broadened NH_3 lines are slightly narrower (by $\sim$10%) than N_2-broadened lines [14, 15, 22, 23]. When the on-peak WMS-2f signal is used for concentration quantification in a gas sensor, a relative inaccuracy of a similar amount as the linewidth change arises on the retrieved concentration, as the WMS-2f signal scales linearly with the absorption peak value. Additionally, there is another smaller contribution due to the fact that the WMS-2f signal also depends on the normalized frequency modulation index $m = \Delta\nu/\Delta\nu_{\text{line}}$, where $\Delta\nu$ is the amplitude of the laser frequency modulation and $\Delta\nu_{\text{line}}$ the width of the absorption line [24]. The modulation index changes if the width of the line varies (at constant laser modulation), which affects the WMS-2f signal. The relative importance of this additional effect depends on the precise value of the modulation index and is minimized for its optimal value ($m = 2.2$ for a Lorentzian lineshape and a sine modulation waveform [9, 24]).

6 Conclusion

It was shown in this work that cross-sensitivity may have a noticeable influence on gas sensing by spectroscopic methods. In principle, this effect does not affect the response of a sensor when the detection method uses the entire absorption profile to determine the gas concentration (e.g., in rapid scan direct absorption or in WMS scan mode), since cross-sensitivity only broadens the line without modifying its strength (integrated area). However, the impact of cross-sensitivity may be significant when only one point of the absorption line (usually at line centre) is considered. This is the case in the simplest and popular implementation of WMS or PAS, which determines the gas concentration from the measured on-peak demodulated signal (2f signal for WMS and wavelength modulation PAS, 1f signal for intensity modulation PAS). Inaccuracy in the retrieved concentration ranging from several percents to a couple of tens of percents may arise in such conditions.

The specific case of H_2O-induced cross-sensitivity on NH_3 was treated in this paper by considering two different NH_3 absorption features at 6612.7 cm^{-1}(line A) and 6596.4 cm^{-1} (line B), respectively. Experimental data obtained at T = 70°C (in order to enable high H_2O mixing ratios to be achieved) have shown that the H_2O-induced broadening was higher than the N_2-broadening by $\sim$20% for line A and by almost a factor two for line B. Therefore, a humidity change from 0 to 20% (absolute concentration) results in an additional broadening of the NH_3 absorption features of $\sim$4% for line A and $\sim$19% for line B. This can lead to a significant inaccuracy of the NH_3 concentration retrieved in a gas sensor based on the on-peak WMS-2f signal in such a large H_2O mixing ratio variation, which can occur in high-temperature applications. However, this effect becomes negligible for NH_3 sensing in standard atmospheric conditions such as 50% relative humidity at 20°C (around 12 mbar H_2O in 1013 mbar of air). In this case, the H_2O broadening contributes to only 1.5% of the linewidth in air for line A and 2.5% for line B, and even a large humidity change down to 0% or up to 100% relative humidity would result in a change of the NH_3 line width of only 0.25% for line A and 1.2% for line B. Therefore, the effect considered in the present study becomes relevant only at higher temperature, where large H_2O concentration variations are susceptible to occur.

Even if the specific case of H_2O-induced cross-sensitivity on NH_3 absorption features was considered in this paper, cross-sensitivity in spectroscopic gas sensors is a more general issue which has to be taken into account for any species when the composition of the diluting gas is subject to significant variations. One of the major potential sources of cross-sensitivity is H_2O for gas sensing application in ambient air, since H_2O is susceptible to significant concentration variations, especially in hot conditions. But other species can also contribute to cross-sensitivity in process control application where specific carrier gases (other than air) are used and are subject to composition variation.

Cross-sensitivity affects the entire absorption spectrum of an analyzed species and is thus unavoidable. This is a major difference compared to interference effects which are also induced by the presence of other species in the mixture but rely on the spectral closeness of interfering lines of these species. A cross-sensitivity results from collision processes between the different molecules (collision broadening) and a line of the analyte is affected whenever the collisional partner has absorption lines in the considered spectral range or not. The only spectral dependence results from the fact that some lines may be more affected than others by cross-sensitivity, since foreign-broadening coefficients can significantly vary between different lines of the same species. As an example, the ratio between the CO_2-broadening coefficient reported by Gamache et al. [31] and the air-broadening coefficient tabulated in Hitran changes by $\sim$50% among a series of H_2O lines in the 2.39–2.52 µm spectral range. For the same lines, the ratio between self-broadening and air-broadening coefficients changes even in a larger proportion (by a factor $\sim$2).

Acknowledgements The author acknowledges Prof. Luc Thévenaz (EPFL, CH) for the realization of the experimental work using his infrastructure (climate chamber, gas mixing unit) and Dr. Renaud

Matthey (CSEM Neuchâtel, CH) for lending the multipass cell used in these experiments and for a careful reading and useful comments on the manuscript.

References

1. S. Schilt, L. Thévenaz, M. Niklès, L. Emmenegger, C. Hüglin, Spectrochim. Acta A **60**, 3259 (2004)
2. L.R. Narasimhan, W. Goodman, C.K.N. Patel, Proc. Natl. Acad. Sci. **98**, 4617 (2001)
3. H. Ahlberg, S. Lundqvist, R. Tell, T. Andersson, Spectrosc. Eur. **6**, 22 (1994)
4. J.-Ph. Besson, M. Gyger, S. Schilt, L. Thévenaz, Individually ventilated microclimate monitoring using photoacoustic spectroscopy, in *4 Konferenz über Optische Analysenmesstechnik in Industrie und Umwelt (OPTAM'04)*, Düsseldorf, Germany, October 7–8, 2004, VDI-Berichte, vol. 1863, p. 99
5. M.E. Webber, T. MacDonald, M.B. Pushkarsky, C.K.N. Patel, Y. Zhao, N. Marcillac, F.M. Mitloehner, Meas. Sci. Technol. **16**, 1547 (2005)
6. R. Claps, F.V. Englich, D.P. Leleux, D. Richter, F.K. Tittel, R.F. Curl, Appl. Opt. **40**, 4387 (2001)
7. G. Totschnig, M. Lackner, R. Shau, M. Ortsiefer, J. Rosskopf, M.C. Amann, F. Winter, Appl. Phys. B **76**, 603 (2003)
8. E. De Tommasi, G. Casa, L. Gianfrani, Appl. Phys. B **85**, 257 (2006)
9. I. Linnerud, P. Kaspersen, T. Jaeger, Appl. Phys. B **67**, 297 (1998)
10. G. Modugno, C. Corsi, Infrared Phys. Technol. **40**, 93 (1999)
11. B.A. Paldus, C.C. Harb, T.G. Spence, R.N. Zare, C. Gmachl, F. Capasso, D.L. Sivco, J.N. Baillargeon, A.L. Hutchinson, A.Y. Cho, Opt. Lett. **25**, 666 (2000)
12. J. Manne, O. Sukhorukov, W. Jäger, J. Tulip, Appl. Opt. **45**, 9230 (2006)
13. A.A. Kosterev, F.K. Tittel, Appl. Opt. **43**, 6213 (2004)
14. C.L. Bell, M. Dhib, G. Hancock, G.A.D. Ritchie, J.H. van Helden, N.J. van Leeuwen, Appl. Phys. B **94**, 327 (2009)
15. J.S. Gibb, G. Hancocka, R. Peverall, G.A.D. Ritchie, L.J. Russell, Eur. Phys. J. D **28**, 59 (2004)
16. L.S. Rothman, I.E. Gordon, A. Barbe, D. Chris Benner, P.F. Bernath, M. Birk, V. Boudon, L.R. Brown, A. Campargue, J.-P. Champion, K. Chance, L.H. Coudert, V. Dana, V.M. Devi, S. Fally, J.-M. Flaud, R.R. Gamache, A. Goldman, D. Jacquemart, I. Kleiner, N. Lacome, W.J. Lafferty, J.-Y. Mandin, S.T. Massie, S.N. Mikhailenko, C.E. Miller, N. Moazzen-Ahmadi, O.V. Naumenko, A.V. Nikitin, J. Orphal, V.I. Perevalov, A. Perrin, A. Predoi-Cross, C.P. Rinsland, M. Rotger, M. Simeckova, M.A.H. Smith, K. Sung, S.A. Tashkun, J. Tennyson, R.A. Toth, A.C. Vandaele, J. VanderAuwera, J. Quant. Spectrosc. Radiat. Transf. **110**, 533 (2009)
17. L. Lundberg-Nielsen, F. Hegelund, F.M. Nicolaisen, J. Mol. Spectrosc. **162**, 230 (1993)
18. M.E. Webber, D.S. Bear, R.K. Hanson, Appl. Opt. **40**, 2031 (2001)
19. L.-H. Xu, Z. Liu, I. Yakovlev, M.Y. Tretyakov, R.M. Lees, Infrared Phys. Technol. **45**, 31 (2004)
20. R.M. Lees, L. Li, L.-H. Xu, J. Mol. Spectrosc. **251**, 241 (2008)
21. H. Jia, W. Zhao, T. Cai, W. Chen, W. Zhang, X. Gao, J. Quant. Spectrosc. Radiat. Transf. **111**, 347 (2009)
22. D.M. O'Leary, J. Orphal, A.A. Ruth, U. Heitmann, P. Chelin, C.E. Fellows, J. Quant. Spectrosc. Radiat. Transf. **109**, 1004 (2008)
23. M.A. Koshelev, M.Y. Tretyakov, R.M. Lees, L.-H. Xu, Appl. Phys. B **85**, 273 (2006)
24. S. Schilt, L. Thévenaz, P. Robert, Appl. Opt. **42**, 6728 (2003)
25. N. Melander, J. Henningsen, AIP Conf. Proc. **463**, 78 (1999)
26. G. Gagliardi, L. Gianfrani, Opt. Lasers Eng. **37**, 509 (2002)
27. S.W. Sharpe, T.H. Johnson, R.L. Sams, P.M. Chu, G.C. Rhoderick, P.A. Johnson, Appl. Spectrosc. **58**, 1452 (2004)
28. G. Durry, V. Zeninari, B. Parvitte, T. Le Barbu, F. Lefevre, J. Ovarlez, R.R. Gamache, J. Quant. Spectrosc. Radiat. Transf. **94**, 387 (2005)
29. V. Vorsa, S. Dheandhanoo, S.N. Ketkar, J.T. Hodges, Appl. Opt. **44**, 611 (2005)
30. R.R. Gamache, J. Quant. Spectrosc. Radiat. Transf. **52**, 481 (1994)
31. R.R. Gamache, S.P. Neshyba, J.J. Plateaux, A. Barbe, L. Régalia, J.B. Pollack, J. Mol. Spectrosc. **170**, 131 (1995)

Measurement of the isotope shift of the $6^3P_1 \leftrightarrow 5^3D_1$ transition of ytterbium by using a diode oscillator fiber amplified laser

L. Lee · H. Park · K.-H. Ko · D.-Y. Jeong

Received: 7 October 2009 / Revised version: 2 February 2010 / Published online: 9 March 2010
© Springer-Verlag 2010

Abstract We demonstrated a Diode Oscillator Fiber Amplification (DOFA) system in order to study the $6^3P_1 \leftrightarrow 5^3D_1$ (1539 nm) transition line of a neutral ytterbium atom that is accessed by the stepwise excitation of the ground state. The frequency of the DOFA system was doubled by a MgO:PPLN crystal for the resonant excitation of the $6^1S_0 \leftrightarrow 6^3P_1$ transition. The frequency of the second harmonic beam was stabilized to the $6^1S_0 \leftrightarrow 6^3P_1$ transition of each isotope with the stability of about 2 MHz. We performed absorption spectroscopy on the $6^3P_1 \leftrightarrow 5^3D_1$ (1539 nm) transition after the velocity selective excitation by the frequency-doubled beam. The isotope shifts in the $6^3P_1 \leftrightarrow 5^3D_1$ (1539 nm) transition were directly measured for the first time. The relative isotope shifts from ^{174}Yb were measured as -105.8 MHz and 109.7 MHz for ^{176}Yb and ^{172}Yb, respectively.

1 Introduction

Stable isotopes have been used in medical, industrial, nuclear engineering and basic science. Ytterbium (Yb) has seven stable isotopes: ^{168}Yb, ^{170}Yb, ^{171}Yb, ^{172}Yb, ^{173}Yb, ^{174}Yb, and ^{176}Yb. Enriched ^{168}Yb and ^{176}Yb can be used as source materials of ^{169}Yb and ^{177}Lu which are useful radioactive isotopes in the medical and industrial applications [1, 2]. The natural abundances of ^{168}Yb and ^{176}Yb are 0.13% and 12.7%, respectively. Therefore, the isotope abundances need to be increased for the applications mentioned above. The laser-based technique is one of the well-known isotope separation methods [3]. In this method, the wavelengths of lasers are tuned to the transition lines of wanted isotope. The excited atoms are then ionized by a strong laser. However, the isotope selectivity is limited due to the linewidth of the pulsed laser and the power broadening [4]. Jeong et al. proposed new isotope separation technique based on optical pumping to overcome above limitations [5, 6]. In this technique, high separation factor in thallium can be achieved in the optical pumping stage where a narrow bandwidth cw-laser is used to pump the population from the ground state to the metastable state. Park et al. extended the optical pumping technique to the ytterbium isotope separation [7]. However, there is a lack of spectroscopic data for the transition lines between the excited states, so the spectroscopic studies are needed for the application.

A tunable diode-laser has been widely used in atomic and molecular spectroscopy due to its narrow linewidth, wide tuning range and compactness. It has been difficult to use the diode laser in the shorter wavelength below 620 nm [8]. Moreover, the optical pumping technique requires a watt-class laser with the single longitudinal mode and tunability. In order to access those transition lines, we established a laser system which consists of a Diode Oscillator Fiber Amplifier (DOFA) and a subsequent frequency-doubling system. Using this system, we performed spectroscopic studies on ytterbium atoms.

In this paper, we describe the DOFA system and subsequent frequency-doubling system for the $6^1S_0 \leftrightarrow 6^3P_1$ transition. The frequency of the frequency-doubled DOFA system was stabilized to the $6^1S_0 \leftrightarrow 6^3P_1$ transition for the study of the $6^3P_1 \leftrightarrow 5^3D_1$ transition. We characterize the DOFA system and its frequency doubling as well as the frequency stabilization to the $6^1S_0 \leftrightarrow 6^3P_1$ transition. Using

L. Lee (✉) · H. Park · K.-H. Ko · D.-Y. Jeong
Quantum Optics Division, Korea Atomic Energy Research Institute, P.O. Box 105, Yuseong, Daejeon, 305-600, Republic of Korea
e-mail: lim2phy@kaeri.re.kr

this laser, we performed saturated absorption spectroscopy on the $6^1S_0 \leftrightarrow 6^3P_1$ transition and absorption spectroscopy on the $6^3P_1 \leftrightarrow 5^3D_1$ transitions. The isotope shifts of the $6^3P_1 \leftrightarrow 5^3D_1$ transition were obtained from the absorption signals.

2 Experiment

Figure 1 shows the partial atomic structure of an ytterbium atom. We consider three states such as the ground state of 6^1S_0, the intermediate state of 6^3P_1, and the highly excited state of 5^3D_1. The transition wavelength and gA value of the $6^1S_0 \leftrightarrow 6^3P_1$ transition are 555.6 nm and 2.2×10^6 s^{-1}, respectively [12]. For the $6^3P_1 \leftrightarrow 5^3D_1$ transition, the transition wavelength is 1539 nm, and the gA value is about 1×10^6 s^{-1} [12]. The lifetimes of the 6^3P_1 state and the 5^3D_1 state are 875 and 380 ns, respectively. The isotope shifts of the $6^1S_0 \leftrightarrow 6^3P_1$ transition were reported [4, 9]. However, the isotope shifts and the hyperfine structure of the $6^3P_1 \leftrightarrow 5^3D_1$ transition have never been directly measured.

The experimental setup is shown in Fig. 2. An External Cavity Diode Laser (ECDL, Toptica, DL 100, Littrow configuration) which was used as an oscillator has the mode hop free tuning range of more than 10 GHz at the center wavelength of 1111 nm. The laser beam with the power of 50 mW and the bandwidth less than 1 MHz was injected on the Ytterbium Doped Fiber Amplifier (YDFA). The amplified infrared laser was incident on a Mg-doped Periodically Poled Lithium Niobate (MgO:PPLN, 5 mol%) crystal to generate the green beam of 555.6 nm. The crystal length was about 50 mm, and the end surfaces were coated to prevent the reflection of the infrared and the visible laser beams. A half-waveplate was used to control the polarization of the incident laser beam, and the periodically poled crystal was mounted on an oven heated to 60.2°C for the phase matching. The frequency-doubled laser beam was split to two beams: one beam was used for Saturated Absorption Spectroscopy (SAS) on the $6^1S_0 \leftrightarrow 6^3P_1$ transition (555.6 nm) and the other was used as an excitation laser beam for absorption spectroscopy on the $6^3P_1 \leftrightarrow 5^3D_1$ transition (1539 nm).

A Distributed Feedback laser (DFB, Toptica, DL100 DFB 01113) was used as a probe laser to study the $6^3P_1 \leftrightarrow 5^3D_1$ transition when the green beam was used to prepare highly excited atoms. The DFB laser had the tuning range of >20 GHz at the center wavelength of 1539 nm and emitted the single-longitudinal-mode laser beam of about 25 mW. The DFB laser was provided to a heat pipe oven to probe the absorption of the $6^3P_1 \leftrightarrow 5^3D_1$ transition. The SAS signals of the $6^1S_0 \leftrightarrow 6^3P_1$ transition were measured as a function of the vapor pressure.

A heat pipe oven was used to maintain the sufficient atomic density of ytterbium. It was filled with Ar gas of 0.3 torr as buffer gas to prevent the window coating by ytterbium atoms. The length of the vapor column was about 30 cm. The vapor pressure of neutral ytterbium is given by the vapor-pressure model as follows [10]:

$$\log_{10} P_{\text{torr}} = 6.94313 - \frac{8655.67}{T} - 0.00046T + 0.63932 \log_{10} T, \tag{1}$$

where P_{torr} is the vapor pressure in torr and T is the temperature in K. At the temperature of 650 K, the vapor pressure and the atomic density were evaluated as 1.6×10^{-5} torr and 2.3×10^{11} atoms/cm^3.

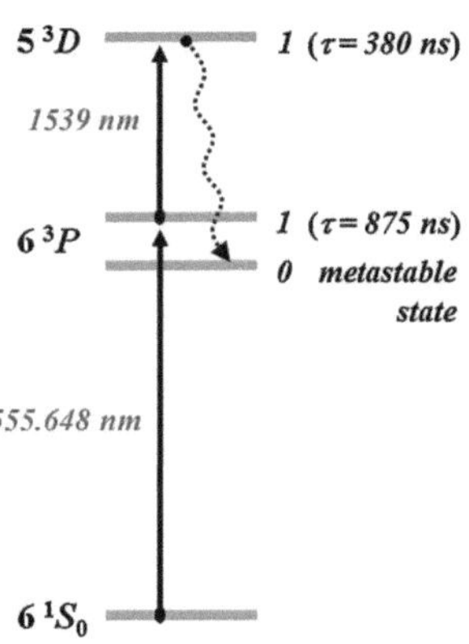

Fig. 1 Partial atomic structure of an ytterbium atom

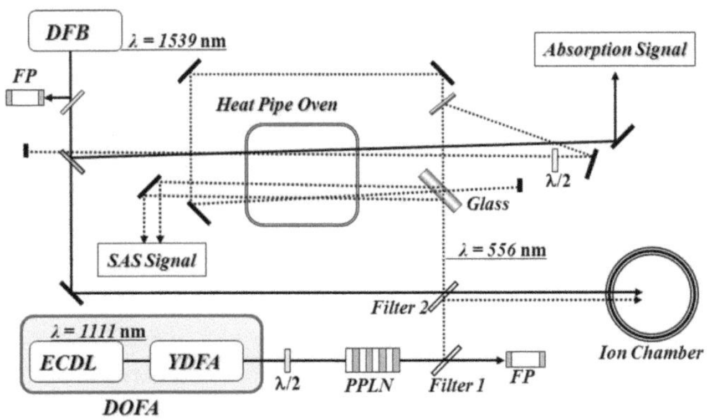

Fig. 2 The experimental setup for the frequency stabilization and spectroscopy by using the DOFA and the frequency-doubling systems

3 Results and discussion

At first, we show the characteristics of the diode oscillator fiber amplifier system at the wavelength of 1111 nm. The input laser wavelength is located at the wing of the gain spectral profile, thus we investigated the amplifying characteristics of the DOFA system. Figure 3(a) shows the spectrum of the amplified laser. The Amplified Spontaneous Emission (ASE) was observed at around 1070 nm. The power ratio of the ASE to the laser beam was measured to be about 1/3.

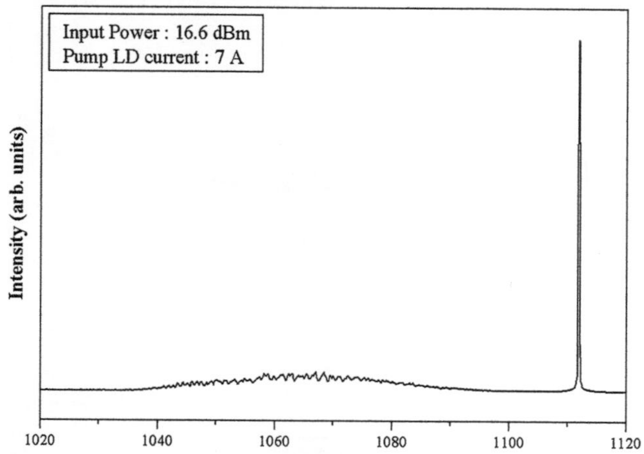

a

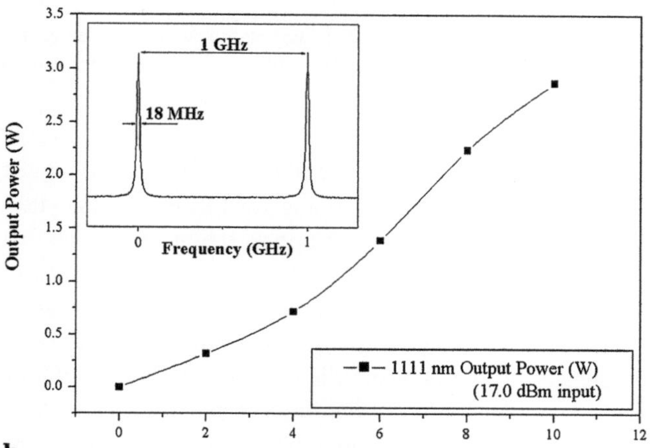

b

Fig. 3 (**a**) The spectrum of the amplified infrared laser. (**b**) The amplified output power versus the currents of the pumping diode laser in the amplifier. The transmission signal through a Fabry–Perot interferometer is shown in the inset

Figure 3(b) also shows the amplified power versus the injected current to the amplifier (currents of the laser diode which was used for pumping the ytterbium doped fiber in the YDFA). The amplified power reached up to about 3.2 W at the input laser power of about 50 mW and the injected laser diode currents of about 11 A. The spectral characteristics of the amplifiers such as continuous tunability and the single longitudinal mode operation were determined by those of the input diode laser. The transmitted laser power through a confocal Fabry–Perot Interferometer (FPI) is shown in the inset, while the injected voltage of the Piezoelectric transducer (PZT) to control the wavelength of a diode laser was scanned continuously. The Free Spectral Range (FSR) of a FPI was about 1 GHz, and the finesse was about 150.

Figure 4 shows the frequency-doubling power versus the input infrared power. As we expected, the second harmonic power increases in a quadratic form of the incident power. The maximum power reached up to about 430 mW at the input power of 3.2 W. The second harmonic power was optimized by varying the crystal temperature and the focus-

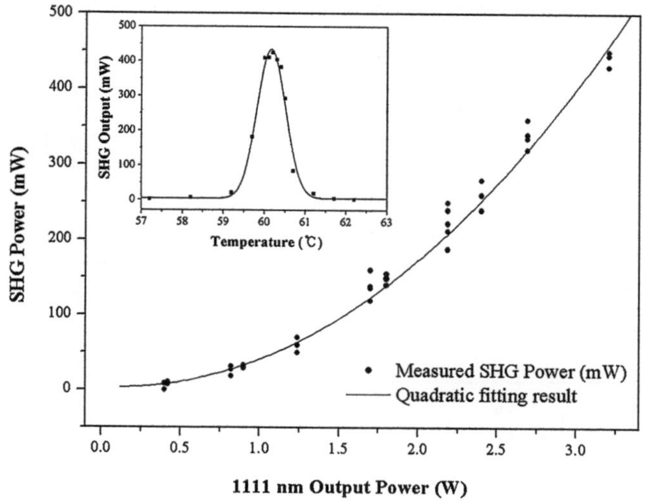

Fig. 4 The second harmonic power versus the fundamental laser power; the temperature tuning curve for the PPLN crystal is shown in the inset

ing lenses. The temperature width was measured to be about 0.7°C from the temperature tuning curve as shown in the inset. When the incident laser power increased, the optimum oven temperature was slightly decreased due to the thermal effect [11]. The green laser with several tens of milliwatts was used for the frequency locking and the absorption spectroscopy, but the high power green beam were to be used for further spectroscopic studies which concerned with the power broadening effect and the optical pumping effect in a thick atomic vapor.

The result of SAS of the $6^1S_0 \leftrightarrow 6^3P_1$ transition is depicted in Fig. 5(a). All isotopes except ^{168}Yb could be resolved and the ^{171}Yb $(1/2 \rightarrow 3/2)$ and ^{173}Yb $(5/2 \rightarrow 5/2)$ peaks could not be identified because the shifts of those peaks are too small (less than 1 MHz) to be identified in our system. The width of the SAS signal was measured to be about 13.8 MHz, while the natural linewidth of the $6^1S_0 \leftrightarrow 6^3P_1$ transition was only 0.182 MHz. The broader width resulted from the collisional broadening such as the ytterbium–ytterbium and ytterbium–argon collisions. The intensities of the probe and pump beam for SAS were 3.2 mW/cm^2 and 19.1 mW/cm^2, respectively.

Figure 5(b) shows the SAS signals at different oven temperatures. Relatively small signals due to low abundances, such as the ^{170}Yb peak and the crossover peak between ^{173}Yb $(5/2 \rightarrow 5/2)$ and ^{173}Yb $(5/2 \rightarrow 3/2)$, increased as the temperature was raised to 670 K. However, the signals decreased above 670 K due to the absorption of the excitation and the probe beam. We could obtain the optimized Signal-to-Noise Ratio (SNR) of the crossover at 670 K and could stabilize the frequency-doubled DOFA system to the peak.

However, the wavelength mismatch between the master oscillator and the fiber amplifier resulted in a large ASE

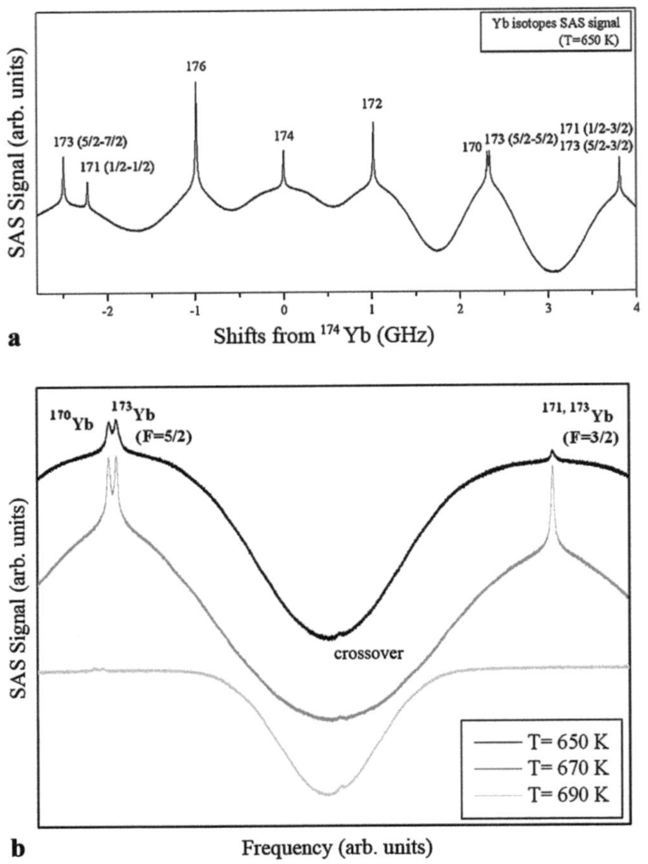

Fig. 5 (**a**) The SAS signal of the $6^1S_0 \leftrightarrow 6^3P_1$ transition; the isotope shifts from ^{174}Yb isotope measured by the developed DOFA system are shown in the figure. (**b**) The SAS signals at different oven temperatures. The peaks corresponding to the isotopes with relatively low abundances increase as the temperature reaches 670 K, and decreases afterwards

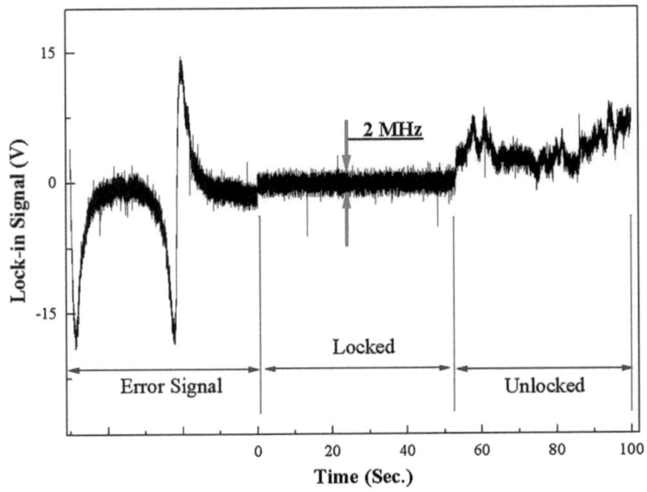

Fig. 6 The error signal of the frequency stabilization. The center frequency was stabilized within 2 MHz, while the laser frequency drifted in the unlocked case

as discussed above. The ASE induced the amplitude noise in the green beam, and thus it made noise in the dispersive spectrum. The noise level became extremely large at the higher frequency range of the ^{173}Yb ($5/2 \rightarrow 5/2$) peak. From the FFT (Fast Fourier Transform) analysis of the noises, the frequency elements such as 10.4 kHz, 15.4 kHz, etc. were conspicuously observed. In order to reduce those frequency elements, we used a low pass filter with the 10 kHz cut-off frequency in the photodiode circuit and modulated the laser frequency at 8.1 kHz by using a PZT. As a result, we obtained a clean dispersive signal and stabilized the laser frequency to each peak of an isotope in the SAS spectrum. We could stabilize the frequency of the frequency-doubled DOFA system within 2.0 MHz to the peaks which have relatively large SNR such as ^{176}Yb peak, within 4.0 MHz to the peaks with small SNR in the range of a large noise level. The stability of 2.4×10^{-6} was achieved for averaging time of 200 seconds. Figure 6 shows the examples of the frequency stabilization to ^{176}Yb peak. It was robust enough to keep on locking over 12 hours long under a noisy experimental environment.

For the spectroscopy on the $6^3P_1 \leftrightarrow 5^3D_1$ transition, the first excitation beam of 555.6 nm was incident on the ytterbium vapor with the linear (π) polarization. While the first excitation beam was locked to the SAS signal of each isotope, linear absorption spectroscopy was performed by using a DFB laser of 1539 nm as a probe beam. The probe beam propagated in the opposite direction of the first excitation laser. The polarization of the probe beam was perpendicular to that of the first excitation beam. The intensities of the excitation and the probe laser beams were measured to be 3.2 mW/cm^2 (2 mm diameter) and 140 μW/cm^2 (1.5 mm diameter), respectively.

Figure 7(a) shows three absorption spectra of the $6^3P_1 \leftrightarrow 5^3D_1$ transition, while the first excitation laser locked to the transition lines of ^{172}Yb, ^{174}Yb and ^{176}Yb isotopes, respectively. The spectra were redrawn in order to easily show the observed relative positions of ^{172}Yb and ^{176}Yb from ^{174}Yb. The linewidth of the peaks were measured to be about 22.7 MHz and the relative positions of peaks from ^{174}Yb were measured to be −450.3 MHz for ^{176}Yb and 470.9 MHz for ^{172}Yb. The spectral positions and widths were also measured by another FPR with FSR of 1.0 GHz and finesse of 150. The high spectral resolution came from the velocity selective excitation by the first excitation laser. The atoms whose velocity was nearly the same in the Doppler broadened atomic medium should be excited by the first excitation laser. The Doppler shifts of the absorption signals are shown in Fig. 7(b). When the first excitation laser is locked to the ^{174}Yb peak, the first excitation laser (555.6 nm) excites the ^{174}Yb atoms with zero velocity. In addition, the excitation laser can excite the ^{176}Yb isotope (relative isotope shift of −954.2 MHz) within the Doppler broadening. In this case, the ^{176}Yb isotope with the velocity $k_1 v_{176} = -\Delta_{(S\rightarrow P)176-174}$ can be excited by the first excitation laser,

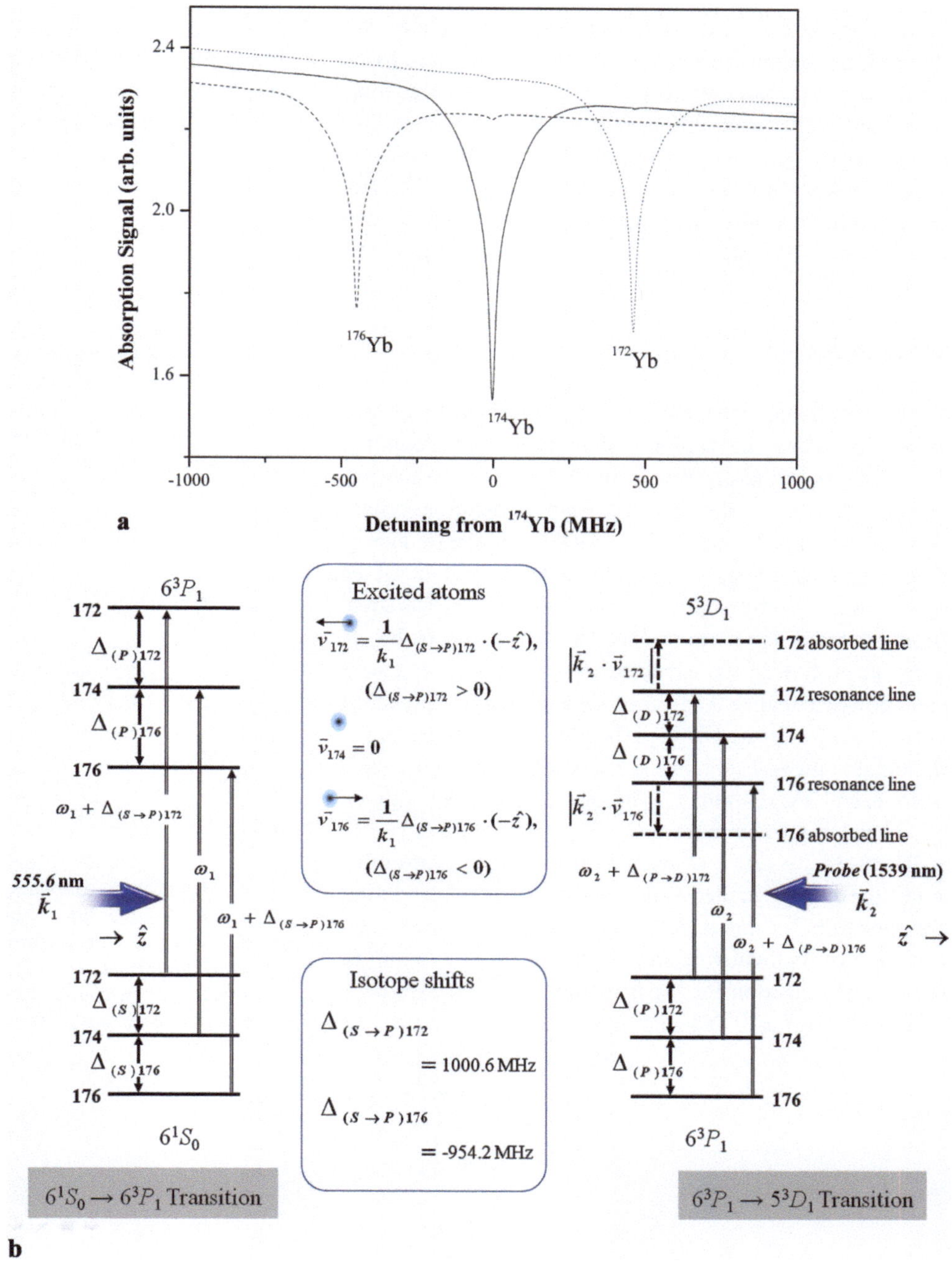

Fig. 7 (**a**) Three absorption spectra of the probe laser (1539 nm light for $6^3P_1 \leftrightarrow 5^3D_1$ transition) when the first excitation laser (555.6 nm for $6^1S_0 \leftrightarrow 6^3P_1$ transition) was locked to the peaks corresponding to ^{172}Yb (*dotted line*), ^{174}Yb (*straight line*) and ^{176}Yb (*dashed line*), respectively. The spectra were redrawn in order to easily show the observed relative positions of ^{172}Yb and ^{176}Yb from ^{174}Yb. (**b**) A schematic diagram to explain the Doppler shifted absorption signals corresponding to the even isotopes in the $6^3P_1 \leftrightarrow 5^3D_1$ transition. One isotope with zero velocity and the other isotope with a specific velocity can be excited to the 6^3P_1 state by the first excitation laser, and they are observed at different positions in the absorption spectrum

where $-\Delta_{(S\to P)176-174}$ is the relative isotope shift from the ^{174}Yb isotope of the $6^1S_0 \leftrightarrow 6^3P_1$ transition, k_i is the wave number of the incident laser ($i = 1$ for 555.6 nm, $i = 2$ for 1539 nm), and v_m is the velocity of the mth isotope. When the ^{176}Yb isotope is probed by the counter-propagated beam with the wavelength of 1539 nm, the Doppler shifted signal of the ^{176}Yb isotope in the $6^3P_1 \leftrightarrow 5^3D_1$ transition can be observed. In this case, the Doppler shift becomes $k_2 v_{176} = k_2(\Delta_{(S\to P)176-174}/k_1)$, where the opposite sign means that the probe beam is propagated in the opposite direction. Therefore, the relative absorption position of ^{176}Yb is located at $\Delta_{(P\to D)176-174} + k_2(\Delta_{(S\to P)176-174}/k_1)$. In the same way, when the first excitation laser is locked to the ^{176}Yb peak, ^{176}Yb isotope around zero velocity can be excited while ^{174}Yb isotope with the velocity of $k_1 v_{174} = \Delta_{(S\to P)176-174}$ can be excited. Thus, the ^{174}Yb peak and the ^{176}Yb peak (from the ^{174}Yb peak of the first case) are observed at $-k_1 v_{174}$ and $\Delta_{(P\to D)176-174}$, respectively. But the relative positions of the ^{176}Yb peak from ^{174}Yb are the same for each case. Considering the Doppler shifts, we obtained the relative isotope shifts of -105.8 MHz for ^{176}Yb and 109.7 MHz for ^{172}Yb.

If the frequency of the first excitation laser was modulated at 8.1 kHz as mentioned above, the absorption signal of the $6^3P_1 \leftrightarrow 5^3D_1$ transition could also be modulated with the same frequency. We obtained the dispersive curve by using a lock in amplifier referenced by a modulation frequency of the first excitation laser. The frequency was also stabilized and the error signal was measured to be 2.5 MHz when the excitation laser was locked to the ^{176}Yb isotope. The long term stability was also measured as 2.0×10^{-6} for the averaging time of 200 seconds.

4 Conclusion

We demonstrated a tunable high power and compact laser system by using the DOFA system at the wavelength of 1111 nm. In addition, the periodically poled crystal was also used for the compact green beam generation system. At first, we characterized both the DOFA and the frequency-doubling system. The frequency characteristics of the DOFA system were determined by the diode oscillator. The green beam power of 555.6 nm reached to about 430 mW, but the green beam power was expected to be increased to a sufficient power to realize the isotope separation as the higher incident laser power was applied. In addition, the frequency of the tunable green beam was successfully stabilized to each isotope, and it was applied to the spectroscopy on highly excited level of ytterbium atoms.

The velocity selective excitation technique is useful to study the highly excited state even in the Doppler broadened medium. In this paper, we performed the direct measurement of the relative isotope shifts of the $6^3P_1 \leftrightarrow 5^3D_1$ (1539 nm) transition for the first time. It was measured as -105.8 MHz for ^{176}Yb and 109.7 MHz for ^{172}Yb from ^{174}Yb, respectively. It is also available to resolve the hyperfine splitting of the odd isotopes, such as ^{171}Yb and ^{173}Yb. The high resolution spectroscopy to obtain the hyperfine splitting of odd isotopes will be performed in the near future.

References

1. H. Yamabayashi, Radioisotopes **43**, 296 (1994)
2. C.S. Cutler, C.J. Smith, G.J. Ehrhardt, T.T. Tyler, S.S. Jurisson, Cancer Biother. Radiopharm. **15**, 531 (2000)
3. P.A. Bokhan, V.V. Buchanov, N.V. Fateev, M.M. Kalugin, M.A. Kazaryan, A.M. Prokhorov, D.E. Zakevskii, *Laser Isotope Separation in Atomic Vapor* (WILEY-VCH, Weinheim, 2006)
4. M. Sankari, M.V. Suryanarayana, J. Phys. B **31b**, 261 (1998)
5. D.-Y. Jeong, K.H. Ko, G. Lim, J. Kim, T. S Kim, S. Rho, C.J. Kim, in *11th International Symposium on Laser Spectroscopy* (2003)
6. K.H. Ko, G. Lim, T.S. Kim, J. Kim, S. Rho, H. Park, C.J. Kim, D.Y. Jeong, in *LAP 2004 International Conference on Laser Probing* (2004)
7. H.M. Park, B.S. Kim, K.H. Ko, Y.H. Cha, J.M. Han, G. Lim, T.S. Kim, D.Y. Jeong, C.J. Kim, in *LAP2008 International Conference on Laser Probing* (2008)
8. Marvin J. Weber, *Handbook of Lasers* (CRC Press, Boca Raton, 2000)
9. H.M. Park, S. Kostritsa, D.H. Kwon, Y.H. Cha, K.T. Lee, S.M. Nam, J.M. Han, Y.J. Rhee, C.J. Kim, J. Korean Phys. Soc. **41**, 322 (2003)
10. A. Javan, P.L. Kelley, IEEE J. Quantum Electron. **QE-2**, 9 (1966)
11. K. Mizuuchi, A. Morikawa, T. Sugita, K. Tamamoto, N. Pavel, I. Shoji, T. Taira, Jpn. J. Appl. Phys. **42**, L1296 (2003)
12. C.H. Corliss, W.R. Bozman, *Experimental Transition Probabilities for Spectral Lines of Seventy Elements—Derived from the NBS Tables of Spectral-Line Intensities*. NBS Monograph, vol. 53 (US G.P.O., Washington, 1962)

Simulation-based comparison of noise effects in wavelength modulation spectroscopy and direct absorption TDLAS

B. Lins · P. Zinn · R. Engelbrecht · B. Schmauss

Received: 15 September 2009 / Revised version: 10 December 2009 / Published online: 14 January 2010
© Springer-Verlag 2010

Abstract A simulative investigation of noise effects in wavelength modulation spectroscopy (WMS) and direct absorption diode laser absorption spectroscopy is presented. Special attention is paid to the impact of quantization noise of the analog-to-digital conversion (ADC) of the photodetector signal in the two detection schemes with the goal of estimating the necessary ADC resolution for each technique. With laser relative intensity noise (RIN), photodetector shot noise and thermal amplifier noise included, the strategies used for noise reduction in direct and wavelength modulation spectroscopy are compared by simulating two respective systems. Results show that because of the combined effects of dithering by RIN and signal averaging, the resolutions required for the direct absorption setup are only slightly higher than for the WMS setup. Only for small contributions of RIN an increase in resolution will significantly improve signal quality in the direct scheme.

1 Introduction

Wavelength modulation spectroscopy (WMS) and direct absorption spectroscopy are two widely used variations of tunable diode laser absorption spectroscopy (TDLAS). Both are used in fields such as environmental monitoring, remote leak detection, combustion analysis or isotopic assessment [1–5].

B. Lins (✉) · R. Engelbrecht · B. Schmauss
Erlangen Graduate School in Advanced Optical Technologies (SAOT), Universitaet Erlangen-Nuernberg, Paul-Gordan-Str. 6, 91052 Erlangen, Germany
e-mail: bastian.lins@lhft.eei.uni-erlangen.de

B. Lins · P. Zinn · R. Engelbrecht · B. Schmauss
Lehrstuhl fuer Hochfrequenztechnik (LHFT), Universitaet Erlangen-Nuernberg, Cauerstr. 9, 91058 Erlangen, Germany

In virtually all present-day setups, the amplified output of a photodetector is digitized by an analog-to-digital converter (ADC) for digital post-processing.

In WMS, the laser wavelength is swept across an absorption feature and additionally modulated in the kHz range. Lock-in detection of frequency components at the first, second or a higher order harmonic is then used to determine the amount of the target gas with the help of calibration or the use of reference spectra. This lock-in based scheme allows the detection to be shifted to the frequency of the respective harmonic and thus reduces the influence of laser intensity noise (RIN), which is typically stronger for low frequencies. For small modulation amplitudes, the detected line shape of the nth harmonic resembles the nth derivative of the original absorption line shape function, which is why this technique is then also referred to as derivative spectroscopy. An advantage of WMS is that the lock-in output signal for the second or any higher harmonic is essentially background-free, resulting in a zero baseline.

Since it is the harmonics of the modulation frequency that are used for detection with the lock-in amplifier (LIA), the DC component and ramping background may be cut off by filtering, e.g., by an AC-coupled frontend of a digital LIA, before the signal is digitized and demodulated. Thus the full resolution of the ADC is available for quantization of the signal that carries the absorption information, and the detection of weak absorption lines is possible even with a limited ADC resolution [6].

The direct absorption scheme, in contrast, uses the direct absorption signature generated by the respective gas absorption line. As for the WMS technique, the laser wavelength is swept across an absorption line by modulation with a ramp current but without the harmonic modulation. Compared to the WMS scheme, the direct absorption measurement offers the potential for calibration-free measurements [7] and eliminates the need for a hardware lock-in amplifier or for a digital implementation of phase-sensitive detection as presented in [8, 9]. The resulting direct absorption signal is generally small compared to the large sloping background of the laser power caused by the ramp tuning of the current. For digital signal processing, the full range of the detector signal, including the ramp and DC background, must be digitized. In this case, only a fraction of the quantization steps of the ADC is used for the comparatively small absorption signal. This leads to the assumption that compared with the WMS technique, much better resolution of the ADC is required for this approach in order to resolve small absorptions.

On the other hand, to reduce noise, a large number of wavelength scans are usually averaged before determining the concentration of the target gas by a calculation of the integrated absorption from the measurement data. There are various noise sources present that dither the signal before quantization and together with the signal averaging this dithering effect may lead to a reduction of the quantization error [10, 11].

Using higher resolution ADCs will increase cost, and usually increased resolution comes at the expense of lower sampling rates. For these reasons, it is worth asking the question which is the necessary resolution for the WMS and the direct absorption scheme to avoid loss of signal-to-noise ratio (SNR) due to quantization errors. To answer this question for our measurement scenario described in the following section, the presented simulation has been carried out. It compares the different noise reduction mechanisms and estimates the required ADC resolution for the two measurement techniques. Though $2f$-WMS has been chosen in this analysis, many results and conclusions will in a similar way be valid for $1f$- or higher order WMS.

2 Simulated spectrometer setups

2.1 Simulated measurement scenario

To compare the two measurement techniques, simulations of their application to the same following measurement scenario have been created. The gas to be measured is ammonia (NH_3) at a path integrated concentration of $c(NH_3) =$ 5 ppm m, a temperature of $T_{gas} = 296$ K and a pressure of $p = 100$ mbar. The used line data is that of an NIR absorption line around 1527 nm taken from [12], resulting in a peak absorption of $A_{max} = 4.6 \cdot 10^{-5}$. In our system, a distributed feedback laser diode (DFB-LD) is used as a source, so the frequency tuning and noise parameters in the simulation were chosen based on measurements on that DFB-LD (NEL NLK1556STB). For both techniques, the average power on the photodetector was $P_0 = 3.5$ mW and the responsivity of the photodetector was set to $R = 1$ A/W, which is close to typical values for InGaAs detectors in the NIR region. Unless stated otherwise, the measurement time has been restricted to $T_{meas} = 1$ s, to make the systems comparable.

2.2 Simulation structure and framework

The described simulation is implemented as a numerical, modular time-domain simulation using *Matlab*. For each device of the spectrometer setups, an individual program module has been created. The DFB-LD, for instance, is simulated in a laserdiode module that takes the laser injection current $i_{LD}(t)$ as an input and calculates the laser output power $P_{opt}(t)$ and the laser output frequency $f_{opt}(t)$ as the outputs. Similarly, all other components of the setups are modeled in individual modules. The parameters that are needed to calculate the respective output variables of one module from the input variables are provided in configuration files for the respective modules. This modular organization makes it very easy to change or rearrange the individual components within the simulation. If, for example, a different laserdiode is to be simulated, all that needs to be done is to change the relevant parameters in the laserdiode configuration file.

All time-dependent variables are represented by a vector containing time-discrete samples of the respective signal. The simulations presented in this work have been carried out with a simulation sample rate of $F_S = 1$ MHz thus covering a simulated bandwidth of $B = 500$ kHz.

2.3 Wavelength modulation spectroscopy setup

The setup underlying the simulation of the $2f$-WMS system is shown in Fig. 1. The DFB-LD is driven by a ramp modulation of frequency $f_{ramp} = 1$ Hz to sweep the optical emission frequency f_{opt} across the desired gas absorption line. This modulation is superimposed with a harmonic modulation with a frequency f_{mod} in the kHz range provided by

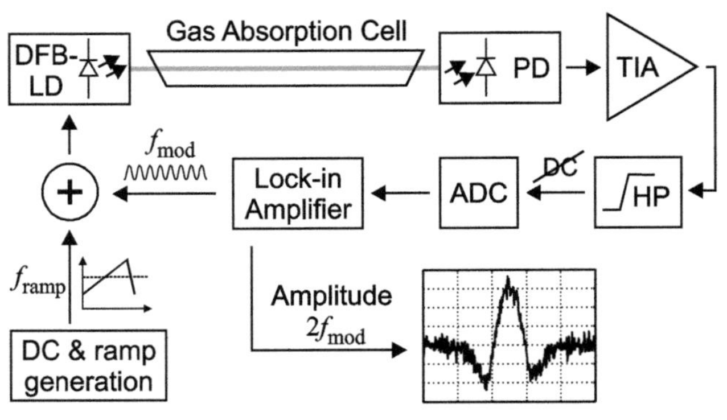

Fig. 1 Simulated $2f$-WMS setup: *DFB-LD*, distributed feedback laserdiode; *PD*, photodiode; *TIA*, transimpedance amplifier; *HP*, highpass filter; *ADC*, analog-to-digital converter

the LIA. The total injection current for the DFB-LD is then given by

$$i_{\mathrm{LD}}(t) = i_0 + i_{\mathrm{ramp}}(t) + \delta I \cos(2\pi f_{\mathrm{mod}} t), \tag{1}$$

with i_0 being the DC current component, $i_{\mathrm{ramp}}(t)$ the ramp current and δI the amplitude of the harmonic modulation. δI is chosen to achieve a modulation index of $m = 2.2$, thus yielding the maximum possible $2f$ signal in the line center, as first derived in [13] and described in a larger context of signal generation in [14]. From the injection current $i_{\mathrm{LD}}(t)$, the simulation calculates the instantaneous optical frequency output $f_{\mathrm{opt}}(t)$ and the optical power output $P_{\mathrm{opt}}(t)$ of the DFB-LD. Since the frequency modulation response of the DFB-LD decreases with increasing modulation frequency, different modulation coefficients M_{DC} and M_{mod} must be used to calculate the frequency response to the (slow) ramp modulation and to the kHz harmonic modulation, respectively [15]:

$$f_{\mathrm{opt}}(t) = f_0 + f_{\mathrm{ramp}}(t) + \delta f_{\mathrm{opt}} \cos(2\pi f_{\mathrm{mod}} t + \phi_{\mathrm{FM}}) \tag{2}$$

with

$$f_{\mathrm{ramp}}(t) = M_{\mathrm{DC}}\, i_{\mathrm{ramp}}(t), \tag{3}$$

$$\delta f_{\mathrm{opt}} = M_{\mathrm{mod}}\, \delta i_{\mathrm{mod}}. \tag{4}$$

While for the ramp modulation the frequency tuning is assumed to be in-phase with the current modulation, a phase shift ϕ_{FM} between the current modulation and the optical frequency modulation can be considered in the simulation. The optical power $P_{\mathrm{opt}}(t)$ is assumed to instantly follow the current modulation in the considered frequency range, so that the optical output power is given by

$$P_{\mathrm{opt}}(t) = P_0 + \eta_{\mathrm{slope}}\left[i_{\mathrm{ramp}}(t) + \delta I \cos(2\pi f_{\mathrm{mod}} t)\right], \tag{5}$$

where P_0 is the optical output power at the DC-current i_0 and η_{slope} denotes the slope of the current–power characteristics of the DFB-LD above threshold. From the output power and emission frequency, the power $P_{\mathrm{gas}}(t)$ after the gas absorption can be calculated by multiplying the transmission $T(f_{\mathrm{opt}})$ at the given frequency with the respective laser output power:

$$P_{\mathrm{gas}}(t) = P_{\mathrm{opt}}(t) \cdot T\big(f_{\mathrm{opt}}(t)\big). \tag{6}$$

To calculate $T(f_{\mathrm{opt}})$, a Voigt line profile, which is a convolution of a Lorentzian and a Gaussian line profile [16], has been used based on the line data given in Sect. 2.1 and the numerical approximation of [17]. The power after the absorption cell is converted to a photocurrent $i_{\mathrm{PD}}(t)$ by a photodiode with responsivity R

$$i_{\mathrm{PD}}(t) = R \cdot P_{\mathrm{gas}}(t) \tag{7}$$

and is then amplified by a transimpedance amplifier (TIA) with gain Z_{TIA} so that the voltage u_{TIA} after the TIA is given by

$$u_{\mathrm{TIA}}(t) = Z_{\mathrm{TIA}} \cdot R \cdot P_{\mathrm{gas}}(t). \tag{8}$$

The voltage $u_{\mathrm{TIA}}(t)$ is then ideally highpass filtered (HP) by subtracting the DC component and ramping background from the measurement signal, which is possible because both are known in the simulation. In a real setup, this filtering may be performed by an AC-coupled TIA with a cut-off frequency high enough to remove the ramp and low enough to let pass the $1f$ and $2f$ measurement signals. The input range of the ADC module is ideally matched to the range of the highpass filtered voltage $u_{\mathrm{TIA,HP}}(t)$ to make use of the full ADC resolution. In a real spectrometer, the ADC may be a separate benchtop or integrated device, or it can be the front-end of a digital LIA. In either case, after the quantization by the ADC, the $2f$-signal is demodulated by the LIA by multiplying the input signal with a $2f$-reference and applying digital lowpass filtering. For details on the general principle of a LIA, see, for example, [18]. The system's equivalent noise bandwidth (ENBW) [19] is given by the noise bandwidth of the LIA's lowpass filter. Here, it has been designed to be 40 Hz, a value being a consequence of the prerequisite of not distorting the $2f$-lineshape. As a final step, the LIA delivers the $2f$-amplitude signal during the 1 s ramp scan as shown in Fig. 1.

2.4 Direct absorption setup

The simulated direct absorption setup is shown in Fig. 2. Since the setup is closely related to the previously described WMS setup, only the differences will be highlighted. As no lock-in detection is performed, the harmonic modulation is omitted and no lock-in amplifier is needed. For the formal description of the direct absorption setup, (1)–(8) were applied for $\delta I = 0$. To enable signal averaging within the same total measurement time of $T_{\mathrm{meas}} = 1$ s, f_{ramp} is increased to

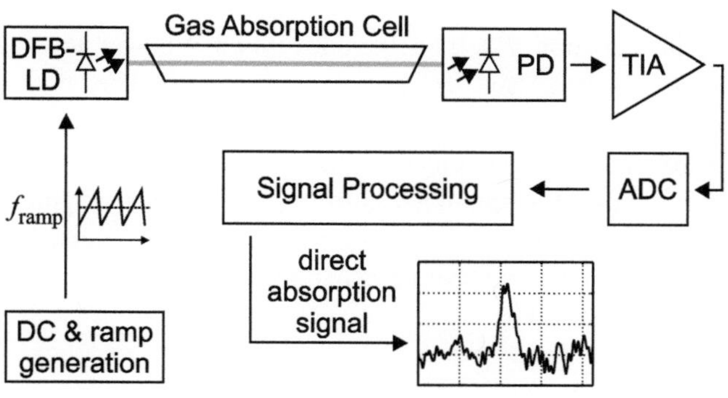

Fig. 2 Simulated direct absorption setup: *DFB-LD*, distributed feedback laserdiode; *PD*, photodiode; *TIA*, transimpedance amplifier; *ADC*, analog-to-digital converter

some tens up to several hundreds of Hz. In the direct absorption scheme, the full voltage $u_{\mathrm{TIA}}(t)$, including the sloping background from the power modulation of the DFB-LD, must be digitized for further signal processing. The range of the ADC is again ideally matched to cover that voltage range. A signal processing stage performs the signal averaging of the multiple line scans and the base line fitting and removal of the background signal to obtain the absorption signal. Additionally, FIR lowpass filtering is performed in this step in order to decrease the total system ENBW after scan averaging.

3 Considered noise effects

To compare the two systems' performances and to evaluate the role of quantization noise, several noise effects are included in the simulation. The respective spectral distributions of the different noise effects have been considered for the full simulation bandwidth of $B = 500$ kHz. To model the noise effects in the simulation, random time series are generated for each noise process according to the respective noise power spectral density of the individual contribution. This way, laser intensity noise, which directly translates into noise in the photodiode current, photodiode shot noise and thermal amplifier noise of the TIA are included.

It should be mentioned that other common noise effects, such as etalons and background drifts, are omitted in this simulation. On the one hand, they are very specific to the respective setup and thus cannot be easily included in a general simulation. On the other hand, the main goal of this simulation is to evaluate the effects of the quantization noise and to enable the proper choice of necessary ADC resolution to ensure that no *additional* noise is acquired by wrong choice of resolution.

Including the noise effects and considering a setup without gas absorption, the expression for the photodiode current $i_{\mathrm{PD}}(t)$ of (7) is modified to

$$i_{\mathrm{PD}}(t) = R P_{\mathrm{opt}}(t) + R P_{\mathrm{RIN}}(t) + i_{\mathrm{shot}}(t) + i_{\mathrm{therm}}(t). \tag{9}$$

In this equation, $P_{\mathrm{opt}}(t)$ is the optical output power according to (5), $P_{\mathrm{RIN}}(t)$ is the fluctuation of the optical power due to intensity noise, $i_{\mathrm{shot}}(t)$ stands for the shot noise contribution and $i_{\mathrm{therm}}(t)$ considers the amplifier noise by means of an equivalent input noise current (see Sect. 3.3). Figure 3 shows the current density spectrum of the noise contributions to $i_{\mathrm{PD}}(t)$, that will be explained in more detail in the following sections.

3.1 Relative intensity noise

Relative intensity noise (RIN) describes intensity fluctuations of the laser output power, that occur even if constant

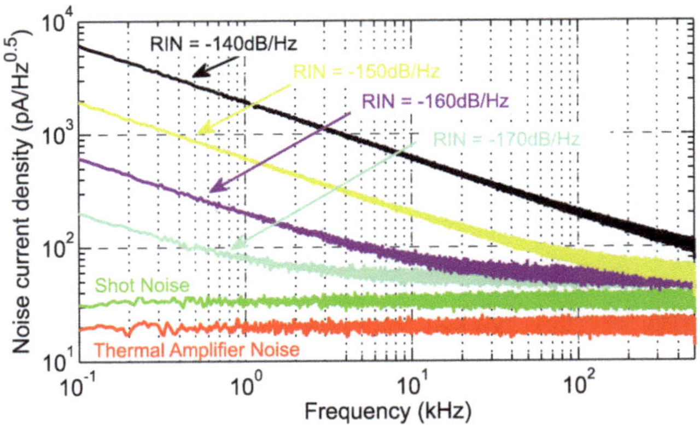

Fig. 3 Spectral noise current density of the photodiode current. Total noise including RIN, thermal and shot noise for four different levels of RIN and pure thermal noise and shot noise as a reference

injection current is applied. RIN is defined by relating the detected electrical noise power in a 1 Hz bandwidth to the DC signal power according to [20]:

$$\mathrm{RIN} = \frac{\langle \Delta i^2 \rangle}{I_{\mathrm{DC}}^2} \tag{10}$$

with $\langle \Delta i^2 \rangle$ as the average noise power in a 1 Hz bandwidth and I_{DC} as the average DC photocurrent. Since the photocurrent is proportional to optical power, (10) is also valid when optical power is used in place of the photocurrent.

In the considered frequency range up to several hundred kHz, the noise spectral power density of the RIN shows $1/f^{\alpha}$-like behavior for DFB-LDs [21] with α being a constant in the range of about 0.5 to 1.6 [22]. To include this noise contribution into the simulation with a realistic order of magnitude of the noise spectral power density, noise measurements on a DFB-LD (NEL NLK1556STB) have been performed. The DFB-LD was driven by a ILX 3724 current controller and the measurements yielded a RIN of -140 dB/Hz at a measurement frequency of 30 kHz and α close to 1, so a $1/f$ distribution has been used for the simulation, although any value for α could easily be included. It is worth noting that the current controller strongly contributed to RIN as was shown by a comparative noise measurement with the laser diode driven by a simple battery-powered current source. The result showed a noise level reduction of about 10 dB, resulting in RIN $= -150$ dB/Hz at the measurement frequency of 30 kHz.

Based on the measured RIN, the simulation generates a random noise time series $P_{\mathrm{RIN}}(t)$ with a $1/f$ spectral power distribution and random phase as described in [23]. $P_{\mathrm{RIN}}(t)$ is then added to the optical output power $P_{\mathrm{opt}}(t)$ as stated in (9). Since the measured RIN of a laserdiode strongly depends on the noisiness of the laser controller and on the laser operating current, further RIN-levels have been included in the simulation, to get the whole picture. The upper four curves in Fig. 3 show the contribution of RIN to the photocurrent noise for different RIN levels. As $1/f$ behavior is assumed for the RIN as defined by (10) using the electrical power ratio, the spectral distribution of the photodiode noise current itself will have $1/\sqrt{f}$ behavior, resulting in a change of a factor of 10 over two decades of frequency as is obvious from Fig. 3.

3.2 Photodetector shot noise

Shot noise is caused by the quantization of the electric current in electrons that arrive at random times at the pn-junction of the photodiode. The response of a photodiode to a constant optical input power P_0 is

$$i_{PD}(t) = i_0 + i_s(t), \tag{11}$$

where $i_0 = R \cdot P_0$ is the average photocurrent and $i_s(t)$ represents the random fluctuation due to shot noise. With the random process following Poisson statistics, the spectral shot noise current density $\bar{i}_s$ is frequency-independent and given by the average photocurrent i_0 and the electron charge q_e [24]:

$$\bar{i}_s = \sqrt{2q_e i_0}. \tag{12}$$

For the simulation, a time series $i_{shot}(t)$ with current spectral density $\bar{i}_s$ is calculated and added to the photocurrent. For the average optical power of $P_0 = 3.5$ mW, resulting in an average photocurrent of $i_0 = 3.5$ mA, the shot noise current density according to (12) is $\bar{i}_{shot} = 33.6\ \mathrm{pA}/\sqrt{\mathrm{Hz}}$. The respective noise spectrum of a generated random noise time series is shown in Fig. 3.

3.3 Thermal noise of the transimpedance amplifier

There are generally many noise mechanisms contributing to transimpedance amplifier (TIA) noise, depending on the exact circuitry of the amplifier itself and on the properties of the connected devices [25, 26]. The accumulated noise of a TIA can be specified by an equivalent input noise current density $\bar{i}_{TIA}$. This enables easy comparison of the amplifier noise with other noise contributions in the measurement system, for instance, the shot noise current at the output of the photodiode. Measurements of the equivalent input noise current density $\bar{i}_{TIA}$ on a Femto DLPCA200 low noise laboratory TIA for a transimpedance gain of $Z_{TIA} = 10^3$ V/A in a frequency range up to 100 kHz showed approximately constant noise current density. This led to the conclusion that thermal noise is dominant in our case and correspondingly the simulation models the amplifier noise as white noise with constant spectral noise current density with its value taken from the measurements:

$$\bar{i}_{therm} = 20.6 \frac{\mathrm{pA}}{\sqrt{\mathrm{Hz}}}. \tag{13}$$

The random time series $i_{therm}(t)$ with spectral noise density $\bar{i}_{therm}$ is added to the photodiode output current in (9).

3.4 Quantization noise

There are plenty of approaches available that model quantization noise as a further noise contribution with a specific spectral distribution, e.g., [27, 28]. However, in this simulation the quantization noise is simply included by emulating the most basic operation that any ADC does, which is the quantization of the input voltage $u_{TIA}(t)$. The sampling rate remains unchanged. For both, the WMS and the direct absorption scheme, the ADC range is adjusted ideally to the respective input voltage range and the N_{ADC} bit that the ADC offers as resolution are distributed to uniform quantization steps within this range.

4 Simulation results and discussion

4.1 Noise performance without quantization

As a validation for the simulation of both systems, the different noise reduction approaches of the WMS and the direct absorption setup have been simulated without quantization by the ADC and will be discussed in this section.

To assess the respective signal quality of the resulting $2f$- or direct absorption signal, a signal-to-noise ratio (SNR) is defined according to

$$\mathrm{SNR} = \frac{\max(Signal)}{3\sigma_{Noise}}. \tag{14}$$

Here *Signal* refers to the LIA output voltage for the WMS system and to the absorption signal after baseline removal for the direct absorption scheme, whereas σ_{Noise} is the standard deviation of the respective $2f$- or direct absorption signal, when no gas is present.

4.1.1 WMS setup

In the $2f$-WMS technique, the harmonic modulation shifts the detection frequency to the second harmonic of the modulation frequency f_{mod}. The lowpass filtering, that takes place in the LIA after the multiplication with the reference, is thus comparable to a bandpass filter centered around the reference frequency. The system noise bandwidth is given by the filter bandwidth of the LIA lowpass filter. The smallest bandwidth that can be realized depends on the ramp frequency f_{ramp}. After the lowpass filtering, the slowly varying $2f$-signal should not be distorted by the filtering process. If smaller bandwidth is desired, the scanning frequency must be reduced, so that the integration time of the LIA can be increased and the lowpass filtering becomes tighter. The

ENBW of the simulated filter that for $f_{ramp} = 1$ Hz fulfills this condition is 40 Hz. That means that for white noise at the filter input, the filter will let through the same noise power as an ideal rectangular (boxcar) filter with a 40 Hz bandwidth. However, due to the noise contribution of the RIN, the spectral noise power density is not constant over frequency. As a result, a different amount of noise will pass the system, depending on whether the filtering with the 40 Hz bandwidth takes place in the baseband interval from DC to 40 Hz or in a 40 Hz interval at a higher frequency.

Consequently, the effect of increasing the modulation frequency in the $2f$-setup can be thought of the 40 Hz detection bandwidth being moved along the frequency axis in Fig. 3 towards higher frequencies. Accordingly, the SNR of the measurement will improve with increasing frequency as the noise power density decays.

Figure 4 shows the simulation results of the SNR over modulation frequency f_{mod} for different levels of RIN. The bottom graph for RIN $= -140$ dB/Hz shows an improvement of SNR with increasing modulation frequency as one would expect. Since for this level of intensity noise, RIN dominates the other noise contributions (Fig. 3), a constant improvement of 10 dB per frequency decade can be observed. The insets on the right of Fig. 4 show the respective $2f$-lineshapes at the SNRs indicated in the graph on the left by (a), (b) and (c). For the graph in the middle, that results from a RIN of -150 dB/Hz, the same behavior can be observed with a slight flattening for higher frequencies. For the top graph (RIN of -160 dB/Hz), this flattening already occurs at a modulation frequency of 20 kHz.

The reason for this flattening gets obvious from Fig. 3, which shows that for a RIN of -160 dB/Hz the $1/f$ contribution of the noise current spectral density reaches the

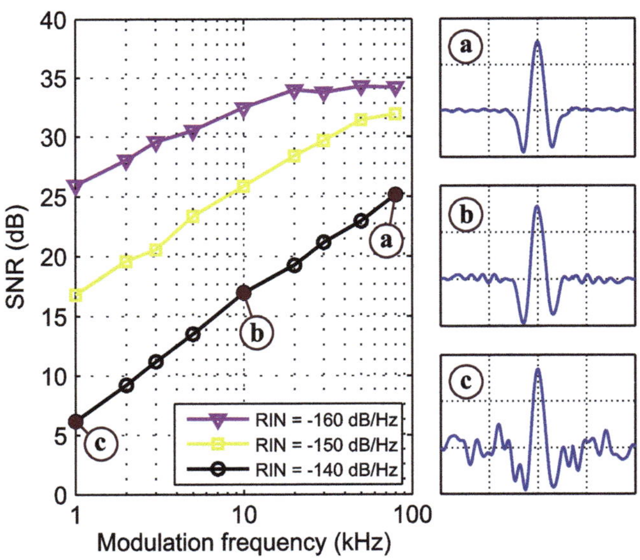

Fig. 4 SNR improvement with increasing modulation frequency for the $2f$-WMS simulation. The different curves apply for different levels of RIN. *Insets* on the right show $2f$-lineshapes for the indicated SNRs

constant noise floor caused by thermal and shot noise in this frequency range. With a modulation at 20 kHz, the $2f$-detection takes place around 40 kHz. Accordingly, even shifting the detection to higher frequencies does not reduce the noise power that passes the system anymore and thus the SNR does not improve any further. For the WMS technique, this situation corresponds to the best possible noise suppression. Further improvement is then only possible by narrowing the detection bandwidth or by additionally applying signal averaging.

4.1.2 Direct absorption setup

As the direct absorption setup does not include the modulation and lock-in scheme, the detection cannot be shifted away from frequencies with high RIN in the same way as for the WMS approach. Instead, averaging as another way of narrowing the detection bandwidth is used. Averaging of subsequent scans will act as a filter with a comb-like transmission spectrum with passbands around the harmonics of f_{ramp}. Thus the filter blocks all components that are not periodic with the scan frequency [29]. The width of these passbands is determined by the number of scans that are used for averaging. For averaging a number N_{av} of scans, the equivalent noise bandwidth of a system is reduced by a factor of N_{av} leading to an SNR improvement by a factor of $\sqrt{N_{av}}$. For instance, the filter effect of averaging $N_{av} = 400$ scans at a scanning frequency of $f_{ramp} = 200$ Hz will result in 0.5 Hz wide passbands around each of the 200 Hz harmonics. Depending on the scan rate f_{ramp} it is sensible to limit the system bandwidth by lowpass filtering, since not all the harmonics of the scanning frequency up to the maximum system bandwidth (in our case the simulation bandwidth $B = 500$ kHz) are needed to determine the absorption signal. The number of harmonics to be included, however, is not a general rule but depends on the width of the scanned line and on the scanned wavelength interval. Different settings for the simulation showed, that in our case it is enough to include about 30 harmonics of the ramp frequency to still achieve correct results.

That means, that if the scanning frequency f_{ramp} is increased, the system bandwidth, as limited by the additional lowpass filter, must be increased, too, so as to match the requirement of including enough harmonics of the ramp frequency. In turn, when using a higher scan rate, more scans in the same measurement time can be performed, and thus the number of averaged scans N_{av} is increased which will narrow the noise bandwidth of the system as described above.

If the spectral distribution of the system noise would be constant, as it is for white noise, it would make no difference, whether the noise bandwidth is decreased by a lowpass filter or by averaging. Looking at the simulated system with the initial noise bandwidth of $B = 500$ kHz, let

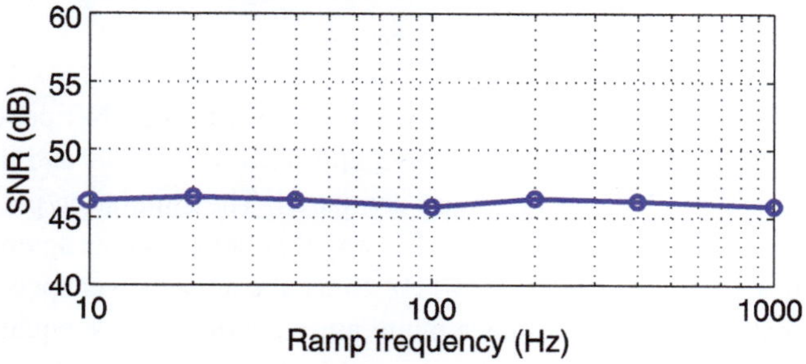

Fig. 5 SNR for different scan rates and averaging off all scans acquired in 1 s combined with FIR lowpass filtering with adjusted cutoff frequencies, if only shot and amplifier noise is present

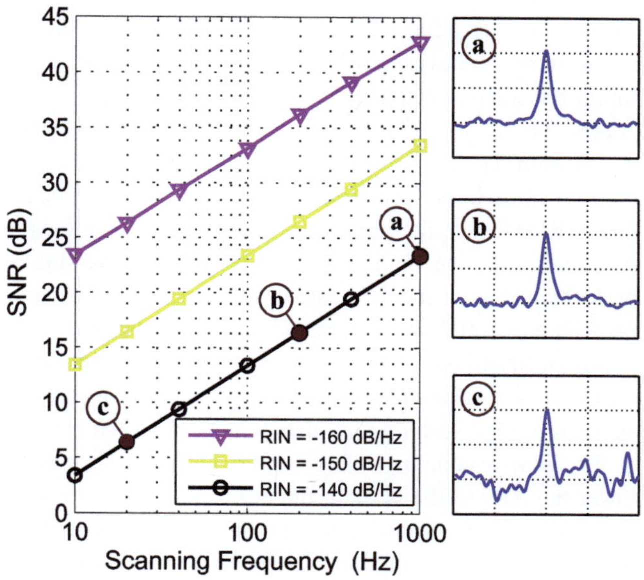

Fig. 6 SNR improvement with increasing scanning frequency/number of scan averages for the direct absorption simulation. For each frequency, all ramps acquired during $T_{meas} = 1$ s have been averaged. The different curves apply for different levels of RIN. *Insets* on the right show absorbance lineshapes for the indicated SNRs

us assume we limit the measurement time to $T_{meas} = 1$ s and we use a slow scan rate of $f_{ramp} = 10$ Hz, yielding $N_{av} = 10$ scans to be averaged. Since we only need a limited number of harmonics of the ramp frequency, we may limit the system bandwidth by using lowpass filtering. For this scanning frequency, we choose an FIR lowpass filter with a cutoff frequency of 270 Hz and a corresponding ENBW of 400 Hz. Averaging of the 10 scans will further reduce the noise bandwidth by the factor of $N_{av} = 10$, resulting in a total equivalent noise bandwidth $\text{ENBW}_{tot} = 40$ Hz. Let us now assume, we instead use a 20 times faster scan rate of $f_{ramp} = 200$ Hz. Then we have to adjust the FIR lowpass filter cutoff frequency by the same factor to include enough harmonics. With the resulting cutoff frequency of 20×270 Hz $= 5.4$ kHz, the corresponding ENBW increases to 8 kHz. In turn, we can now average $N_{av} = 200$ scans, that we collected during the measurement time of 1 s, which again brings us down to $\text{ENBW}_{tot} = 40$ Hz. Thus, if only white noise were present, exactly the same noise power would pass the signal processing, as when using slow scanning. Simulation results for this somewhat unrealistic situation are shown in Fig. 5, where only shot noise and amplifier noise were considered. The cutoff frequency of the FIR filter has been adjusted for each scanning frequency as described above. It is obvious that the achieved SNR does not depend on the scanning frequency, in this case.

However, with the large contribution of RIN to the system noise, the spectral noise distribution is far from being constant, but is dominated by the $1/f$ characteristic of RIN. In this case, the mere consideration of ENBW is not sufficient, as now it is important in which spectral region noise passes the filtering process. The first passband of the comb-like transmission spectrum of the averaging process occurs around the scanning frequency f_{ramp}. Increasing f_{ramp} means that the passband is moved to a higher frequency at the same time. There is now an advantage compared to the white noise case, as RIN will be lower at this higher frequency and we can average a larger number of scans due to the faster scanning.

Figure 6 shows the simulation results for the SNR achieved with different scanning frequencies f_{ramp} for different levels of RIN. Again, the system noise bandwidth has been limited by FIR lowpass filtering with the cutoff frequencies matched to the respective scanning frequency in order to include the same number of harmonics of f_{ramp} each time. For each scanning frequency, all scans taken during a measurement time of $T_{meas} = 1$ s have been averaged. In contrast to Fig. 5, higher scanning frequency now leads to an improvement of SNR. As RIN has been modeled with a $1/f$ spectral noise distribution as described in Sect. 3.1, the observable improvement of SNR is 10 dB per decade of the scanning frequency. Figure 6 includes curves for RIN between -140 dB/Hz and -160 dB/Hz. For all these RIN levels, the $1/f$ spectral characteristic dominates in the frequency range of the considered scanning frequencies (compare Fig. 3). A further reduction of RIN would lead towards the situation when white noise is dominant and thus approach the case of Fig. 5.

4.1.3 Comparison of WMS and direct absorption

A general comparison of the two variations of TDLAS remains difficult, due to the differences in the applied techniques. A universal statement on which approach is better is hard to give. For a given system, it will depend on the noise and especially on its spectral distribution, whether a $2f$-approach or a direct measurement will be more promising.

Nevertheless, a few issues should be commented based on the simulation results. When comparing Fig. 4 and 6, one can find that for a RIN level of -140 dB/Hz, the direct system reaches an SNR of about 23.5 dB for the maximum considered scanning frequency (and number of averages) of $f_{\text{ramp}} = 1000$ Hz. To reach similar SNR in the wavelength modulation scheme, a modulation frequency of f_{mod} of 50 kHz is needed, as visible from Fig. 4. For reduced RIN levels, the required scanning frequency for the direct absorption technique that is needed to achieve similar performance as in the $2f$-system, decreases. For instance, for RIN $= -160$ dB/Hz, the wavelength modulation system yields an SNR just below 35 dB, even for high modulation frequencies, a value that is exceeded by the direct absorption simulation for a scanning frequency of $f_{\text{ramp}} = 200$ Hz.

The general conclusion that may be drawn here is that the lower the RIN of the system, the more it is worth thinking about using direct absorption measurements. Or in reverse, if direct absorption techniques are applied, it is worth making an effort to achieve the lowest possible RIN.

Finally, it should be pointed out, that although Fig. 6 suggests a steady increase in SNR for increasing scanning frequency, in real systems there are limits to the choice of the scanning frequency f_{ramp}, that are currently not modeled by the simulation. One issue is the scanning frequency dependent tuning behavior of DFB-lasers. With increasing scanning frequency f_{ramp}, the wavelength tuning of the laser diode decreases [7, 21], which at some point will make it difficult to scan over the complete absorption line. This limitation of tunability holds for the harmonic modulation in the $2f$-approach in a similar fashion, since there will be a maximum modulation frequency f_{mod}, for that the required modulation index m can be achieved (compare Sect. 2.3).

Further constraints may be given by timing jitter that may make averaging of the scans challenging at high scan rates or by the minimum required number of samples per scan for an accurate determination of gas concentrations which will be limited by the maximum available sample rate of the analog-to-digital conversion.

4.2 Impact of quantization noise

To assess the impact of quantization noise to the SNR performance of both measurement approaches, a WMS and a direct absorption system are compared. The simulation parameters were chosen in such a way that both systems again have the same ENBW $= 40$ Hz. Since the noise power spectral density does not have white noise characteristics, equal ENBW does not necessarily make the systems comparable. For this reason, the remaining simulation parameters have been adjusted in such a fashion, that similar SNR is achieved for both systems for an intensity noise of RIN $= -140$ dB/Hz, when no quantization by the ADC is considered. Thus for the WMS system, a modulation frequency of 12 kHz has been chosen, yielding an SNR of 19.2 dB. For the direct absorption system, the scanning frequency was set to $f_{\text{ramp}} = 320$ Hz with signal averaging performed over 320 scans resulting in a comparable SNR value of 18.8 dB. The simulated measurement time of both systems is 1 s.

Starting from this initial configuration, quantization with a varying ADC resolution N_{ADC} has been simulated. As in the previous simulations, different levels of RIN have been included. Figures 7 and 8 show the simulation results for the SNR as a function of the ADC resolution N_{ADC} for the $2f$-WMS and for the direct absorption system, respectively.

When considering the case of RIN $= -140$ dB/Hz (bottom curve), the results for both systems are similar in the respect that no significant improvement of SNR is achieved when the resolution is increased above 11 bit. This result strikes as surprising, especially for the direct absorption approach. A quantization with 11 bit corresponds to a smallest normalized resolvable value of $1/2^{11} = 4.88 \cdot 10^{-4}$ so that it should hardly be possible to measure even the peak absorption of $A_{\max} = 4.6 \cdot 10^{-5}$.

The explanation is that the direct measurement system, in fact, implies three signal processing steps that in combination lead to an increase of the effective ADC resolu-

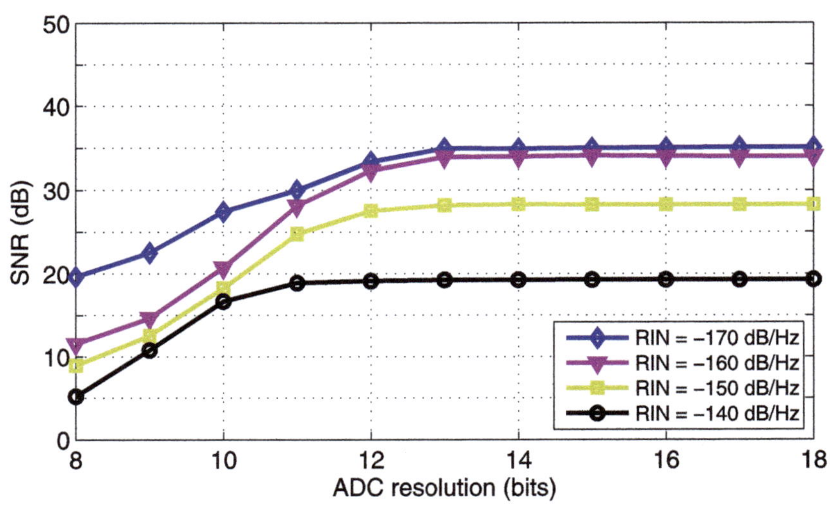

Fig. 7 Influence of ADC resolution to the SNR for the $2f$-WMS system. Different curves apply for different levels of RIN

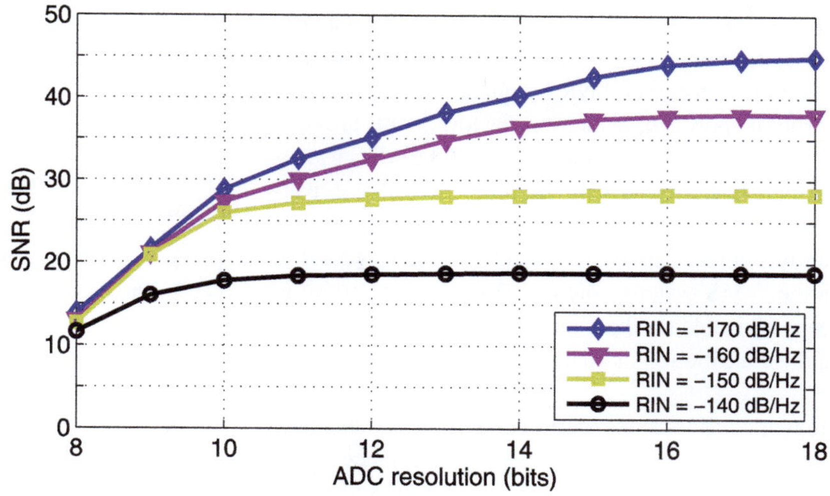

Fig. 8 Influence of ADC resolution to the SNR for the direct absorption system. Different curves apply for different levels of RIN

tion, namely oversampling, dithering and averaging [30, 31]. The intensity noise present in the system dithers the voltage that is sampled by the ADC, thus randomly toggling the least significant bit (LSB). In combination with the averaging procedure, which can be considered a kind of filtering, an increased effective resolution is achieved, making the absorption signal measurable. Thus, even with a resolution of 11 bit, an SNR of 18 dB is achieved (Fig. 8, bottom curve). Considering the peak absorption of $A_{\max} = 4.6 \cdot 10^{-5}$, this value corresponds to a detection limit of about $5.8 \cdot 10^{-6}$ with SNR $= 1$. Although this detection limit is unlikely to be achieved with a real setup, the reason for that is not the quantization error of the ADC but rather additional perturbations by etalons and drifts, which were not included in the simulation.

Due to the dithering effect of RIN, in this case the impact of quantization noise is low. The situation changes slightly for lower RIN when the effect of quantization is no longer buried in intensity noise and RIN is not strong enough to toggle the LSB. For the $2f$-setup (Fig. 7) even for a RIN level as low as -170 dB/Hz, a resolution of 13 bit results in an SNR value that is as good as for ideal quantization. For the direct absorption system, in contrast, the required resolution depends on the present RIN level. While for RIN $= -150$ dB/Hz 13 bits are enough to avoid SNR degradation, higher resolutions are required for lower RIN. For the case of RIN $= -170$ dB/Hz (Fig. 8, top curve) an improvement of SNR is achieved even when increasing resolution from 16 bit to 17 bit. These somewhat stricter requirements for the resolution for the direct absorption setup are in agreement with the idea that the full background signal must be quantized, as explained in Sect. 2.

A further issue that can be found from the comparison of Fig. 7 and 8 is that for very low RIN of -170 dB/Hz the SNR that can be achieved in the direct absorption technique is higher than that for WMS, if sufficient ADC resolution is provided. This difference is comprehensible, as for such low levels of RIN the total spectral noise distribution approaches that of white noise as visible from Fig. 3. Since the two simulated systems have been designed to possess equal equivalent noise bandwidth, their output will now contain equal noise power. With the direct absorption signal being stronger than the $2f$-signal and equal noise for both systems, the SNR is larger for the direct absorption scheme. In such a system, the advantage of the WMS technique to escape intensity noise lapses and a direct measurement appears more appropriate. Intensity noise levels as low as that, however, are far from the normal case, but can be achieved by balanced detection schemes for noise canceling [32], making very sensitive absorption measurements without wavelength modulation feasible [33].

Finally, it should be mentioned that the SNRs shown in Fig. 7 and 8 are very unlikely to be reached in any real system. With additional disturbances such as background drift and etalons, the achievable SNRs will be significantly lower. Nevertheless, as the goal of this investigation is to avoid *additional* noise by the quantization, the simulation results serve as a way to assess the impact of quantization noise.

5 Conclusion

In conclusion, a simulation of WMS and direct absorption TDLAS systems has been developed. With the inclusion of laser relative intensity noise, thermal amplifier noise and photodetector shot noise, the simulation illustrates how noise is reduced by different approaches in the two systems. For a given measurement scenario and known noise characteristics, the simulation helps determining whether a WMS or direct absorption measurement approach is more promising. By considering quantization of the measurement signal by an ADC, it was shown for the considered systems that as long as RIN is the dominating noise effect, the requirements for the ADC resolution in $2f$-WMS and direct absorption spectroscopy setups are similar. Then comparatively low resolutions of 11 bit or 12 bit suffice in order to not additionally deteriorate SNR by quantization. In the direct absorption approach, resolutions as low as this are only possible due to a resolution enhancement of the ADC by signal dithering caused by RIN in combination with the averaging of several scans. For lower RIN levels, SNR does not significantly improve for ADC resolutions above 13 bit to 14 bit in the WMS system, whereas resolutions of 16 bit to 18 bit seem advisable for the direct absorption setup to make sure that the contribution of quantization noise remains negligible.

Future work will look at experimental comparisons of the two approaches in order to assess their performance in the presence of background or etalon noise. Furthermore, the simulation may be extended to include chemometric algorithms in order to compare the performance on the basis of the variance of the determined gas concentrations.

Acknowledgement The authors gratefully acknowledge funding of the Erlangen Graduate School in Advanced Optical Technologies (SAOT) by the German Research Foundation (DFG) in the framework of the German excellence initiative.

References

1. M.W. Sigrist, R. Barlome, D. Marinov, J.M. Rey, D.E. Vogler, H. Wächter, Appl. Phys. B **90**, 289 (2008)
2. P.A. Martin, Chem. Soc. Rev. **31**, 201 (2002)
3. H. Teichert, T. Fernholz, V. Ebert, Appl. Opt. **42**, 2034 (2003)
4. M.R. McCurdy, Y. Bakhirkin, G. Wysocki, R. Lecicki, F.K. Tittel, J. Breath Res. **1**, 014001 (2007)
5. L. Gianfrani, G. Gagliardi, M. van Burgel, E.R.Th. Kerstel, Opt. Express **11**, 1566 (2003)
6. M. Andersson, L. Persson, T. Svensson, S. Svanberg, Rev. Sci. Instrum. **78**, 113107 (2007)
7. S. Hunsmann, K. Wunderle, S. Wagner, U. Rascher, U. Schurr, V. Ebert, Appl. Phys. B **92**, 393 (2008)
8. T. Fernholz, H. Teichert, V. Ebert, Appl. Phys. B **75**, 229 (2002)
9. T. Svensson, M. Andersson, L. Rippe, S. Svanberg, S. Andersson-Engels, J. Johansson, S. Folestad, Appl. Phys. B **90**, 345 (2008)
10. L. Schuchman, IEEE Trans. Instrum. Meas. **12**, 162 (1964)
11. P. Carbone, D. Petri, IEEE Trans. Instrum. Meas. **43**, 389 (1994)
12. M.E. Webber, D.S. Baer, R.K. Hanson, Appl. Opt. **40**, 2031 (2001)
13. R. Arndt, J. Appl. Phys. **36**, 2522 (1965)
14. P. Kluczynski, J. Gustafsson, A.M. Lindberg, O. Axner, Spectrochim. Acta A **56**, 1277 (2001)
15. R. Engelbrecht, Spectrochim. Acta A **60**, 3291 (2004)
16. U. Platt, J. Stutz, *Differential Optical Absorption Spectroscopy: Principles and Applications* (Springer, Berlin, 2007)
17. J.J. Olivero, R.L. Longbothum, J. Quant. Spectrosc. Radiat. Transfer **17**, 233 (1977)
18. M. Johnson, *Photodetection and Measurement* (McGraw-Hill, New York, 2003)
19. W.C. Van Etten, *Introduction to Random Signals and Noise* (Wiley, Chichester, 2005)
20. D. Derickson, *Fiber Optic Test and Measurement* (Prentice Hall, Upper Saddle River, 1998)
21. K. Petermann, *Laser Diode Modulation and Noise* (Kluwer Academic, Dordrecht, 1991)
22. E. Milotti, arXiv:physics/0204033 (2002)
23. J. Timmer, M. König, Astron. Astrophys. **300**, 707 (1995)
24. G.P. Agrawal, *Fiber-Optic Communication Systems* (Wiley, New York, 2002)
25. G.B. Clayton, S. Winder, G. Clayton, *Operational Amplifiers* (Newnes, Oxford, 2003)
26. S. Franco, *Design with Operational Amplifiers and Analog Integrated Circuits* (McGraw-Hill, New York, 2002)
27. J. Candy, O. Benjamin, IEEE Trans. Commun. **29**, 1316 (1981)
28. K. Hejn, A. Pacut, IEEE Trans. Instrum. Meas. **45**, 41 (1996)
29. P.C.D. Hobbs, *Building Electro-Optical Systems: Making it All Work* (Wiley, New York, 2000)
30. J. Potzick, Rev. Sci. Instrum. **70**, 2038 (1999)
31. L. Krause, Measurement **39**, 681 (2006)
32. P.C.D. Hobbs, Appl. Opt. **36**, 903 (1997)
33. R. Claps, F.V. Englich, D.P. Leleux, D. Richter, F.K. Tittel, R.F. Curl, Appl. Opt. **40**, 4387 (2001)

Temperature and current coefficients of lasing wavelength in tunable diode laser spectroscopy

M. Fukuda · T. Mishima · N. Nakayama · T. Masuda

Received: 11 September 2009 / Revised version: 1 December 2009 / Published online: 14 January 2010
© Springer-Verlag 2010

Abstract The factors determining temperature and current coefficients of lasing wavelength are investigated and discussed under monitoring CO_2-gas absorption spectra. The diffusion rate of Joule heating at the active layer to the surrounding region is observed by monitoring the change in the junction voltage, which is a function of temperature and the wavelength (frequency) deviation under sinusoidal current modulation. Based on the experimental results, the time interval of monitoring the wavelength after changing the ambient temperature or injected current (scanning rate) has to be constant at least to eliminate the monitoring error induced by the deviation of lasing wavelength, though the temperature and current coefficients of lasing wavelength differ with the rate.

1 Introduction

Optical sensing using tunable diode laser spectroscopy (TDLS) has been widely used in various engineering fields and researched for new users ranging from micro-scale to large-scale applications [1]. For example, optical sensing is being used for bio-sensing, environmental monitoring, etc. Laser diodes are key devices for those applications and are indispensable for building sensing equipment. Edge- and surface-emitting laser diodes are used for those purposes. Edge-emitting types with single lasing modes, such as distributed feedback (DFB) lasers, have been conventionally used in sensing equipment because of their high power and single lasing modes. Surface-emitting types are also widely used in sensing systems because they are easy to handle.

The lasing wavelength of these laser diodes is ordinarily controlled by varying the ambient temperature and injected current in TDLS. These variables corresponding to bandgap change due to ambient temperature and the band-filling effect induced by carrier injected into the active layer of the lasers. In addition, refractive index change of the active layer, which is induced by temperature and injected carrier density, takes the important role of changing the lasing wavelength. The changing rate of these physical properties determines the conventionally used temperature- and current-coefficient of lasing wavelength. However, the correlation between the physical properties and the temperature- and current-coefficient has not been discussed in detail. This paper discusses the physical mechanisms of temperature and current control of the lasing wavelength for TDLS.

2 Samples

Several kinds of edge-emitting InGaAsP/InP Fabry–Perot and DFB laser diodes lasing at 1300, 1550, and 2006 nm were used in this study. The active layers of these laser diodes were composed of bulk, quantum well, and strained quantum well grown by metal organic vapor phase epitaxial (MOVPE) method. The cavity length of these lasers was about 0.3 mm and was defined by cleavage from wafers. After the cleavage, the front and rear facets were respectively coated with an anti- and high-reflecting dielectric

M. Fukuda (✉) · T. Mishima · N. Nakayama
Toyohashi University of Technology, 1-1, Hibarigaoka, Tempaku-cho, Toyohashi, Aichi 441-8580, Japan
e-mail: fukuda_mitsuo@eee.tut.ac.jp

T. Masuda
Optoelectronic Industry and Technology Development Association (OITDA), Sumitomo Edogawabashiekimae Bldg., 7F, 1-20-10, Sekiguchi, Bunkyo, Tokyo 112-0014, Japan

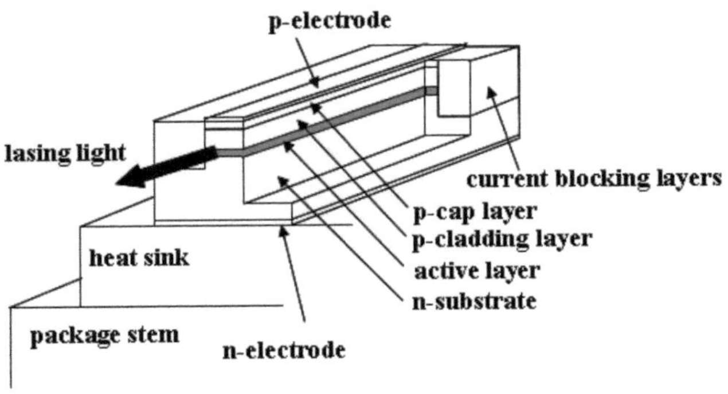

Fig. 1 Schematic diagram of a buried heterostructure laser diode used in this experiment and its mounting configuration

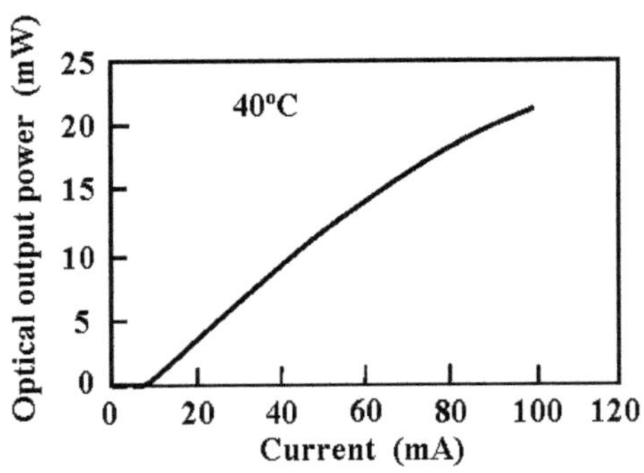

Fig. 2 Typical current–light output power characteristics of 2000-nm-band DFB laser diode

film. These laser diodes were mounted in a junction-up or junction-down configuration on silicon or ceramic heat sinks as shown in Fig. 1 and then installed in TO-can packages.

Typical current-light output power characteristics are shown in Fig. 2 for a 2000-nm-band laser diode. The threshold current was about 6 mA. The output power was more than about 20 mW at an injected current of 100 mA and at temperatures below 40°C. The 2000-nm-band as well as the other-band laser diodes showed stable operation with a single longitudinal mode at temperatures ranging from below 0 to above 60°C. The side-mode suppression ratio (SMSR) of all the DFB laser diodes used were more than 40 dB for 1300- and 1550-nm-band laser diodes and more than 25 dB for 2000-nm-band laser diodes at 10 mW and 25°C. Figure 3 shows an example of a lasing spectrum of 2000-nm-band laser diodes at 40°C. A SMSR of more than 25 dB was maintained even at 40°C. The lasing spectral linewidth is not a critical characteristic for spectroscopy and 1300- and 1550-nm-band laser diodes used showed sufficiently narrow linewidth, for example, less than a few MHz at 10 mW for 1550-nm-band laser diodes. This value corresponds to less than 0.1 pm in the 1550-nm wavelength band.

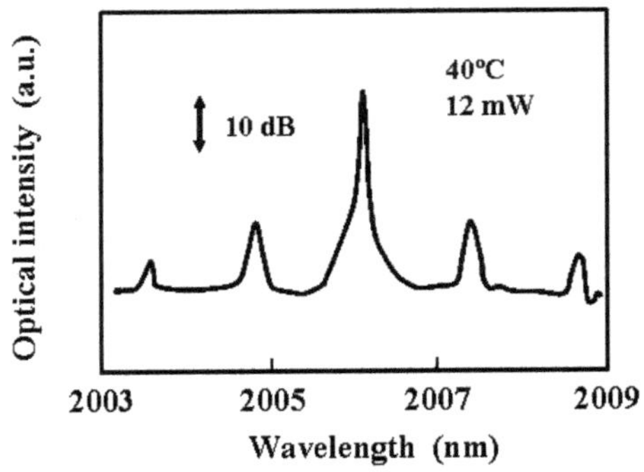

Fig. 3 Typical spectral characteristics of 2000-nm-band DFB laser diode at 40°C

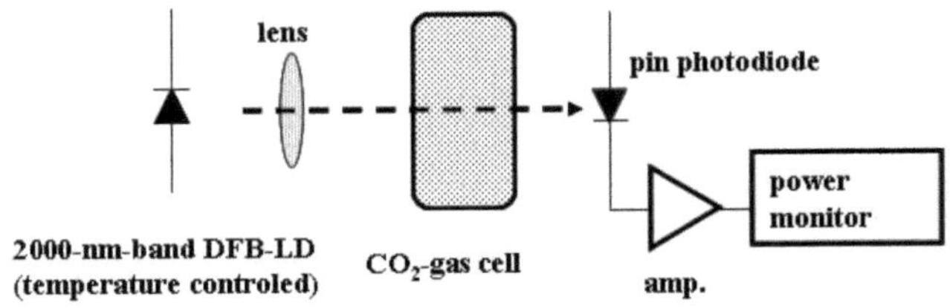

Fig. 4 Schematic diagram of the experimental setup

3 Experimental setup

3.1 Lasing wavelength scanning system

To analyze the influence of the ambient temperature and injected current on the lasing wavelength of laser diodes, the temperature and current coefficient of lasing wavelength were monitored with the system shown in Fig. 4. The optical output power from the laser diode was reshaped into a parallel beam with a lens and passed through a 4-cm-thick gas cell filled with CO_2 gas at atmospheric pressure and then received by a pin photodiode or an optical spectrum analyzer to monitor the wavelength spectra. The lasing wavelength was automatically scanned by controlling the temperature of the laser diode with a thermoelectric cooler or the injected current of laser diode. Here, the change rates of the temperature and current were controlled with a computer.

3.2 Junction temperature monitoring system

As described in the next section, the change in lasing wavelength by controlling ambient temperature or injected current mainly resulted from the temperature change in the active layer of the lasers. To analyze the thermal conductance of each sample, a junction temperature monitoring system was used [2]. The concept of the pn-junction-temperature estimation is depicted in Fig. 5. This method has often been used in the estimation of thermal resistance of diodes and transistors. The active layer temperature (or pn-junction temperature), T_j can be expressed with the equation

$$T_j = T_a + R_{th} I_F V_F, \quad (1)$$

where T_a is the ambient temperature, R_{th} is the thermal resistance in units of °C/W, I_F and V_F are the operating current and voltage, respectively. If operating current is quickly changed from I_F to a small current, I_f, at which Joule heating can be ignored, the junction voltage is lowered by ΔV_f, which is determined by Joule heating due to the operating current, I_F, and then reaches a constant value under the small current, I_f (see Fig. 5). The monitoring delay time, t_d should be small to eliminate monitoring error. Ideally, ΔV_f is obtained by $V_f - V_M$ with $t_d = 0$. Consequently, the temperature dependence of the pn-junction voltage is estimated from the voltage change, ΔV_f. If the temperature dependence of the junction voltage is expressed with $K = -dV_f/dT_a$, the temperature change, ΔT_j can be given by

$$\Delta T_j = T_j - T_a = \Delta V_f / K, \tag{2}$$

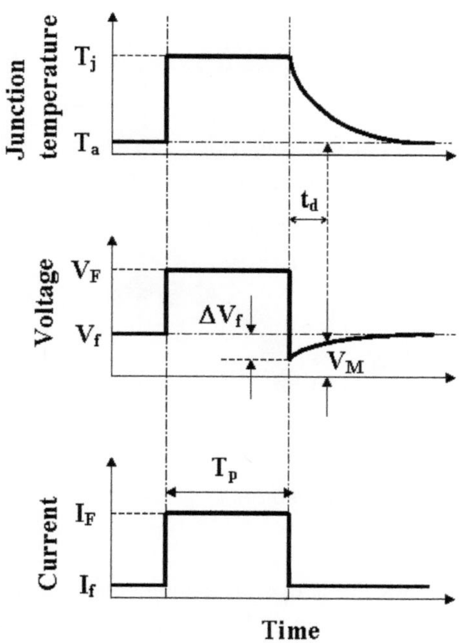

Fig. 5 Concept of active-layer- (pn-junction-) temperature estimation method

where K can be calculated as the slope of the junction-voltage vs. ambient temperature under very small forward current injection which scarcely influences the temperature rise of the junction. The behavior of the active layer temperature can be estimated with this method. If the width of pulse current, T_p increases, the heat generated at the active layer diffuses to the laser chip, heat sink, package stem, package, etc. in response to the thermal conductance of each part. This behavior can be monitored with the systems shown in Fig. 5.

4 Lasing wavelength

Several factors govern the change in lasing wavelength of laser diodes as shown in Fig. 6. A decrease (an increase) in the refractive index of the active region originates from an increase (a decrease) in threshold carrier density and shortens (lengthens) the lasing wavelength of each Fabry–Pelot (FP)-mode in FP-lasers. This phenomenon is induced by the plasma effect related to carrier density in semiconductors [3]. In DFB lasers, the lasing mode is shortened (lengthened) with a decrease (an increase) in effective grating pitch introduced by the decrease (increase) in the refractive index. The increase (decrease) in the refractive index is introduced by rising (lowering) temperature. In addition, the rising (lowering) temperature shifts the envelope of FP-modes (gain envelope) to the longer (shorter) range. This is due to a reduction (an increase) of the bandgap energy.

Before lasing, the peak wavelength of FP-modes shortens due to the band-filling effect, and that of DFB-mode also shortens as the injected carrier density increases through the refractive index reduction. After lasing, the main factor is the thermal effect because threshold carrier density is fixed at the threshold value after lasing. Joule heating is generated and light output power changes in response to injected current under the constant carrier density.

These are basic mechanisms for changing lasing wavelength in laser diodes. Among them, the change in lasing

Fig. 6 Basic mechanisms of lasing wavelength variation in laser diodes

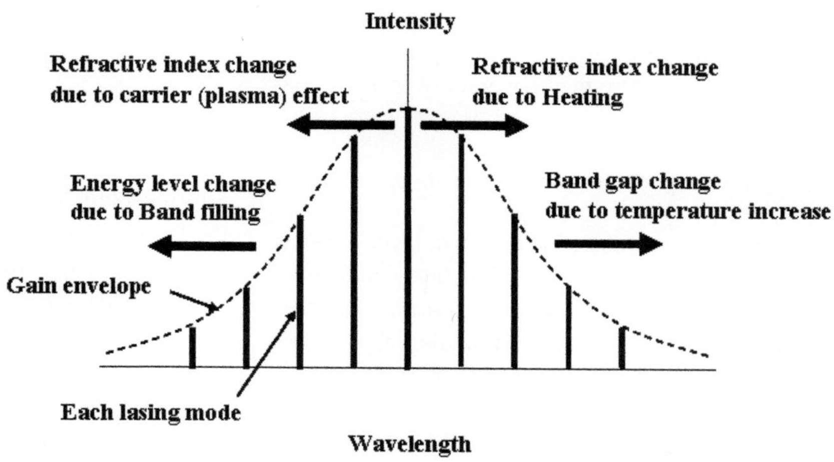

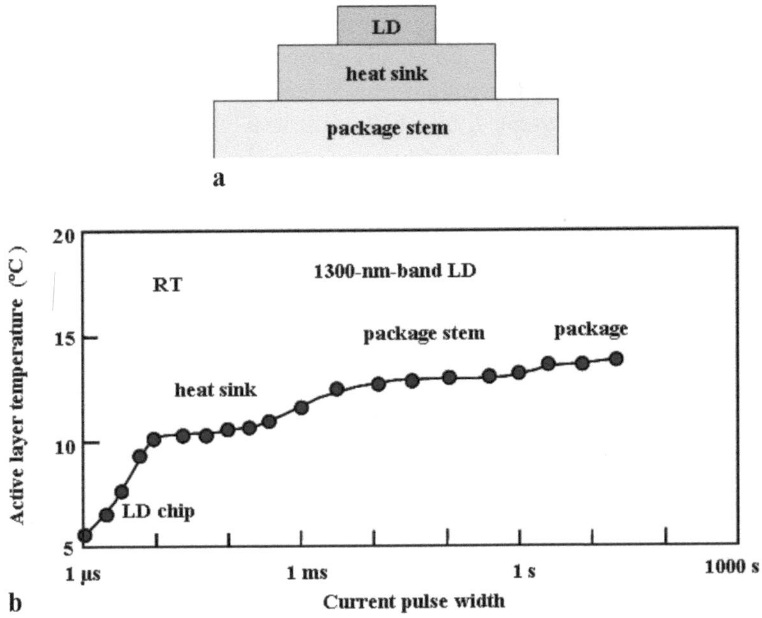

Fig. 7 (**a**) Sample configuration and (**b**) the estimated temperature rise in active layer as a function of pulse width. The sample is a 1300-nm-band FP laser diode. The labels of LD chip, heat sink, package stem, and package in (**b**) indicate the responsible parts of heat conduction for the heat generated at the active layer. These are estimated by the time constant of the heat conduction

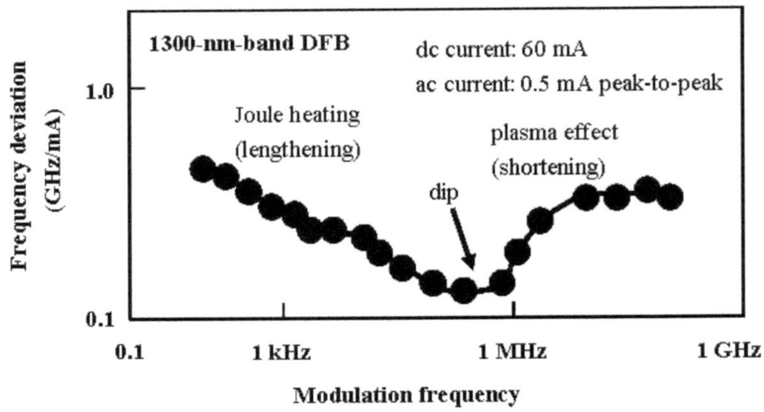

Fig. 8 Lasing frequency (wavelength) deviation for a 1300-nm-band DFB laser diode as a function of frequency. The modulation current was a 0.5-mA peak-to-peak sinusoidal wave and the DC bias was set at 60 mA

wavelength by controlling ambient temperature under a constant current is mainly generated by bandgap change in FP-lasers and refractive index change in DFB lasers. Controlling the lasing wavelength with the magnitude of injected current also occurs by the bandgap change due to Joule heating at the active layer (or pn-junction) because the injected carrier density is nearly constant after lasing. The temperature- and current-coefficient of lasing wavelength is analyzed in DFB lasers from the viewpoint of thermal conductivity in the following sections.

4.1 Wavelength change due to current injection

The rate of temperature change in the active layer depends on a transient phenomenon determined by heat conduction. The Joule heating generated at the active layer gradually diffuses from the active layer to the surrounding region, and thus the change rate in lasing wavelength strongly depend on the mounting configuration and packaging structure. This behavior was analyzed with the system indicated in Fig. 5. Figure 7 shows the active layer temperature increase as a function of the current pulse width for a 1300-nm-band FP laser diode. The pulse current, I_F, and the monitoring current, I_f, were set at 100 and 1 mA, respectively. The temperature of each point in the figure was calculated by $V_f - V_M$ with $t_d = 1$ μs. The temperature coefficient of the junction voltage, K, was about 1 mV/°C. The Joule heating due to current injection diffused within the laser chip and then towards the outside of the active layer, heat sink, package stem, package, and equipment, as pulse width was widened. This heat conduction transient phenomenon governs the temperature of the active layer and is influenced by the laser-chip mounting configuration (configuration of the heat-conducting path).

These behaviors are closely related to the rate and range of wavelength change under current modulation. In Fig. 8, the horizontal axis indicates modulation frequency and the vertical axis corresponds to the frequency deviation, which is equal to the wavelength variation. As modulation fre-

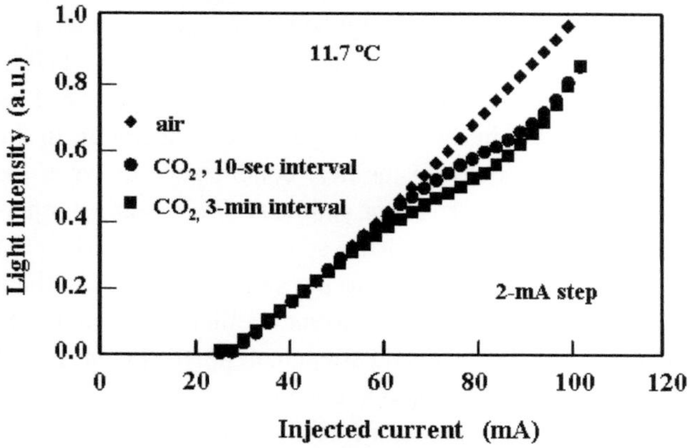

Fig. 9 Typical current-light output power relation under CO_2-gas absorption. The ambient temperature was set at the wavelength just before a CO_2-gas-absorption peak. The current was increased with a 2-mA step

quency increases from 100 Hz, the frequency deviation decreases because the response to heat conduction is gradually small. This behavior is also recognized in Fig. 7, in which the current pulse width corresponds to the modulation frequency of the laser diode from the viewpoint of heat conduction. A dip appears after 100 kHz in Fig. 8. After the dip, the plasma effect is dominant and the lasing wavelength tends to be shortened (blue shift). This frequency deviation is called FM-response or chirping in the optical fiber communication field [4]. The frequency range used for TDLS is below the dip frequency and the frequency at which the influence of heat is dominant (red shift).

4.2 Wavelength change due to ambient temperature

Heat is inversely transmitted from the ambient to the active layer of the laser diode when the ambient temperature or package temperature is changed. The heat conductance of the package, package stem, and heat sink is the same for the case of the diffusion of Joule heating at the active layer, and a certain time interval is needed until the temperature of the active layer is equal to the ambient temperature as shown in Fig. 7.

4.3 Current coefficient and temperature coefficient

The temperature and current coefficients of lasing wavelength were monitored at a specific interval after changing the package temperature and the injected current. The spectral characteristics were scarcely degraded and responded well to CO_2-gas absorption under these conditions.

The current coefficient of wavelength and absorption peak wavelength varied with the time interval of monitoring after changing the current as shown in Fig. 9. The current was changed with a 2-mA step under a constant package

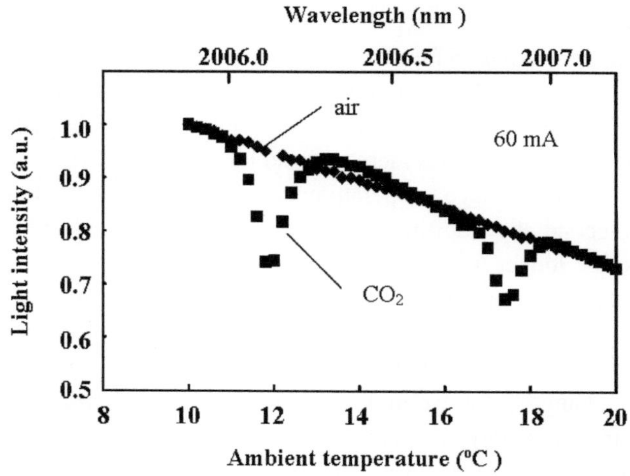

Fig. 10 Typical CO_2-gas absorption spectrum. The laser diode was biased at 60 mA, and the ambient temperature was changed with a 0.2°C step. The time interval between steps was set at 1 and 3 min, respectively

temperature (11.7°C). The lasing wavelength was also monitored at the all monitoring points in Fig. 9, and in the extreme case, the current coefficient was estimated to be about 0.008 and 0.006 nm/mA at 3 min and at 10 s, respectively, just after changing the current under a constant package temperature. The difference in the coefficient was caused by the temperature difference at the time interval of 3 min and 10 s which corresponds to the pulse width in Fig. 7. When the package temperature was not controlled, the current coefficient increased about 10% because the package temperature increased. The current-light output characteristics under CO_2-gas absorption therefore differed from the change rate of injected current. When the injected current is quickly changed, the increase in the temperature is not sufficient and the increase is not saturated. The temperature difference between the active layer and the package temperature becomes large or small in response to the magnitude of injected current when the package temperature is set at a constant temperature. The current coefficient is, therefore, not constant and varies with the rate of current increase. Their coefficients were, however, kept at fixed values if the time interval of monitoring were fixed at the constant values and strongly influenced by the materials used and the chip-mount configuration.

The temperature coefficient of wavelength and absorption peak wavelength will vary depending on the time interval of monitoring after changing the ambient temperature (see Fig. 7). If the change rate of package temperature is set at values of more than 1 second, the temperature coefficient is the same because the change in package temperature diffuses to the active layer. Figure 10 shows examples of the CO_2-gas absorption spectra monitored at several time intervals after 0.2°C-step change in the package temperature. The absorption spectra monitored at the time intervals of 1 and

3 min after 0.2°C-step change nearly coincided with each other as shown in Fig. 10. The temperature rise at the active layer was nearly constant in this time range, and the coefficient was estimated to be about 0.13 nm/°C. When the package temperature was increased and the change rate (time interval) of package temperature was set at values of less than 1 second, the absorption peak tended to be small and shifted to the longer wavelength range against the horizontal axis (package temperature).

These coefficients are governed by the change of each factor discussed in Sect. 4. When ambient temperature is changed, for example, threshold current density and bandgap energy vary simultaneously and lasing wavelength changes complicatedly. The coefficients result from the overall change in the factors. Consequently, the coefficients will vary with the material used, the mounting configuration, the monitoring time interval, etc. It can be said that the change rate of the injected current and ambient (package) temperature has to be constant during TDLS to eliminate wavelength error, although the coefficient differs with the change rate.

5 Conclusion

The physical mechanisms determining temperature and current coefficients of lasing wavelength, the band-filling effect, the plasma effect, and the heating effect in relation to the dynamic property of the junction (active layer) temperature have been discussed in detail and clarified for TDLS. The wavelength change after lasing is mainly governed by the temperature in the active layer. The temperature in the active layer depends on the heat conductance between the active layer and package (or ambient). Joule heating generated at the active layer quickly spreads within the laser chip and then gradually diffuses to the surrounding region (heat sink and package), and the current coefficient differed with the monitoring time interval after changing the magnitude of injected current. The thermal conductance also has a strong influence on the temperature of the active layer when the ambient (package) temperature changes. From these results, it can be said that the monitoring time interval after changing the ambient temperature or injected current (scanning rate) should be strictly fixed to eliminate monitoring error related to lasing wavelength, although the coefficients will be different with the scanning rate.

Acknowledgement This work was partially supported by the Ministry of Economy, Trade, and Industry, Japan.

References

1. M.W. Sigrist (ed.), Appl. Phys. B **90**(2) (2008)
2. M. Fukuda, *Optical Semiconductor Devices* (Wiley, New York, 1999)
3. F.R. Nash, J. Appl. Phys. **44**, 4696 (1973)
4. Y. Yamamoto, IEEE J. Quantum Electron. **QE-19**, 34 (1983)

Quantitative gas measurements using a versatile OPO-based cavity ringdown spectrometer and the comparison with spectroscopic databases

S. Persijn · F. Harren · A. van der Veen

Received: 7 October 2009 / Revised version: 16 November 2009 / Published online: 14 January 2010
© The Author(s) 2010. This article is published with open access at Springerlink.com

Abstract A versatile OPO-based cavity ringdown spectrometer is reported for quantitative and sensitive gas measurement down to nmol/mol levels. The system is based on cavity ringdown spectroscopy (CRDS) in combination with a continuous wave optical parametric oscillator tunable from 2693 to 3505 nm. Using a single set of CRDS mirrors, spectra were recorded of methane, ethane, benzene, propane, water, acetone and formaldehyde. Gas mixtures were gravimetrically prepared in cylinders or via dynamic generation using diffusion tubes (formaldehyde). Results were compared with data from the Hitran, PNNL and NIST databases. Good agreement was found with PNNL and NIST data for most molecules while agreement with Hitran was less for ethane and formaldehyde.

1 Introduction

Gas monitoring systems based on laser spectroscopy have become standard equipment in industry and research environments due to their sensitivity, selectivity, and speed of measurement. Fields that are served include environmental monitoring, purity analysis of gases in the semiconductor industry, and leak detection in the oil and gas industry. Existing gas monitoring systems are in general limited to measurement of a single component as they are equipped with standard continuous wave diode lasers that lack wide tunability. Analysis of multiple components requires the use of a light source covering a wide wavelength range. In the near-infrared a multitude of inexpensive fibered lasers with a limited tuning range have been combined [1]. In the mid-infrared range, recent developments with external cavity quantum cascade lasers are promising. Nowadays, tuning ranges exceeding 100 cm^{-1} in combination with an output power greater than 100 mW can be achieved [2]. Difference frequency generation sources based on widely available near-infrared lasers form another alternative. Employing waveguide instead of bulk PPLN crystals results in high conversion efficiencies with output powers up to tens of mW ([3] and references therein). Wide tuning can also be achieved by optical parametric oscillators (OPOs). Tuning ranges in the order of 1000 cm^{-1} in combination with output powers of more than 1 watt have been achieved [4–6]. Their high output power renders them ideal as a source for photoacoustic spectroscopic techniques [7, 8] while their narrow line width makes them also a useful source for high-resolution spectroscopy [9, 10]. Recently, several reviews have appeared that focus on developments in the design, operation, and spectroscopic applications of tunable OPOs [11, 12].

Accurate spectroscopic gas detection requires reliable spectroscopic data and good reference gas mixtures. When one only relies on spectroscopic data, knowledge of accurate absorption strengths and their uncertainty is of key importance. The Hitran and GEISA databases are two of the best known and most extensive resources containing several million line strengths of tens of molecules [13, 14]. Data are based on a combination of experimental work and theoretical calculations. They have proven to be useful resources for applications such as planetary observations or radiative transfer calculations. Care should be taken not to rely on

S. Persijn (✉) · A. van der Veen
VSL, Thijsseweg 11, 2629 JA Delft, The Netherlands
e-mail: spersijn@vsl.nl

F. Harren
Life Science Trace Gas Facility, Institute for Molecules and Materials, Radboud University, 6525 ED Nijmegen, The Netherlands

these databases as a sole source for calibration of a gas detection system. If possible, calibration of the system with an appropriate set of certified reference gas mixtures should be performed. In some cases, this may not be possible such as in differential absorption LIDAR, combustion research or satellite measurements. However, one should be aware that this may lead to erroneous results. For example, using near-infrared spectroscopy, the SCIAMACHY instrument onboard ESA's ENVISAT satellite observed unexpectedly large tropical emissions of methane [15]. Recently however, a retrieval error was identified related to inaccuracies in water vapor spectroscopic parameters, causing an overestimation by 30% of methane amount correlated with high water vapor abundances [16]. Another example is formed by the formaldehyde line intensities in the strong 3.6 μm band; Perrin and co-workers found out these to be on the average 28% stronger than those quoted in the HITRAN 2004 database. Several studies are based on these incorrect line strengths [17]. Accurate determination of line strengths is complex and very time-consuming. The best results up to now reach an accuracy of 0.1% and have been obtained on well-isolated lines of small molecules such as CO_2 [18].

Many molecules do not readily lend themselves to a line-by line description such as species possessing very dense spectra and numerous low-frequency modes of vibration. In this case, one has to rely on databases which encompass FTIR absorption spectra that are normally recorded at atmospheric pressure and temperature. Unfortunately, many of them suffer from low resolution, lack of sample quantification, and use of impure samples which limits the usefulness of these databases. Two databases which we do found useful are the NIST library and the PNNL database [19, 20]. Data consists of composite spectra obtained from measurements on at least 9 different partial pressures of the compound. For both databases an extensive description of the used methods and accuracy budget is available. For most compounds, the relative expanded uncertainty ($k = 2$) of the NIST database is between 2% and 2.3% and it is lower than 3.1% for the PNNL database. The instrumental resolution is 0.125 and 0.112 cm^{-1}, respectively. It should be noted that both databases just like any FTIR library have limitations regarding applications outside their intended temperature and pressure ranges.

Here, we report on a spectrometer which can quantitatively detect tens of different gases down to nmol/mol levels. The system is based on cavity ringdown spectroscopy (CRDS) [21] in combination with a continuous wave optical parametric oscillator tunable from 2693 to 3505 nm. Using a single set of CRDS mirrors, spectra over typically several wavenumbers were recorded of a variety of compounds including ethane, benzene, propane, acetone and formaldehyde. Gas mixtures were gravimetrically prepared in cylinders or via dynamic generation using diffusion tubes (formaldehyde). Results were compared with data from the Hitran, PNNL and NIST databases. Finally, the system was applied to the purity analysis of carbon monoxide and methane.

2 Experimental approach

A schematic of the experimental set-up is shown in Fig. 1 which encompasses a continuous wave optical parametric oscillator in combination with cavity ringdown spectroscopy. As a pump source for the optical parametric oscillator a 7.8 watt, single-frequency DFB fiber laser centered at 1063.9 nm is operated (Koheras). The laser can be tuned thermally over 45 GHz or via applying a tensile strain to the fiber using a multi-layered piëzo over 53 GHz (fast). The combined tuning range is 90 GHz. The stated line-width of the laser is better than 70 kHz. After passing through a free-space isolator (Electro-Optics Technology), the pump laser is tightly focused in a MgO-doped (5 mol%) congruent periodically-poled lithium niobate (PPLN) crystal (50 mm × 8 mm × 0.5 mm) with poling periods ranging from 28.5 to 31.5 μm (HC Photonics). The crystal has 1° wedges on both sides to reduce etalon effects and suppress feedback into the pump laser. The crystal is placed in an oven made of peek (polyetheretherketones), which can be machined much easier compared to the more commonly used ceramic materials. The crystal is placed in a bow-tie optical cavity which is resonant for the signal wavelength. A 400 μm thick solid uncoated YAG etalon (free spectral range is 7 cm^{-1}) is placed in the cavity to suppress mode hops.

The OPO cavity is enclosed in a thick aluminum housing which is flushed with dry nitrogen to improve the stability

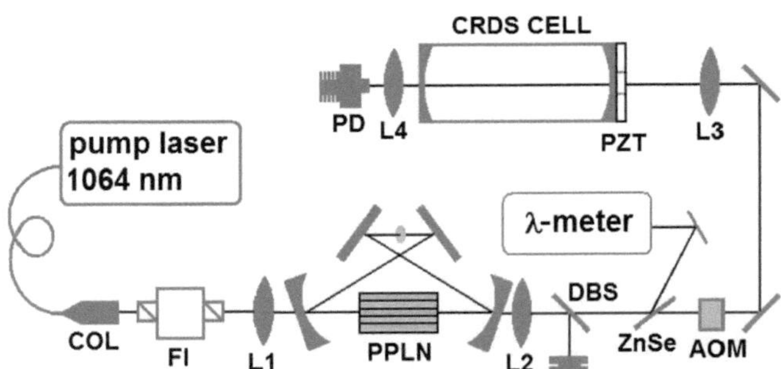

Fig. 1 Light from the pump laser is coupled into the OPO via a collimator (COL), Faraday isolator (FI) and AR-coated focusing lens (L1). The output of the OPO is collimated using an uncoated CaF_2 lens (L2). Signal and residual pump are separated from the idler using a dichroic beams splitter (DBS). Part of the idler beam is directed to a wavelength meter using a ZnSe window placed near Brewster angle. The idler is guided through an acousto-optic modulator (AOM), and via mode-matching optics (L3) coupled into the ringdown cavity. Length of the cell is modulated at 35 Hz using a ring piëzo (PZT). Light emerging from the cell is focused with a lens (L4) on a fast photodetector (PD)

and output power at certain wavelengths that are absorbed by gases in the atmosphere. The idler wavelength can be tuned from 2693 to 3505 nm with a maximum idler output power of 1.2 watt around 3270 nm. The idler wavelength is measured using a wavelength meter (model 621A, Bristol instruments). The stability of the idler frequency (better than 30 MHz over a period of 10 seconds) and power are good taking into consideration that no active stabilization is used.

A standard detection scheme is used to record the ringdown events. The 40 cm long cavity consists of two highly reflective mirrors from Los Gatos Research Inc. with a maximum reflectivity of 99.987% at 3150 cm^{-1} (corresponding to a decay time of 10.5 μs). The cell volume is 150 ml. The cavity length is modulated over one free spectral range at a frequency of 35 Hz by means of a ring piezo-electric transducer attached to one of the mirrors. Light emerging from the cell is focused on a Peltier-cooled (HgCdZn)Te photodetector (PVI-2TE-5, Vigo Systems S.A.). Once sufficient intracavity light intensity has built up, the idler beam is rapidly extinguished using an acousto-optic modulator as a fast switch. Spectra are recorded by measuring the ringdown time as function of the frequency. For spectra spanning up to 53 GHz, only pump laser tuning is used which takes about 90 seconds per scan. For spectra up to 210 GHz, this is combined with scanning of the etalon, while for even wider spectra temperature tuning of the PPLN crystal is needed [22]. Typically ten scans are averaged. The minimum detection limit is 6×10^{-8} cm^{-1} in the short-term (1 second) and 5×10^{-9} cm^{-1} in the long-term (5 minutes) at 3150 cm^{-1}.

The gas flow from the cylinders to the cell is regulated using mass flow controllers (flow rate between 5 and 30 l/h). At the exit of the cell, a combination of a pressure regulator and a membrane pump are used to maintain a constant pressure in the cell during operation. The ringdown cell is equipped with a pressure sensor (Edwards Barocel) and thermistor (Tempcontrol) for temperature measurements. Measurements in this work were recorded at atmospheric pressure and a temperature of (20 ± 1)°C.

All gas mixtures used in this work were prepared in house. Formaldehyde was generated dynamically following ISO 6145. All other compounds were prepared as cylinder mixtures following ISO 6142 which describes the preparation of calibration gas mixtures using gravimetric method. Keys to the accurate preparation are the use of high purity compounds, suitable cylinders and good facilities for weighing gas cylinders. Relative measurements are done against a reference cylinder to compensate external variables such as the changes of the temperature, humidity and buoyancy. The nitrogen which was used for preparing the gas mixtures was of high purity and was analyzed separately on impurities. Amount of substances ranged from 10 nmol/mol up to 5 μmol/mol with typical uncertainties of less than 1 percent. Mixtures were made in 5 liters cylinders with a pressure of about 100 bars.

The accuracy of the derived measurements depends on several factors including uncertainties in pressure and temperature, interferences by other compounds, and photodetector non-linearity. In comparison to near-infrared photodetectors, mid-infrared photodetectors have a markedly larger non-linearity [23]. The non-linearity results in an overestimation of the decay times both of the sample and background (high purity nitrogen). Overall the calculated absorption coefficients will be too small. The non-linearity of the photodetector used in this work was measured to be $\leq 0.5\%$ over the intensity range used. The influence of the time response of the photodetector (20 ns) is negligible as typical decay times are between 5 μs and 10 μs. For strong absorptions, the expanded uncertainty ($k = 2$) of the measured absorption coefficients is about between 2% and 3%. In case of strong interferences (often water vapor) or at the edge of the ringdown mirror reflectivity bandwidth, this value is higher. Note that we assumed that the concentration of the compound in the ringdown cell equals that of the gravimetrically determined value. This might not be correct for all compounds due decomposition or polymerization of the compound (e.g. formaldehyde) or adsorption in the flow system or gas cylinder (especially an issue for polar molecules like acetone).

3 Experimental results ethane (C_2H_6)

Figure 2 depicts a measurement of 10 nmol/mol ethane and of pure nitrogen. A good signal to noise is obtained at this low concentration level. The detection limit for ethane is below 1 nmol/mol (SNR = 3). In nitrogen the increasing decay time at higher frequencies is due to a small yet significant increase in mirror reflectivity over this short wavelength range.

Figure 3a shows a recorded spectrum of 50 nmol/mol ethane together with spectra from Hitran and PNNL data-

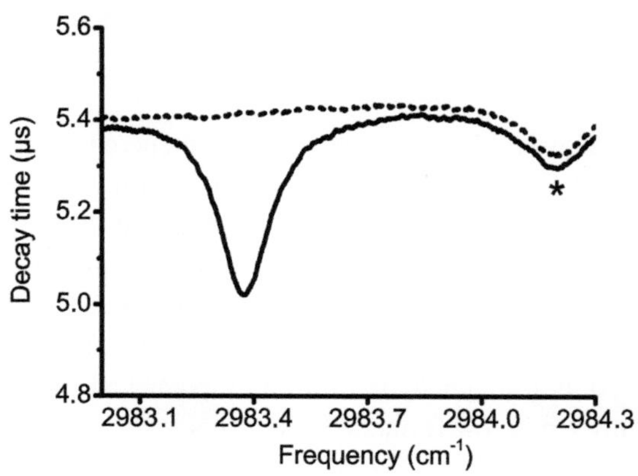

Fig. 2 Observed decay time for a mixture of 10 nmol/mol ethane (*solid line*) in nitrogen and of pure nitrogen (*dashed line*). An *asterisk* marks a water absorption line

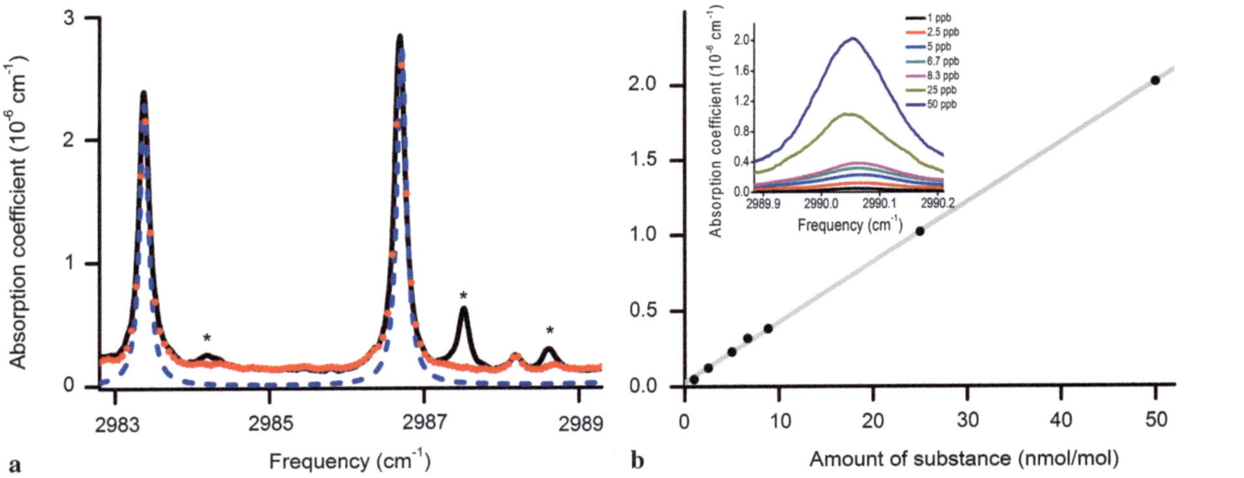

Fig. 3 **a** Measurement of 50 nmol/mol ethane (*solid black line*) and the comparison with PNNL (*red dots*) and Hitran (*dashed blue line*). Peaks in the measurements marked by an *asterisk* are due to water vapor. **b** Linearity of the peak signal measured at 2990 cm^{-1} for amount of substances ranging from 1 to 50 nmol/mol ethane. The corresponding slope is $(3.99 \pm 0, 03) \times 10^{-8}$ cm^{-1}/(nmol/mol). The *inset* shows the corresponding spectra

bases. Our measurements agree within their combined uncertainty with PNNL. While the peak heights in Hitran match those of our measurements and of PNNL many of the weaker lines (giving rise to the absorption in between the strong peaks) are missing in Hitran. In addition, the position of the peak at 2986.7 cm^{-1} in Hitran is shifted in agreement with observations by NASA [24]. The incompleteness of Hitran is due to the complexity of the ethane spectrum at 3 μm. Many interacting vibrational modes and the presence of hot bands make it difficult to accurately predict the spectrum. This is also recognized in the most recent Hitran paper which states about the ν_7 band at 3.3 μm [13]: '... this band is sorely in need of improvement...'. Figure 3b shows the good linearity of measured absorption coefficients. Amount of substances below 25 nmol/mol were obtained by dynamic dilution using mass flow controllers.

4 Experimental results dynamically generated formaldehyde (CH_2O)

Formaldehyde (CH_2O) mixtures were prepared from paraformaldehyde powder (Merck). The paraformaldehyde was kept in glass diffusion tubes in an oven heated to 60°C. The formaldehyde amount of substance was determined by periodic weighing of the diffusion tube and measuring the flow rate (Fig. 4). High flow rates of typically several l/min were required as formaldehyde is strongly polar and in order to get sufficiently low formaldehyde levels.

After subtraction of water vapor and methane interference the measurement results agreed excellently with PNNL (Fig. 5). Also agreement with Hitran 2008 is good except near 2950 cm^{-1} where Hitran line strengths are too high. In

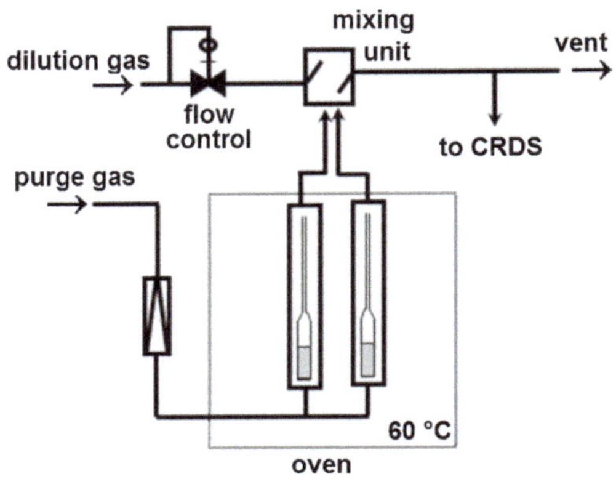

Fig. 4 Set-up for dynamic generation of formaldehyde by heating diffusion tubes containing paraformaldehyde to 60°C

Hitran 2004 most peaks are 30% too weak while the peak at 2950 cm^{-1} is much too strong. Many of the smaller peaks are completely absent in this version of Hitran. Formic acid, which might also be formed, was not detected.

To test the linearity of the system different amount of substances were made using a set of different diffusion tubes and dilution flows (Fig. 6). The spectral region lying between 2949.8 and 2951.3 cm^{-1} was selected for the CRDS measurements as absorption of water and methane are very weak in this wavelength region.

Using the absorption coefficients from PNNL 2.3% higher amount of substances were found than the values obtained from the weighing. The expanded uncertainty in the dynamic preparation of formaldehyde in this experiment is large with 7%. This has been attributed to use of diffusion tubes with a short neck which led to instabilities in the mass loss rate. In later experiment using diffusion tubes with a

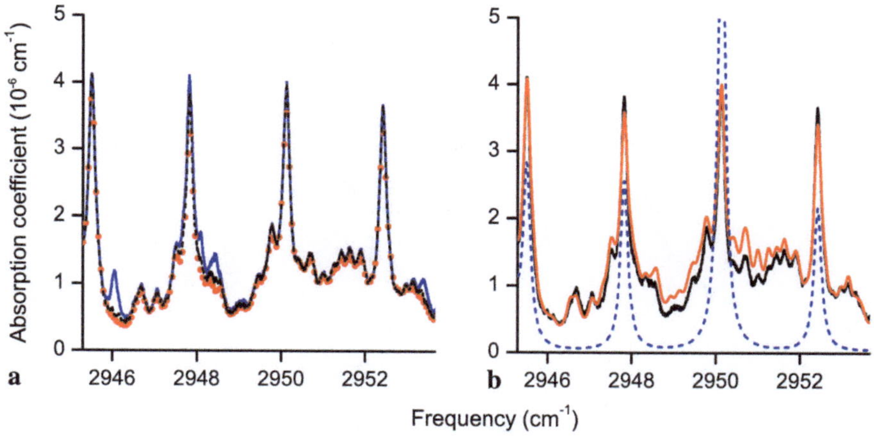

Fig. 5 **a** Measurement of 1.3 µmol/mol formaldehyde (*blue line*) and after subtraction of water and methane interference (*black dashed line*). Good correspondence is observed with PNNL (*red dots*). **b** Comparison of the cleaned measurement spectrum (*black line*) with Hitran 2004 (*blue dashed line*) and 2008 (*red line*)

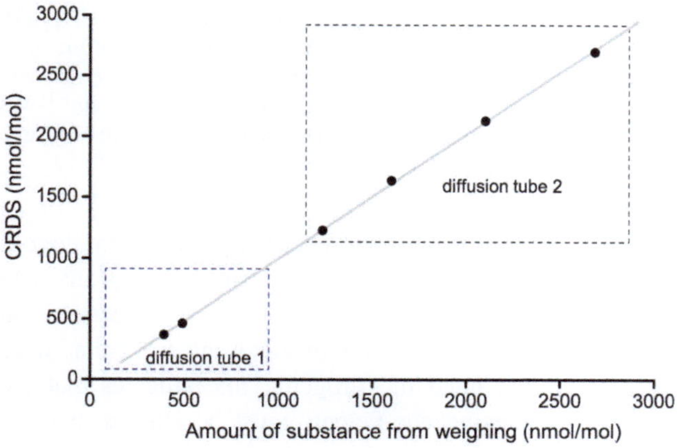

Fig. 6 Comparison of the measured (using the PNNL absorption coefficients) and set amount of substances. The slope of the linear fit is 1.023 ± 0.008

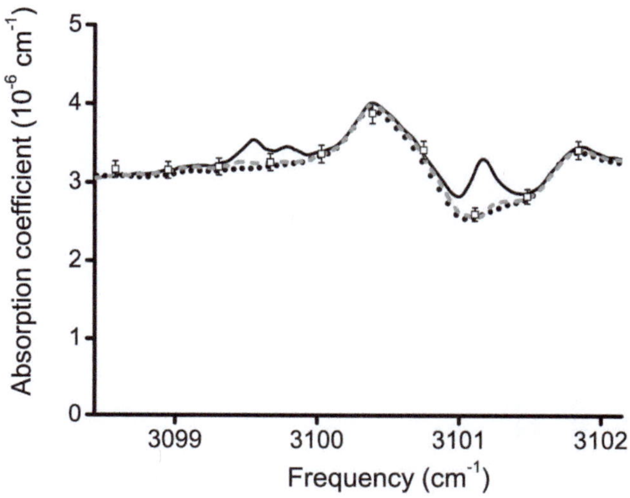

Fig. 7 Measurement of 1 µmol/mol benzene (*black line*) and after subtraction of water vapor interference (*dashed grey line*). Spectra from PNNL (*black dots*) and NIST (*black squares* including expanded uncertainty) database are also shown

longer neck, the expanded uncertainty in the amount of substance could be reduced to 3.5%.

5 Experimental results benzene (C_6H_6), acetone (C_3H_6O), and propane (C_3H_8)

At atmospheric conditions, the spectra of acetone, propane and benzene in the 3 µm region do not exist out of well-isolated lines but of a very broad absorption band spanning from 100 to 200 cm^{-1}. Recording the entire spectra would be very time-consuming with the current set-up as it requires multiple changes in the crystal temperature. Therefore we chose to focus on the Q-branch regions which are much narrower but still specific for each molecule.

Figure 7 shows a measurement of 1 µmol/mol benzene around 3100 cm^{-1}. This feature is relatively narrow and was therefore chosen instead of the 30% stronger but broader absorbing feature centered at 3047 cm^{-1}. After subtraction of water vapor interference, good agreement is obtained with both NIST and PNNL databases.

For acetone a 5 µmol/mol mixture was available. The absorption by this sample is too high at the peak of the absorption and therefore the mixture was diluted with pure nitrogen using mass flow controllers to an amount of substance fraction of 1.67 µmol/mol (Fig. 8).

Qualitatively good agreement is obtained with both PNNL and NIST databases. However the values we measured are consistently 5% higher, most clearly visible at the peak of the absorption. Measurements with new mixtures are planned to sort out the difference. From the inset in Fig. 8, it can be seen that the maximum absorption coefficient that can be measured is about 10^{-5} cm^{-1} at this mirror reflectivity; above this the recorded spectrum becomes unreliable.

For propane a 1 µmol/mol mixture was available. As propane absorbs very strongly at 3 µm, the mixture was further diluted using mass flow controllers to yield a concentration of 200 nmol/mol. This resulted in an increased uncertainty in the amount of substance of $(2.00 \pm 0.03) \times$

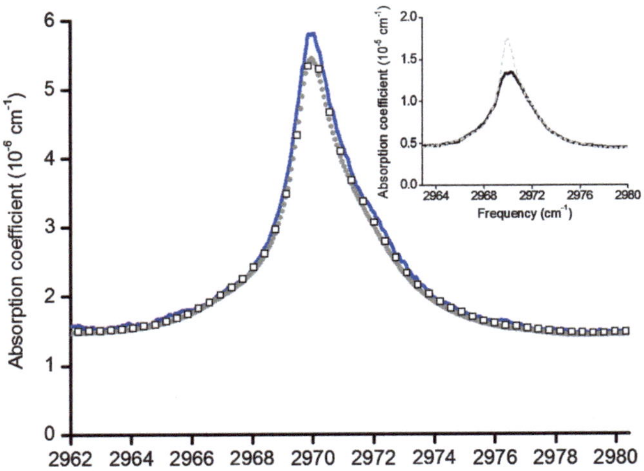

Fig. 8 Measurement of 1.67 μmol/mol acetone (*solid blue line*) and the comparison with PNNL (*grey dots*) and NIST (*squares*) database. The *inset* shows a measurement of 5 μmol/mol (*black line*) and a scaled measurement of 1.67 μmol/mol (*grey dashed line*, multiplied by 5/1.67)

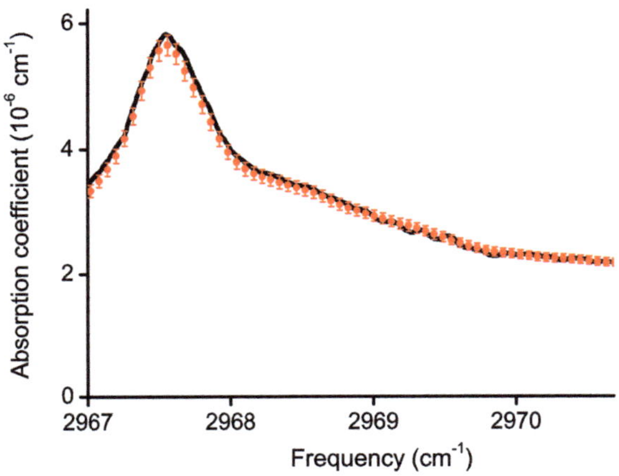

Fig. 9 Measurement of 200 nmol/mol propane (*solid line*) and the comparison with PNNL database (*red dots* including expanded uncertainty)

10^{-7} mol/mol. Figure 9 shows the results. Absorption coefficients match within their combined uncertainty (<3% for the PNNL data and 3.4% for our data).

6 Application to purity analysis

Pure reference materials play an essential role in the traceability chain for chemical measurements as one has to identify the entities concerned. The high sensitivity of our set-up renders it very suitable to quantify the often low levels of impurities present. Not only infrared-inactive molecules (e.g., nitrogen and oxygen) but also many infrared active molecules can be analyzed for their purity due to the wide tuning capability of the OPO.

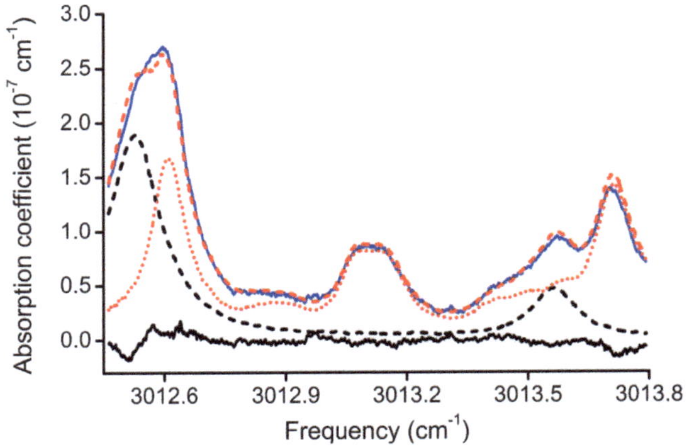

Fig. 10 Measurement of CO with purity 4.7 (*solid blue line*) and the corresponding fit (*dashed red line*) which is the sum of the absorption of 1.6 μmol/mol H_2O (*dashed black line*) and 12 nmol/mol CH_4 (*dotted red line*). The *solid black line* represents the residual of the measurement and the fit. Individual H_2O and CH_4 spectra were recorded in nitrogen. A small background absorption of 3×10^{-8} cm^{-1} was subtracted from the measurement data (due to the presence of a large absorbing molecule or far wing absorptions by CO)

Figure 10 depicts a measurement of pure carbon monoxide (CO) which reveals several absorption peaks that can be attributed to H_2O and CH_4. The measured spectrum can be well fitted by the spectra of H_2O and CH_4 that were recorded before hand in nitrogen. This indicates that the pressure broadening coefficients must be similar in these two matrices.

Figure 11a shows the measurement of two pure methane samples containing different traces of water vapor. The measured spectrum consists of methane and water absorption on top of a broad background absorption of about 3.3×10^{-6} cm^{-1}. Such high background absorption was found in methane from different suppliers. It is due to far wing absorptions by methane itself. Also shown is the methane spectrum as calculated from Hitran 2008. No broad background is present, as in our calculation we did not take into account the contributions from distant methane lines which are up to 6 orders of magnitude higher. Band absorption data for methane have been developed as methane dominates part of the infrared spectra of the giant outer planets and Titan. From the data of Irwin and co-workers a three times higher background absorption of 1×10^{-5} cm^{-1} can be derived at this wavelength range [25]. Nevertheless, a quite good correspondence between Hitran 2008 and our measurements is observed taking into account that the absorption is weak (maximum line strength 1.6×10^{-26} cm^{-1}/(molecule cm^{-2})). The quality of the methane data in this wavelength range has been drastically improved in this version of Hitran as compared to Hitran 2004, which suffered from incorrect line positions and intensities.

Figure 11b shows the difference spectrum of the wet and dry methane samples. The shape of this spectrum is signif-

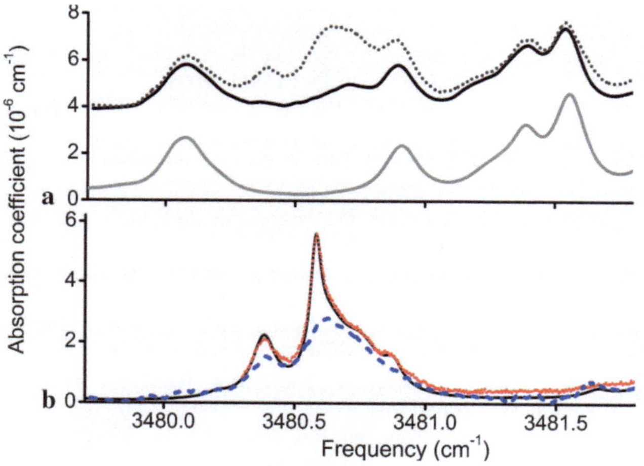

Fig. 11 **a** Recording of two methane samples containing a high (*grey dots*) and low (*solid black line*) amount of water vapor. Also the CH_4 spectrum (*grey solid line*) calculated from Hitran is shown. **b** Difference spectrum of the high and low humidity CH_4 samples (*dashed blue line*). For comparison the measured (*solid black*) and Hitran (*light red dots*) spectrum of 25 µmol/mol H_2O in N_2 is shown

icantly different from the H_2O spectrum as recorded in nitrogen, indicating different pressure broadening coefficients. The amount of substance of water vapor is obtained by integration of the area under the spectra yielding 21 µmol/mol for the wet sample and less than 2 µmol/mol for the dry sample.

7 Discussion and conclusions

A versatile OPO-based CRDS spectrometer was presented for quantitative gas measurements. The wide tuning range of the OPO enables detection of tens of different compounds. Compounds included several molecules with broad absorption patterns such as acetone and benzene which could be specifically measured by recording the Q-branch spectra. Good agreement was obtained with PNNL and NIST databases while agreement with Hitran for in particular ethane and to lesser extent also formaldehyde was less satisfactory. This work shows that results from CRDS are quantitative and are comparable with the FTIR data from PNNL and NIST (which were recorded at much higher concentration levels). With the current state of the art it is not advisable for most molecules to solely rely on Hitran as a source for quantitative gas measurements.

The presented system can be significantly improved in a number of ways. Currently, spectra spanning up to 7 cm^{-1} can be readily recorded using a combination of pump and etalon tuning. Access to wider spectral ranges requires time-consuming temperature tuning of the PPLN crystal. To overcome this, one could replace the pump laser by an external cavity tunable low power seed laser in combination with a high power fiber amplifier. The latter are becoming rapidly cheaper and continuous tuning over hundreds of wavenumbers should be feasible [26]. An additional advantage is that (nearly) all measurements can be done in a single period of the crystal while currently tweaking of the idler beam may be required upon changing the crystal period. To increase the sensitivity of the system further we are currently testing new cavity ringdown mirrors with a higher reflectivity. In addition, we plan to lock the cavity to the OPO wavelength to increase the number of ringdown events that are sampled per second by a factor of 10 to 1000 [27]. The combined effect of these two measures should lead to 10–100 lower detection limits for most molecules. Finally, better mode-matching and temperature control of the cell are needed to push the expanded uncertainty in the measurements to less than 1%.

Acknowledgements The authors thank the colleagues from the department of calibration and reference materials for preparing the gas mixtures, and Annarita Baldan for her contribution to the formaldehyde experiments. We also acknowledge Patrick Irwin (Clarendon Laboratory) for performing the band calculations for methane. This work was financially supported by the Dutch Ministry of Economic Affairs and the European Community's 7th Framework Programme, ERA-NET^{+}, under the iMERA^{+} project—grant agreement No. 217257.

Open Access This article is distributed under the terms of the Creative Commons Attribution Noncommercial License which permits any noncommercial use, distribution, and reproduction in any medium, provided the original author(s) and source are credited.

References

1. Y. Ding, P. Macko, D. Romanini, V. Perevalov, S.A. Tashkun, J.L. Teffo, S.M. Hu, A. Campargue, J. Mol. Spectrosc. **226**, 146 (2004)
2. R. Maulini, I. Dunayevskiy, A. Lyakh, A. Tsekoun, C.K.N. Patel, L. Diehl, C. Pflügl, F. Capasso, Electron. Lett. **45**, 107 (2009)
3. D. Richter, A. Fried, P. Weibring, Laser Photon. Rev. **3**, 343 (2009)
4. A. Henderson, R. Stafford, Opt. Express **14**, 767 (2006)
5. M. Vainio, J. Peltola, S. Persijn, F.J.M. Harren, L. Halonen, Opt. Express **16**, 11141 (2008)
6. J. Kiessling, R. Sowade, I. Breunig, K. Buse, V. Dierolf, Opt. Express **17**, 87 (2009)
7. F. Kühnemann, K. Scheider, A. Hecker, A.A.E. Martis, W. Urban, S. Schiller, J. Mlynek, Appl. Phys. B **66**, 741 (1998)
8. A.K.Y. Ngai, S.T. Persijn, I.D. Lindsay, A.A. Kosterev, P. Gross, C.J. Lee, S.M. Cristescu, F.K. Tittel, K.J. Boller, F.J.M. Harren, Appl. Phys. B **89**, 123 (2007)
9. H. Verbraak, A.K.Y. Ngai, S.T. Persijn, F.J.M. Harren, H. Linnartz, Chem. Phys. Lett. **442**, 145 (2007)
10. E.V. Kovalchuk, D. Dekorsy, A.I. Lvovsky, C. Braxmaier, J. Mlynek, A. Peters, S. Schiller, Opt. Lett. **26**, 1430 (2001)
11. B.J. Orr, Y. He, R.T. White, Spectroscopic applications of tunable optical parametric oscillators, in *Tunable Laser Applications*, 2nd edn. (CRC Press, New York, 2008)
12. A.K.Y. Ngai, S.T. Persijn, M.M.J.W. van Herpen, S.M. Cristescu, F.J.M. Harren, Photoacoustic spectroscopy using continuous wave optical parametric oscillators, in *Mid-Infrared Coherent Sources and Applications* (Springer, Dordrecht, 2008)

13. L.S. Rothman, I.E. Gordon, A. Barbe, D. Chris Benner, P.F. Bernath, M. Birk, V. Boudon, L.R. Brown, A. Campargue, J.P. Champion, K. Chance, L.H. Coudert, V. Dana, V.M. Devi, S. Fally, J.M. Flaud, R.R. Gamache, A. Goldman, D. Jacquemart, I. Kleiner, N. Lacome, W.J. Lafferty, J.Y. Mandin, S.T. Massie, S. Mikhailenko, C.E. Miller, N. Moazzen-Ahmadi, O.V. Naumenko, A. Nikitin, J. Orphal, A. Predoi-Cross, V. Perevalov, A. Perrin, C.P. Rinsland, M. Rotger, M. Šimečková, M.A.H. Smith, K. Sung, S. Tashkun, J. Tennyson, R.A. Toth, A.C. Vandaele, J. Vander Auwera, J. Quant. Spectrosc. Radiat. Transfer **110**, 533 (2009)
14. N. Jacquinet-Husson, N.A. Scott, A. Chédin, L. Crépeau, R. Armante, V. Capelle, J. Orphal, A. Coustenis, C. Boonne, N. Poulet-Crovisier, A. Barbe, M. Birk, L.R. Brown, C. Camy-Peyret, C. Claveau, K. Chance, N. Christidis, C. Clerbaux, P.F. Coheur, V. Dana, L. Daumont, M.R. De Backer-Barilly, G. Di Lonardo, J.M. Flaud, A. Goldman, A. Hamdouni, M. Hess, M.D. Hurley, D. Jacquemart, I. Kleiner, P. Köpke, J.Y. Mandin, S. Massie, S. Mikhailenko, V. Nemtchinov, A. Nikitin, D. Newnham, A. Perrin, V.I. Perevalov, S. Pinnock, L. Régalia-Jarlot, C.P. Rinsland, A. Rublev, F. Schreier, L. Schult, K.M. Smith, S.A. Tashkun, J.L. Teffo, R.A. Toth, Vl.G. Tyuterev, J. Vander Auwera, P. Varanasi, G. Wagner, J. Quant. Spectrosc. Radiat. Transfer **109**, 1043 (2008)
15. C. Frankenberg, J.F. Meirink, M. van Weele, U. Platt, T. Wagner, Science **308**, 1010 (2005)
16. C. Frankenberg, P. Bergamaschi, A. Butz, S. Houweling, J.F. Meirink, J. Notholt, A.K. Petersen, H. Schrijver, T. Warneke, I. Aben, Geophys. Res. Lett. **35**, L15811 (2008)
17. A. Perrin, D. Jacquemart, F. Kwabia Tchana, N. Lacome, J. Quant. Spectrosc. Radiat. Transfer **110**, 700 (2009)
18. G. Casa, R. Wehr, A. Castrillo, E. Fasci, L. Gianfrani, J. Chem. Phys. **130**, 184306 (2009)
19. P.M. Chu, F.R. Guenther, G.C. Rhoderick, W.J. Lafferty, J. Res. Natl. Inst. Stand. Technol. **104**, 59 (1999)
20. S.W. Sharpe, T.J. Johnson, R.L. Sams, P.M. Chu, G.C. Rhoderick, P.A. Johnson, Appl. Spectrosc. **58**, 1452 (2004)
21. G. Berden, R. Peeters, G. Meijer, Int. Rev. Phys. Chem. **19**, 565 (2000)
22. A.K.Y. Ngai, S.T. Persijn, G. von Basum, F.J.M. Harren, Appl. Phys. B **85**, 173 (2006)
23. E. Theocharous, J. Ishii, N.P. Fox, Appl. Opt. **43**, 4182 (2004)
24. http://mark4sun.jpl.nasa.gov/report/c2h6_spectroscopy.pdf
25. P.G.J. Irwin, L.A. Sromovsky, E.K. Strong, K. Sihra, N.A. Teanby, N. Bowles, S.B. Calcutt, J.J. Remedios, Icarus **181**, 309 (2006)
26. M.E. Klein, D.-H. Lee, J.-P. Meyn, K.-J. Boller, R. Wallenstein, Opt. Lett. **24**, 1142 (1999)
27. G. von Basum, D. Halmer, P. Hering, M. Mürtz, S. Schiller, F. Müller, A. Popp, F. Kühnemann, Opt. Lett. **29**, 798 (2004)

Results of TDLS application for ammonia monitoring in a process of high-purity arsine and phosphine production

A.P. Kotkov · V.A. Ivanov · N.D. Grishnova · A.V. Shirayev · A.G. Berezin · A.I. Nadezhdinskii · Y.Y. Ponurovskii · I.P. Popov · Y.P. Shapovalov · D.B. Stavrovskii · I.E. Vyazov

Received: 16 September 2009 / Revised version: 23 November 2009 / Published online: 14 January 2010
© Springer-Verlag 2010

Abstract Using the TDLS method it has been found that ammonia (NH_3) is the main impurity in arsine (AsH_3) and phosphine (PH_3), produced by hydrolysis of magnesium arsinide and phosphinide, respectively. NH_3 behavior is abnormal in solutions of these hydrides: NH_3 reveals the properties of a more volatile impurity in relation to arsine and phosphine, although its boiling temperature is higher than that of AsH_3 and PH_3. The observable anomaly is connected with the fact that in solutions of arsine and phosphine NH_3 shows properties differing from the properties of pure ammonia. It was supposed that the influence of intermolecular interaction between ammonia molecules, when diluted by arsine or phosphine, decreases. During the purification of arsine or phosphine one must continuously monitor the NH_3 concentration in the extraction of the light fraction in order to define the point at which to terminate the purification process.

1 Introduction

High-purity hydrates such as arsine (AsH_3) and phosphine (PH_3) are used for the production of epitaxial layers in A_3B_5 semiconductors. The quality of semiconductors manufactured depends on the presence of impurities [1]. Most important, from the point of view of the quality of the epitaxial layers, is the presence of volatile inorganic hydrides of IV and VI group elements. Germane, silane and hydrogen sulphide are used for intentional doping of epitaxial A_3B_5 layers in CVD processes by n-type impurities. To achieve a current carrier concentration below $10^{15}\ cm^{-3}$, the impurity concentration in hydrides has to be below 10 ppb.

The impurity content of hydrides is formed during their synthesis [2–4]. A universal method for the synthesis of volatile inorganic hydrides, including arsine and phosphine, is acid hydrolysis from an alloy of the hydride generating element with magnesium (either aluminium, or zinc):

$$Mg_3P_2 + 6HCl \rightarrow 2PH_3 + 3MgCl_2 \quad (1)$$

This method is used in the industrial production of arsine and phospine. The universality of this method of synthesis presents a potential danger of pollution of arsine and phosphine by other hydrides, including ammonia.

The above method was used for the synthesis of phospine, while for arsine synthesis a more selective method based on arsenic trichlorine reduction by sodium boron hydride was applied:

$$AsCl_3 + 3NaBH_4 + 9H_2O \rightarrow AsH_3 + 3NaCl + 3H_3BO_3 + 9H_2 \quad (2)$$

Existing techniques for the analysis of volatile hydrides during the synthesis of other hydrides by gas chromatography (see, for example, [5]) are time consuming and do not allow for the conduction of real time multichannel monitoring of the purification processing. Therefore, working out new, more sensitive and convenient techniques is a current and very important task.

Preliminary studies showed that the presence of ammonia in arsine at a 3–5% level influences the electrophysical para-

A.P. Kotkov · V.A. Ivanov · N.D. Grishnova · A.V. Shirayev
Scientific-Industrial Enterprise "Salut", Larina Str. 7, Nizhnii Novgorod 603950, Russia

A.G. Berezin (✉) · A.I. Nadezhdinskii · Y.Y. Ponurovskii · I.P. Popov · Y.P. Shapovalov · D.B. Stavrovskii · I.E. Vyazov
A.M. Prokhorov General Physics Institute of Russian Academy of Sciences, Vavilov Str. 38, Moscow 119991, Russia
e-mail: anber@nsc.gpi.ru

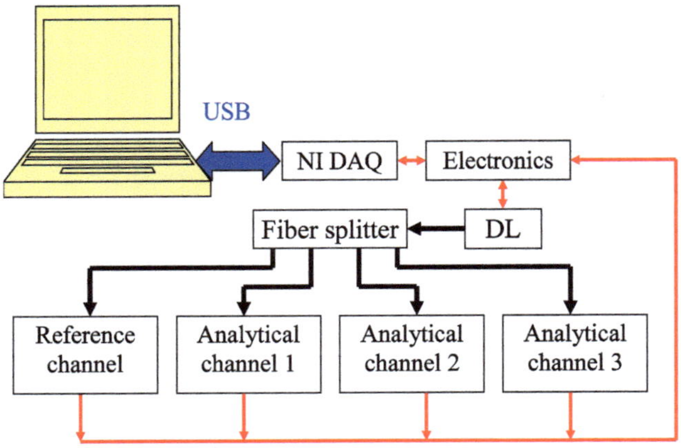

Fig. 1 Block-diagram of experimental set-up. *Black solid lines* correspond to optical fiber connections, narrower *red lines* to electrical connections

meters of the GaAs epitaxial layers: layers with p-type conductivity and with a carrier concentration of 4×10^{16} cm^{-3} and a mobility of 100 cm^2/V s at 77 K were obtained. Since the presence of ammonia in the other hydrates is important, it stipulates the purpose of this work: investigation of the possibility of applying TDLS to the monitoring of the behavior of ammonia impurities in arsine and phosphine in the course of their synthesis and purification.

2 Experimental set-up

The block-diagram of the experimental set-up is shown on Fig. 1. This diagram is similar to the one employed in [6] except for the fact that a multiple analytic channel system was used. This scheme allowed for the simultaneous detection of ammonia in the lower, middle and upper part of the purification column.

2.1 Diode laser

The NEL DFB 503074 (1.512 μm) pigtailed diode laser produced by NTT Electronics company [7] was used as the spectroscopic source for the instrument.

2.2 Electronics

The ammonia concentration measurements were performed by the well-known diode laser frequency scanning method (see, for example, [8]). In this method the initial scanning uses a diode laser, whose frequency was kept close to the absorption line by fine temperature tuning. Acquisition of the selected absorption line was accomplished by varying the DL current excitation. The diode laser was excited by current pulses of a trapezoidal waveform, allowing for DL frequency scanning.

Operation of the diode laser was performed using a NI 16-bit USB-6215 NI (250 kHz sample rate) electronic board instead of a 12-bit card, used in previous experiments. Two DAC (digital-to-analogue converters) of this board were used for generation of the DL current pulses and temperature control. Five ADC (analogue-to-digital converters) acquired signals from the temperature sensor, and from three analytical and one reference photodetectors (PD). Three additional ADC acquired signals, proportional to the pressure in the analytic cells. The device operation was controlled by a specialized program written in LabVIEW-8. This program performed signal processing (subtraction of the slope and the pedestal caused by the change in the laser intensity), low-pass filtering, averaging and concentration calculation.

The concentration was determined using the correlation function F_A for one of the analytic channels and the self-correlation function F_R for the reference channel:

$$F_A(\tau) = \int_0^T f_A(t+\tau) f_R(t)\, dt \tag{3}$$

$$F_R(\tau) = \int_0^T f_R(t+\tau) f_R(t)\, dt \tag{4}$$

Here f_A is the time-dependent signal in the analytic channel, f_R the time-dependent signal in the reference channel, T the pulse duration. Using the linear relationship

$$F_A = aF_R + b \tag{5}$$

The coefficients a and b were found by standard fitting. The concentration of ammonia in the analytic channel C_A could be found from the simple proportion

$$C_A = a\frac{L_R}{L_A}C_R \tag{6}$$

here C_R is for the concentrations in the reference channel, L_A and L_R are the optical paths in corresponding cells. Determination of the concentration with the help of correlation function serves as a filter since it takes into account only signals with the special waveform (the one in the reference channel). In order to get the correct concentration values the line waveform in the analytic channel should be the same or at least be close to the one in the reference channel.

2.3 Optical scheme

The DL output radiation was collimated within optical fibers and was divided into four channels with the help of two beam splitters: one divided radiation into reference and joint analytic channel in a proportion approximately 1:10 and the second split the joint analytic radiation into three analytic channels. In the analytic channels one-pass cells with different lengths (3, 5, 70 and 140 cm) were used. In short cells

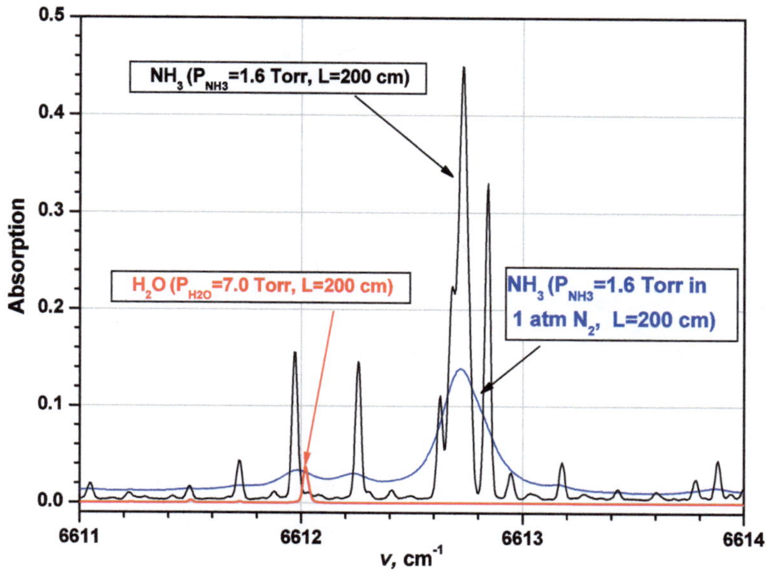

Fig. 2 Ammonia absorption lines near 1.51 μm: *black curve*—at the pressure 1.6 Torr, *blue curve*—1.6 Torr in nitrogen at atmospheric pressure. *Red line*—absorption of 7 Torr of water. The optical path in all cases was 200 cm

radiation was directly sent to a detector located at the opposite side of a cell; for long cells this was performed with a help of a collimator.

The analytical cells were made of electro-polished stainless steel tubes with quartz glass windows at their ends. The gas samples in the analytical cells were kept at lowered pressures. The pressure of ammonia in the reference cell, the sealed molybdenum glass tube of 30 mm long, was kept at the pressure of 50 Torr.

2.4 Experimental

The spectra of ammonia were recorded within the DL temperature tuning range and a group of absorption lines at 6612.7 cm^{-1} (1.512 μm) was chosen for the detection (Fig. 2). These absorption lines were tested in order to find out if they might overlap with absorption lines of the other molecules present in the investigated system, first of all with arsine, phosphine, carbon dioxide and water. It was found that arsine, phosphine and carbon dioxide did not absorb in this region and water has a weak absorption line ($\nu_1 + 2\nu_2$ band, transition 4_{41}–5_{50}, wavenumber 6612.018 cm^{-1} according to HITRAN database [9, 10]), shown by a red arrow on Fig. 2. This line is rather far from the ammonia absorption line chosen for detection. This situation was also analyzed in [11] and it was shown that the influence of this water line on ammonia detection is small, even at high water pressures.

This ammonia group of lines are resolved only at very low pressures, less than 10 Torr. At higher pressures this group of lines behaved as one line, as can be seen from Figs. 2 and 3.

Figure 3 shows the detected signals in the reference channel. The higher curve presents the signal, obtained with a Fabry-Pérot interferometer. This interferometer comprises a plane-parallel plate of ZnSe ($n = 2.43$, 10 mm thickness) having a free dispersion range of 0.206 cm. Taking that into account, the range of whole DL current scan can be calculated (0.6 cm^{-1}). The lower curve represents the signal, recorded with the reference cell. Having a calibrated abscissa scale (in cm^{-1}), it is possible to find the pressure dependencies of the line width (FWHM, full width at half maximum) and line shift. This was done for pure ammonia by fitting a Lorentz function (in Origin 8), and the results are presented in Fig. 4. At pressures higher than 300 Torr the absorption was too high and the line became too broad and fitting was incorrect. As can be seen, there is a strong nonlinear dependence of the line width and line shift upon pressure, and for correct detecting of ammonia with the help of a reference correlation function it is necessary to keep the pressure in the analytic cells low and close to that in the reference channel.

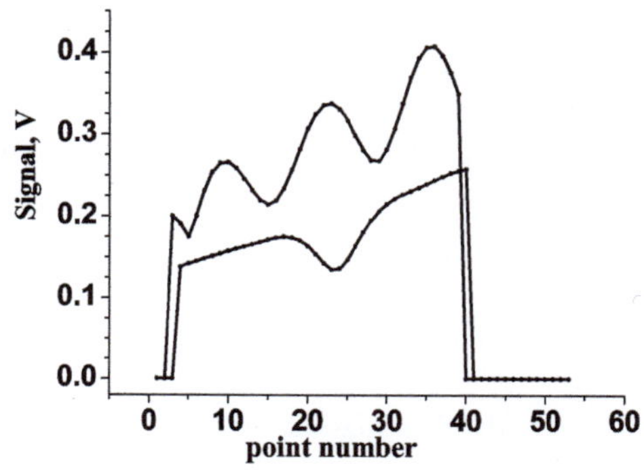

Fig. 3 Detected signals in reference channel: *higher curve*—etalon fringes in Fabry-Pérot interferometer (10 mm ZnSe plate), *lower curve*—reference cell 30 mm long with 50 Torr of ammonia

Experimentally, however, it was not convenient to keep the pressure in the cells so low and stable since the gas samples have to be changed often. So the calibration curve of the

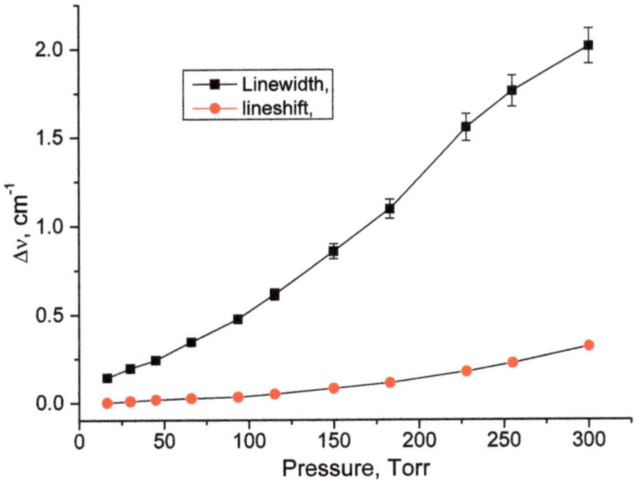

Fig. 4 Pressure dependence of the line width and line shift for the pure ammonia 1.512 μm absorption line

concentration upon pressure dependence was experimentally measured and these data were used to find the ammonia concentration at pressures less than 300 Torr. The pressures in the cells were measured independently and were recorded as a function of time along with other parameters.

3 Results and discussion

The basic method of arsine and phosphine purification is the method of low-temperature distillation. For the purification the column with the middle reservoir was used. In the column there were three monitoring points: in the upper, lower and in the middle part of the purification column, the content of ammonia was monitored by the TDLS method (with the aid of the one-pass cells of different length). Parallel gas samples were taken into the gas chromatograph ("Tzvet-800", thermoconductivity detector), or into the cell of a Fourier spectrometer (Bruker "Tenzor-27", cell length 20 cm, resolution 1 cm^{-1}).

First the gases, contained in balloons with raw materials and in balloons obtained after the purification process, were analyzed. Ammonia was not found in arsine, obtained by the reduction of trichlorine arsenic by sodium boron hydride (2), which confirmed the selectivity of this method of synthesis. At the same time, NH_3 was present in arsine and phosphine, synthesized by the universal method - the hydrolysis of magnesium arsenide or phosphide (1). The source of ammonia is, most probably, magnesium nitride, which is formed as a result of interaction of magnesium with atmospheric nitrogen while producing magnesium arsenide or phosphide [12]. The content of ammonia in synthesized hydrides depends on the conditions of synthesis and initial raw material. In phosphine it reaches a value 0.05%, while in arsine the values are of order of 1%.

Table 1 Contents of ammonia in arsine and phosphine, synthesized by different methods

	Arsine		Phosphine
	From $AsCl_3$	From Mg_3As_2	from Mg_3P_2
Raw material	<0.005%	~0.05%	~0.05%
Light fractions	<0.005%	0.5–5%	0.5–5%

In the balloons with the light fractions, in which more volatile components than arsine and phosphine are concentrated in the process of arsine and phosphine purification, a high content of NH_3 was observed: up to tens of percent in arsine and up to the units of percent in phosphine. In the balloons with the heavy fractions, in which less volatile impurities are concentrated, ammonia was not detected.

The measured ammonia concentration in the light fraction is in contradiction with to the physical properties of ammonia, arsine and phosphine. The boiling point of ammonia (−33°C) is considerably higher than that of arsine (−62°C) and phosphine (−85°C). From this point of view ammonia had to concentrate in the heavy fractions of arsine or phosphine and should be absent in the light fractions. Table 1 shows that it is not so.

Arsine and phosphine have a critical temperature higher than room temperature and are kept in the balloons in the form of two-phase equilibrium system - liquid and gas phase. The measurements of the ammonia content in the gas and liquid phase of arsine and phosphine showed that ammonia has a higher content in the gas phase (Table 2).

Using data of ammonia concentrations in gas and liquid phase, the dependence of ammonia concentration in gas phase versus its concentration in liquid phase was plotted. The concentrations in the gas phase were measured when a balloon was in normal position, from the valve on its top. The concentrations in the liquid phase were measured when a balloon was turned upside-down. A small portion of liquid taken from the same valve in upside-down position was fully evaporated and the concentration was measured in this gas phase. The results are given in Fig. 5 along with the Raoult's law prediction [13] for an ideal gas. The saturation pressures, required for the calculation of the Raoult's law dependence, were taken at 8 and 14 atm for ammonia and arsine, respectively. The deviations of the experimental results from an ideal gas model are clearly seen.

Ammonia has the lowest saturated vapor pressure in comparison with arsine and phosphine, including that at room temperature. Based on this, ammonia in the balloon with arsine and phosphine should be concentrated in the liquid phase. The observed anomaly is connected with the fact that in the solutions of arsine and phosphine NH_3 reveals properties different from the properties of pure ammonia. In the solution the effect of intermolecular interaction between the molecules of ammonia apparently decreases.

Table 2 Contents of ammonia in the gas and liquid phases in the balloons with arsine and phosphine

Hydride	Balloon number (volume, l)	NH_3 content, % vol		
		in gas phase, Cg	in liquid phase, Cl	Cg/Cl
AsH_3 from As_2Mg_3	075(2)	15.5	6.2	2.5
	0011(4)	17.1	7.6	2.3
PH_3 from P_2Mg_3	058(4)	0.043	0.030	1.4

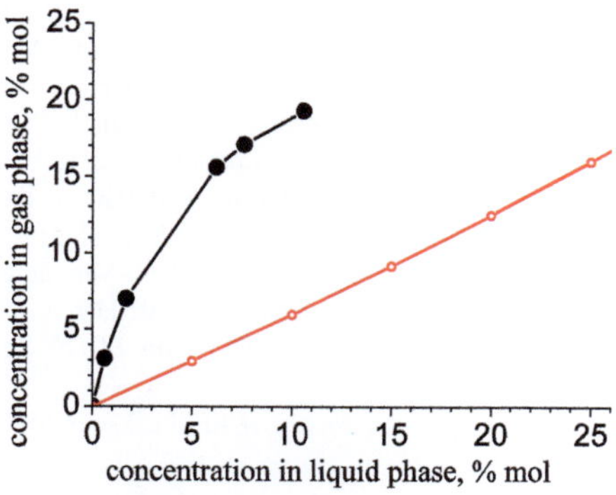

Fig. 5 Ammonia concentration in gas phase vs. its concentration in liquid phase. *Red line*: Raoult's law for ideal gas

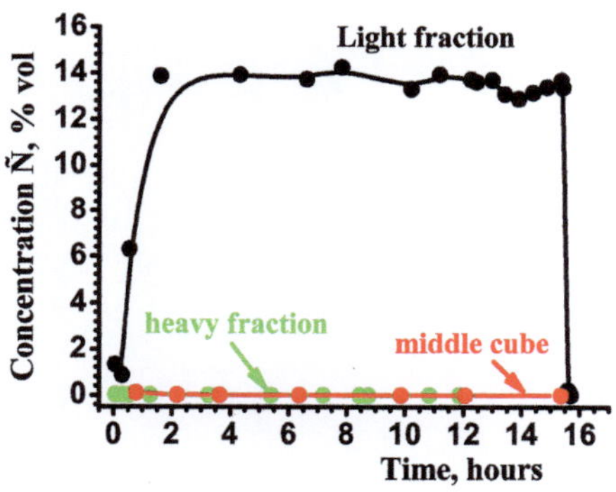

Fig. 6 Time dependence of HN_3 concentration measured in three TDLS channels during the distillation process

The time dependencies of ammonia content in different parts of the distillation column were also analyzed. Figure 6 shows the behavior of the ammonia concentration in the arsine purification process. In the course of purification an increase of ammonia concentration to the value of ~14 vol.% was observed in the upper part and it remained constant for the 16 hours of purification. Simultaneously the contents of ammonia in the heavy fraction and in the middle cube were monitored. In the lower and in the middle parts of the distillation column, the ammonia concentration was below the detection limit (0.01%).

The experimental results again confirmed that in the solution of arsine ammonia behaves anomalously. It is concentrated in the light fraction; in this case its apparent boiling point is lower than in arsine, i.e., lower, than −62°C.

The arsine and phosphine purification processes were carried out with continuous monitoring of the NH_3 content

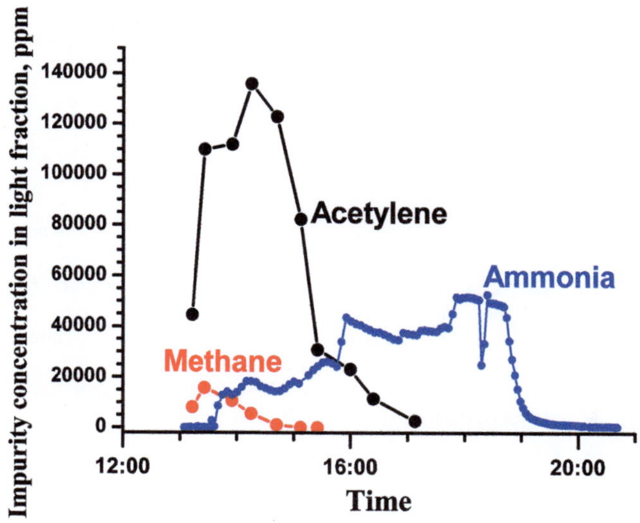

Fig. 7 Monitoring of impurities in light fraction of PH_3 from sampling time (time at which sample is removed)

in the samples of the light fractions by the TDLS method, together with Fourier spectrometer and gas chromatograph methods. In the samples of the light fraction of arsine, besides NH_3 such impurities as acetylene, phosphine and hydrocarbons were detected, and in the samples of the light fraction of phosphine, methane and acetylene were detected. Figure 7 shows a change in the impurity content in the light fraction sample of phosphine according to the data from the Fourier spectrometer and gas chromatography from the sampling time (the time in the process at which samples were extracted). It is evident that at first methane was taken away, then acetylene, and only after a rather long time (6 hours) ammonia was taken off.

In the process of the removing the arsine light fraction a separation process was also observed. At first more volatile phosphine was detected, then acetylene and then ammonia. Thus, it was found that ammonia is one of the basic impurities in arsine and phosphine, synthesized by hydrolysis of As_2Mg_3 and P_2Mg_3 in the solution of hydrochloric acid.

4 Conclusion

With the aid of the TDLS method the continuous monitoring of the NH_3 content in the sampling of the light fraction, middle cube and lower part of the distillation column in the process of AsH_3 and PH_3 low-temperature purification was successfully realized and tested.

It was found that, in the solutions of arsine and phosphine, NH_3 manifests the properties of more volatile impurity with respect to these hydrides, although judging by their boiling points, NH_3 must be less volatile. During the purification of arsine and phosphine, ammonia is taken off longer than all highly volatile impurities. Therefore the continuous monitoring of the NH_3 content in the light fraction is necessary in order to determine the moment of the end of ammonia removal.

References

1. J.V. DiLorenzo, D.D. Khandelwal (eds.), *GaAs FET Principles and Technology* (Artech House, Norwood, 1982)
2. G.G. Devyatykh, A.D. Zorin, *High-Purity Volatile Inorganic Hydrides* (Nauka, Moscow, 1974) (in Russian)
3. A.F. Zhigach, D.S. Stasinevich, *Hydrides Chemistry* (Chimia, Leningrad, 1976) (in Russian)
4. N.T. Ivanova, N.A. Vislykh, V.V. Voevodina, High-Purity Subst. **5**, 198 (1990) (in Russian)
5. S. Monroe, J. IEST **41**(1), 21 (1998)
6. A.G. Berezin, A.I. Nadezhdinskii, Ya.Ya. Ponurovskii, D.B. Stavrovskii, I. Vyazov, A.P. Kotkov, V.A. Ivanov, N.D. Grishnova, A.V. Shirayev, V.A. Sidorov, D.A. Kotkov, Appl. Phys. B **90**, 317 (2008)
7. www.nel-world.com
8. A. Berezin, O. Ershov, A. Nadezhdinskii, Appl. Phys. B **75**, 203 (2002)
9. www.cfa.harvard.edu/HITRAN/
10. L.S. Rothman, I.E. Gordon, A. Barbe, D. Chris Benner, P.F. Bernath, M. Birk, V. Boudon, L.R. Brown, A. Campargue, J.-P. Champion, K. Chance, L.H. Coudert, V. Dana, V.M. Devi, S. Fally, J.-M. Flaud, R.R. Gamache, A. Goldman, D. Jacquemart, I. Kleiner, N. Lacome, W.J. Lafferty, J.-Y. Mandin, S.T. Massie, S.N. Mikhailenko, C.E. Miller, N. Moazzen-Ahmadi, O.V. Naumenko, A.V. Nikitin, J. Orphal, V.I. Perevalov, A. Perrin, A. Predoi-Cross, C.P. Rinsland, M. Rotger, M. Simeckova, M.A.H. Smith, K. Sung, S.A. Tashkun, J. Tennyson, R.A. Toth, A.C. Vandaele, J. Vander Auwera, J. Quant. Spectrosc. Radiat. Transf. **110**, 533 (2009)
11. S. Schilt, in *Abstracts of 7th Int. Conf. on TDLS*, Zermatt, July 2009, p. 12
12. H.L. Glinka, *General Chemistry* (Chimia, Leningrad, 1988) (in Russian)
13. P. Perrot, *A to Z of Thermodynamics* (Oxford University Press, Oxford, 1998). ISBN:0198565569

Measurements of the temperature and water vapor concentration in a hot zone by tunable diode laser absorption spectrometry

M.A. Bolshov · Y.A. Kuritsyn · V.V. Liger · V.R. Mironenko · S.B. Leonov · D.A. Yarantsev

Received: 25 September 2009 / Revised version: 11 December 2009 / Published online: 27 January 2010
© Springer-Verlag 2010

Abstract A tunable diode laser absorption spectroscopy (TDLAS) technique and appropriate instrumentation was developed for the measurement of temperature and water vapor concentrations in heated gases. The technique is based on the detection of the spectra of H_2O absorption lines with different energies of low levels. The following absorption lines of H_2O were used: 7189.344 cm^{-1} ($E'' = 142$ cm^{-1}), 7189.541 cm^{-1} ($E'' = 1255$ cm^{-1}), 7189.715 cm^{-1} ($E'' = 2005$ cm^{-1}). Spectra were recorded using fast frequency scanning of a single distributed feedback (DFB) laser. A unique differential scheme for the recording of the absorption spectra was developed. An optimal technique for fitting the experimental spectra was developed.

The validated TDLAS technique was applied for detection of temperature and H_2O concentration in the post-combustion zone of a supersonic ($M = 2$) air-fuel flow. Hydrogen and ethylene were used as the fuel. The combustion process was ignited and sustained by a pulsed electric discharge.

Presentation of the transient absorption spectra as 2D images was used as the first step of data processing. The estimated precision of the temperature measurement was ± 40 K. The high signal-to-noise ratio enabled the reconstruction of the temporal behavior of temperature with a resolution of ~ 1 ms.

M.A. Bolshov (✉) · Y.A. Kuritsyn · V.V. Liger · V.R. Mironenko
Institute for Spectroscopy, Russian Academy of Sciences,
5 Fizicheskaya str., 142190 Troitsk, Moscow reg., Russia
e-mail: bolshov@isan.troitsk.ru
Fax: +7-496-7510886

S.B. Leonov · D.A. Yarantsev
Joint Institute for High Temperatures, Russian Academy of Sciences, 13/2 Izhorskaya str., 125412 Moscow, Russia

1 Introduction

Tunable diode laser (DL) absorption spectroscopy (TDLAS) is a widely used spectroscopic technique for the detection of various parameters of heated zones [1–7]. This technique provides remote, non-perturbing measurements of the parameters of a hot zone with time resolution in the μs-ms range depending on the specific experimental conditions. The technique is usually based on the measurements of the ratio of the absorption line intensities of a test molecule. If a Boltzmann distribution of the energy levels is established, the ratio of the line intensities depends only on the kinetic temperature of the object.

The technically easiest and most straightforward version of TDLAS is the detection of direct absorption. It works well in the case of high signal-to-noise ratio (SNR). For weak lines the well-developed wavelength modulation (WM) technique can be applied for reducing low-frequency noise (flicker noise) in the baseline and correcting for non-specific absorption [8].

For direct absorption, different approaches have been developed aiming to increase temporal resolution and accuracy of the temperature measurement. One can measure the ratio of the amplitudes of different absorption lines using two DLs, tuned to the selected lines. This approach provides high temporal resolution but is less accurate. More accurate is the measurement of the ratio of the integral line intensities. This can be done by scanning a single DL wavelength across the spectral interval with the lines of interest. In both cases the key problem is the correct estimation of the baseline in the detected experimental absorption spectra [9, 10].

The parameters of the probed medium are obtained as a result of experimental spectra fitting. In most cases the individual absorption lines are fitted using Voigt profiles. This approach does not require the exact mechanism of line broadening. The temperature is inferred from the ratio of the integrals (or amplitudes) of the selected lines using the ratio calculated from spectroscopic databases. The alternative is the fitting of a whole spectral interval, which includes the selected lines. In this case the spectroscopic parameters of the lines from the databases are used. One should correctly account for different mechanisms and fit experimental line profiles using selected models of line broadening. The problem of optimal fitting is especially serious if the baseline is irregular.

In this paper we present the results of temperature measurements by TDLAS with the main emphasis on the optimization of the algorithm for fitting the experimental spectra in the case of an irregular baseline. Molecular water was used as a test molecule. Water vapor is one of the major combustion products and key indicators of the extent of combustion and is therefore widely used as a tracer of combustion processes in mixed gas flows. Initially, the optimal TDLAS strategy was developed in laboratory experiments with stable conditions in an evacuated cell filled with the air. The optimized and validated version of the TDLAS technique was then applied to the measurement of the temperature, total gas pressure and H_2O concentration in a plasma-assisted supersonic combustion flow.

2 Basic principles

The basic equation of absorption spectroscopy is the Beer-Lambert law. The general exponential form of the law can be expanded for the case of low absorption to the form

$$\begin{aligned} I_\nu &= I_{\nu,0}\exp\big[-S(T)g(\nu-\nu_0)NL\big] \\ &\approx I_{\nu,0} - I_{\nu,0}S(T)g(\nu-\nu_0)NL, \end{aligned} \qquad (1)$$

where I_ν is the intensity of the monochromatic radiation with a frequency ν (cm^{-1}) transmitted by an absorbing medium of length L (cm), ν_0 (cm^{-1}) is the frequency of the line center, S (cm/mol) is the line strength (integral intensity), g (cm) is the normalized line-shape function, N (cm^{-3}) is the number density of H_2O molecules. The line strength S depends on temperature while the line-shape function depends on temperature, pressure, gas composition and the mechanisms of line broadening. For pressures of about an atmosphere the line shape is usually approximated by a Voigt profile. The typical absorption line width of the simple molecular components in mixing gas flows and flames is of the order of 0.1 cm^{-1} (3 GHz). Thus, single-mode DLs with typical line-widths ∼10–50 MHz can be treated as monochromatic light sources and (1) can be used for evaluation of the experimental data with high precision.

The line strength $S(T)$ depends on temperature in the form:

$$\begin{aligned} S(T) &= S(T_0)\frac{Q(T_0)}{Q(T)}\exp\left[-\frac{hcE''}{k}\left(\frac{1}{T}-\frac{1}{T_0}\right)\right] \\ &\quad\times\left[\frac{1-\exp(hc\nu/kT)}{1-\exp(hc\nu/kT_0)}\right], \end{aligned} \qquad (2)$$

where $Q(T)$ is the partition function, E'' is the lower state energy of the quantum transition, k is Boltzmann's constant, $S(T_0)$ is the line strength at a reference temperature T_0, h is Planck's constant and c is the speed of light.

The ratio of the line strengths of two different transitions depends only on the temperature and, assuming the second bracket in (2) is equal to unity, can be written in the form:

$$R = \left(\frac{S_1}{S_2}\right)_T = \left(\frac{S_1}{S_2}\right)_{T_0}\exp\left[\frac{hc\Delta E}{k}\left(\frac{1}{T}-\frac{1}{T_0}\right)\right], \qquad (3)$$

where ΔE is the energy difference between the lower levels of the selected transitions. Thus, from simultaneous measurement of the absorption at different lines one can deduce the temperature of a medium and from the measurement of the absolute absorption line intensity one can deduce the concentration of the absorbing species, provided that the spectroscopic parameters of the selected transitions are known.

For uniform media (1)–(3) give the actual temperature and concentrations. For non-uniform media path-integrated temperature and concentration can be calculated from the absorption spectra. The line strengths and broadening parameters for many transitions at ambient temperatures are well known [11–13]. However, the spectroscopic parameters of the selected transitions for different temperatures have to be validated [14, 15].

3 Laboratory experiments

3.1 Experimental cell

A simplified scheme for the experimental cell used for preliminary measurements is shown in Fig. 1. The cell was constructed from a quartz tube (50 cm long, inner diameter 32 mm) inserted in an electrically heated cylindrical oven. The absorbing air layer was formed by a quartz cylinder (10 cm long, outer diameter 22 mm) pressed between quartz windows attached to two cylindrical inserts (50 cm long, outer diameter 22 mm). The planes of both end surfaces of the cylinder were roughly ground and tilted slightly with respect to the optical axis. In this way the volume inside the cylinder was connected to the volume of the gap between the inserts and the main tube. The inserts were fixed inside the

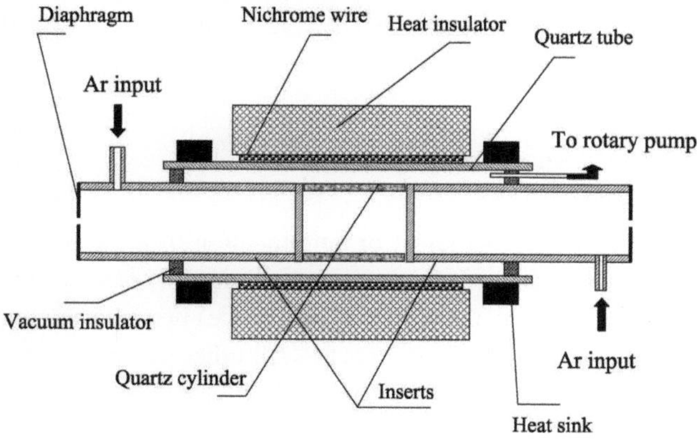

Fig. 1 Experimental cell (see explanation in the text)

main tube by vacuum insulators. The space between the inserts and tube could be evacuated by a rotary pump through a metallic needle. In order to provide a uniform temperature profile along the absorbing layer, the oven was 3 times as long as the central quartz cylinder, 30 cm.

To minimize DL absorption by laboratory air outside the central part of the cell the cylindrical inserts were continuously flushed with a stream of pure argon. The temperature of the absorption air layer was measured by a commercial thermocouple PS2007 (Instrument Specialists Inc., USA) fixed inside the main tube close to the central cylinder containing the absorbing air. The air pressure of the absorbing layer was maintained or changed by a rotary pump and an inlet valve (not shown in Fig. 1) and measured by a pressure sensor G16K (BOC Edwards, UK).

3.2 Diode laser

A pig-tailed distributed feedback (DFB) diode laser (NLK1E5GAA, NEL, Japan) with output in the 1.39 μm range was used in these experiments. The temperature and injecting current controllers were TED350 and LDC202 (Thorlabs), respectively. Fast tuning was provided by modulating the injection current. For a 1 kHz modulation frequency and a DL temperature of 25°C the tuning range was $\sim$1.2 cm^{-1} for a current saw-tooth ramp from 10 to 120 mA. Fast wavelength tuning within $\sim$1 cm^{-1} was accompanied by strong amplitude modulation. The injection current variation over this current range causes a gradual increase of the DL intensity from 3 to 30 mW.

Absolute wavelength calibration was done by recording the H_2O room temperature absorption spectra over the entire tuning range of the DL. The absorption lines were identified using the HITRAN database [9]. Most experiments were performed using the following H_2O lines: 7189.344 cm^{-1} ($E'' = 142$ cm^{-1}), 7189.541 cm^{-1} ($E'' = 1255$ cm^{-1}), 7189.715 cm^{-1} ($E'' = 2005$ cm^{-1}). These lines were selected because of the reasonable difference between the low energy levels and because they were reasonably well resolved. Henceforth, these lines will be denoted as 1, 2 and 3, respectively.

The saw-tooth current modulation was provided by a home-made generator, which was synchronized with the oscilloscope used for data acquisition.

We have checked the linearity of the DL wavelength tuning within the selected spectral interval ($\sim$1.2 cm^{-1}). The Fabry-Pérot etalon made of uncoated glass cylinder ($L = 100$ mm) was placed in the beam about 1.5 m from the DL. The wavelength was tuned linearly within about 2/3 of the scanning range. The F-P fringe spacing was constant in the central part of the scan but varied at the scan extremes. To simplify the recorded spectra processing the DL temperature and limits of the DL injection current modulation were adjusted to keep the selected absorption lines within the linear part of the scan.

3.3 Optical scheme

A simple optical scheme was used for the laboratory experiments. The DL radiation was routed to an entrance diaphragm fixed at the right insert (Fig. 1) via a single-mode fiber (core ∅ 9 μm). The transmitted DL beam was focused by a lens (focal length 5 mm, diameter 7 mm) onto an InGaAs photodiode (Hamamatsu, G8370-2) with 2 mm sensitive area. A lens with high numerical aperture was selected to increase signal stability. The lens was placed just behind the diaphragm fixed at the exit of the left insert (Fig. 1) to minimize the free path of the DL beam outside the experimental cell.

To minimize the influence of baseline fluctuations and strong amplitude modulation of DL intensity, a differential recording scheme was developed. The DL pig-tail was connected to a commercial fiber multiplexer (Poisk-Tr. Ltd, Moscow), which divided the beam into two equal parts. One fiber (1.5 m long single-mode) routed half of the beam to the cell; the second fiber routed the second half to the reference channel. The output of each fiber was coupled with a commercial gradient collimator. The diameters of both output beams were 1.8 mm with a divergence of $\sim 10^{-3}$ (the beam diameter increased to 5 mm at 5 m distance). An identical InGaAs photodiode (Hamamatsu, G8370-2) was used in the reference channel. The output collimator of the reference channel was placed in close proximity to the photodiode to minimize the free path of the reference beam in laboratory air.

3.4 Recording system

The differential scheme which was developed was based on the principle proposed in [16]. The differential signal is the photocurrent generated at the common point of an operational amplifier. At this point the currents from the signal and reference photodiodes are combined, the polarity of the currents being opposite. To compensate for the variations of the signal photocurrent, which are not caused by H_2O absorption (intensity ramp, low-frequency random amplitude modulation, technical noise etc.) the currents in both channels must be well equalized. This is done by tuning the differential current to zero when there is no H_2O absorption in the signal channel. A pair of matched MAT04 transistors (Analog Devices, Inc.) automatically equalize both channels. As a result, any variations of the photocurrent in the signal channel, which are not caused by H_2O absorption, are automatically compensated by the reference channel and the output of the differential scheme is equal to zero. The efficiency of the compensation of low-frequency amplitude modulation in the signal channel depends on the similarity of the characteristics of both channels. With modern electronic components the difference of the two channels can be easily reduced to 40 dB.

Besides this differential output the electronic scheme has a second output. The current of this output is proportional to the instantaneous value of the DL intensity. This signal is used to normalize the absorption signal according to (1). The output currents of the differential and second channels are digitized for further processing. Both signals were recorded by a dual channel 60 MHz, 1 GS/s digital oscilloscope (Agilent DSO3062A).

The bandwidth of the photodetector-amplifier tandems were 470 kHz. It was broad enough to reproduce linearly the shapes of the absorption lines.

3.5 Measurement procedure

The temperature of the absorbing layer was changed by regulation of the electric current of the oven. The strategy for temperature measurements by TDLAS was validated over an interval of 300–1200 K with increments 100 K. After each step of the current increase, the temperature stabilized in about 20 minutes. The actual temperature and its stability were measured by the thermocouple.

The DL was switched on and continuously scanned across the selected spectral range. The duration of each scan was ~850 µs, the saw-tooth profile of the injection current provided a DL wavelength tuning over an interval of 7189.0–7190.0 cm^{-1}. To a first approximation the variation of DL intensity linearly depends on the injection current. During one measurement cycle the transmitted DL intensity was recorded, averaged and stored in the oscilloscope memory. An average of 16 or 32 scans was typically used.

3.6 Optimization of the fitting model

All absorption lines detected and processed in our experiments were weak enough to use the linear approximation in (1). The differential signal Y_i at a frequency ν_i detected by the recording system can be rewritten in the form:

$$Y_i = \alpha I_{0i} \sum_j S_j(T) g_j(\nu_i - \nu_{0,j}) N L + b_i + \varepsilon_i, \tag{4}$$

where α is the coefficient of amplification of the electronics, ν_{0j} is the center of the absorption line j, b_i presents the baseline and ε_i is the residual (difference) between the experimental and simulated spectra. All other parameters are denoted for (1).

Two different fitting methods were compared. The first one is the common profile fitting method in which each absorption line is fitted separately using different kinds of modeling line profile. We tried the following theoretical line shapes: Voigt function with independent Gaussian and Lorentzian parts (VP), Voigt profile with equal Gaussian width for both lines and different Lorentzian widths (VPEG), and a Rautian profile (RP) [17], which takes into account Dicke narrowing. Parameters characterizing line profiles (center, area, width and so on) were adjustable. The baseline was approximated by the polynomial $b_i = \Sigma k_n Q_{in}$, with k_n for the fitting parameters, and Q_{in} for the orthogonal polynomial. A polynomial rank of 2 or less was used. The temperature was determined from the ratio of the areas under two fitted lines by comparing this ratio with the theoretical one.

The second approach was spectrum fitting (SF), in which the experimental spectrum was fitted to a theoretical one calculated using spectral databases. The fitting was started with the matching of the maxima of the selected lines to the tabulated reference values. Linear scanning of the DL wavelength with time during a scan was assumed. At the next step physically reasonable values of the initial temperature T and total pressure P were selected and a theoretical spectrum in the region of interest was calculated using spectroscopic parameters from the HITRAN 2006 database [11] and partition function Q(T) from the SPECTRA database [13]. Then, using the nonlinear least-squares Nelder-Mead method [18], the values of temperature and total pressure were found from the best fit of experimental spectra, e.g., from the minimal value of $\sqrt{\langle \varepsilon_i^2 \rangle}$.

An example of the experimental spectra fitting is presented in Fig. 2. Part of the spectra with two lines #2, 3 for an air temperature ~1000 K is shown in the upper trace, the residuals for two models—Spectrum Fitting (SF) and Voigt Profile Fitting with equal Gaussian width (VPEG) are shown below.

In Fig. 3 temperatures derived from DL spectra using different fitting procedures are plotted versus temperatures measured by the thermocouple. It is noteworthy that the minimal residual is obtained for the VPEG (Fig. 2c), but

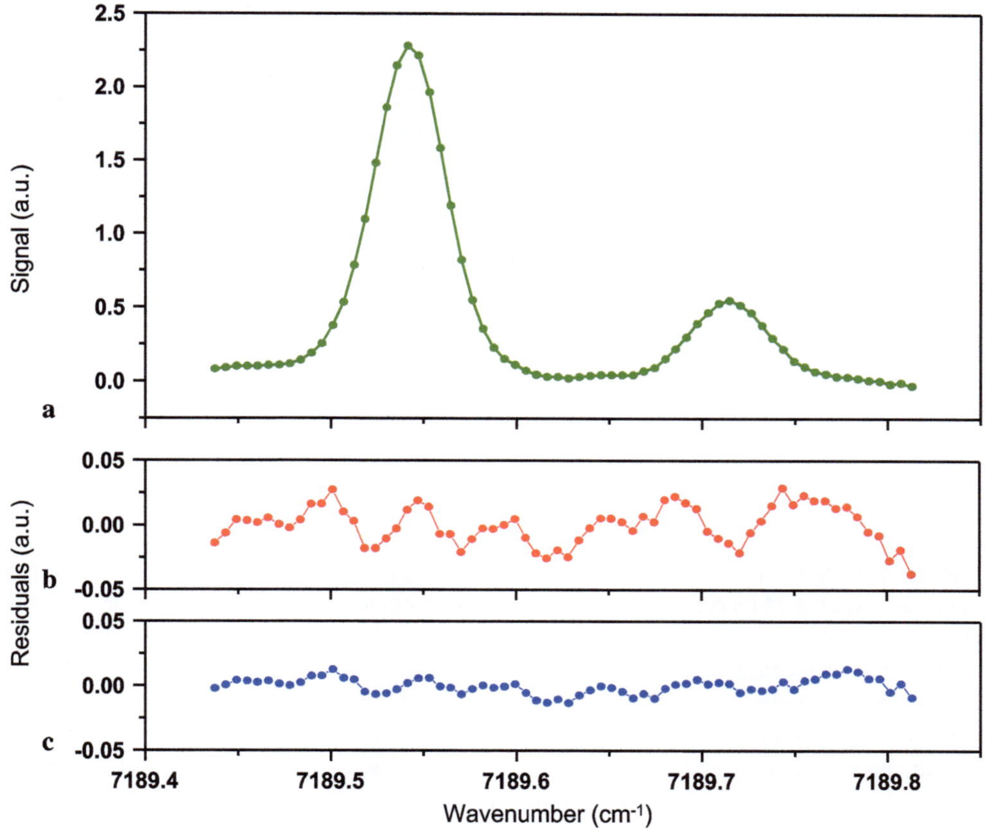

Fig. 2 Example of the experimental line fitting. (**a**) H_2O absorption lines ##2, 3; (**b**) residual for SF model; (**c**) residual for VPEG model

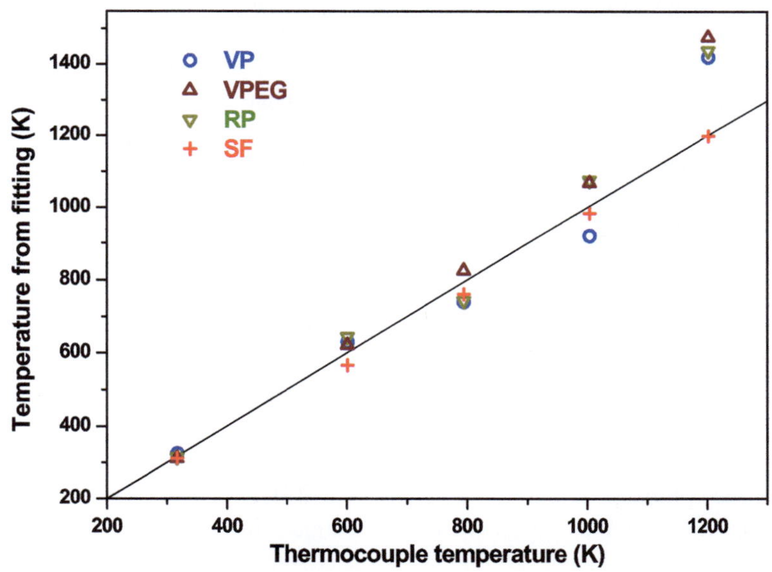

Fig. 3 Temperature evaluated by TDLAS vs temperature measured by the thermocouple

this model gives a large deviation of the evaluated temperatures from temperatures measured by the thermocouple, particularly for temperatures above 1000 K (Fig. 3). For these temperatures the variations of the baseline were more pronounced from scan to scan as compared to the temperature range (300–800) K. It can be explained by the increased fluctuations of the DL beam inside the cell caused by higher temperature gradients. We have not tried to stabilize the baseline because the developed technique and algorithm of data processing must be robust enough to be applied in real experimental conditions of high acoustic and electrical noise in a test combustion chamber.

The SF model shows larger minimal residual, but provides better accuracy (Fig. 3). This result can be explained by the influence of the baseline variations on the weak line. These variations have modified the actual profile of the weak

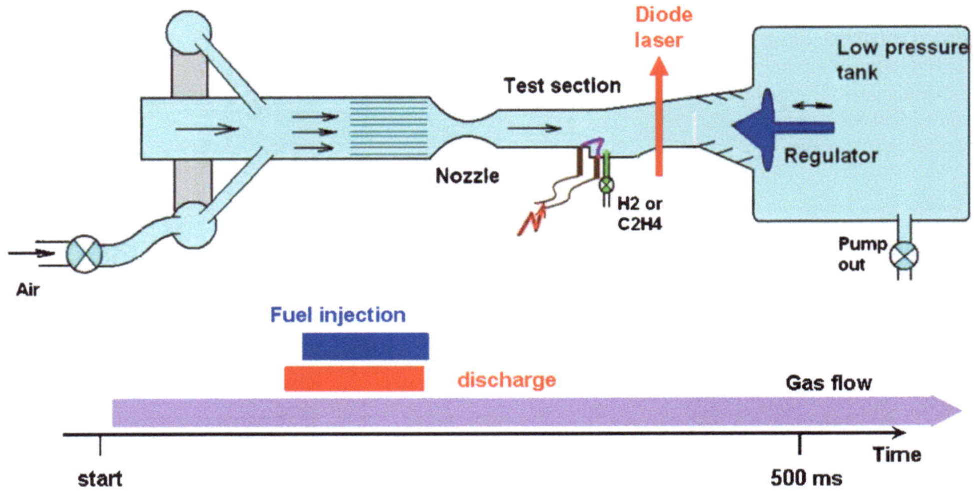

Fig. 4 Test chamber of the aerodynamic tube—*upper figure*; time scale of a run—*lower figure*

line. As a result the program fits weak lines quite precisely but at the expense of increasing the parameter of line broadening and, hence, by increasing the integral line intensity. The SF fits the entire spectral interval using actual line broadening parameters from HITRAN database. The SF calculates the Gaussian part from the fitting parameter T; the Lorentzian part is calculated using the total pressure and water vapor concentration as the fitting parameters. In this way broadening of the water lines by air and self-broadening are taken in consideration. Figure 3 shows that this self-consistent scheme provides a better accuracy for temperature evaluation.

The systematic error (accuracy) of temperature evaluation by SF model varied for different temperatures from 5 to 40 K. Basing on the statistical processing of the set of experimental spectra we estimated the precision (statistical errors) of temperature evaluation by SF model to ~30 K. The main part of this error is defined by irregular variations of the baseline from scan to scan. We did not invest much effort to minimize these variations because the developed technique has to work in the harsh environment of a supersonic facility, where such variations as caused by electronic noises, thermal and acoustic shocks were expected to be much stronger.

The results of the laboratory experiments allowed for the selection of the optimal procedure for temperature and H_2O concentration measurements by the TDLAS technique. In our opinion, when the baseline exhibits pronounced variations from scan to scan, the most accurate values for temperature are given by the SF model.

4 Measurements of the parameters of the hot zone in the plasma-assisted combustion

4.1 Experimental chamber

The test chamber was located at the Joint Institute for High Temperature RAS. Experiments were conducted in a closed-type section of the supersonic aerodynamic tube, shown in Fig. 4. The details of the facility are described elsewhere [19]. The length of the section from the exit of the shock tube to the diffuser is $X = 600$ mm, the cross section varies from $Y \times Z = 60 \times 72$ mm at the entrance to $Y \times Z = 72 \times 72$ mm at the exit in downstream direction. The section has three equidistant pairs of windows (100 mm diameter). The air flow parameters are: Mach $M = 2$, static pressure 150–300 Torr, kinetic temperature 300 K.

The pulsed electric discharge, which ignites and sustains the combustion in supersonic flow, is generated at a flat ceramic wall between the electrodes. The electrodes are fixed at the wall in an anode-cathode-anode scheme. The line direction is perpendicular to the direction of the flow. The discharge current varied from 2–8 A, the applied power varied within 3–9 kW. The applied electric power does not exceed 5% of the kinetic power of the gas flow. The flow of gaseous hydrogen is injected a few cm downstream from the discharge. The duration of a single cycle including air flow injection, plasma ignition, fuel injection, combustion and cooling is about 500 ms. The discharge is switched on after the gas flow achieves a supersonic steady-state regime; the duration of the discharge pulse is 80–100 ms. The fuel injection starts 20–30 ms after plasma ignition and ends about 20–30 ms after the electric pulse is off. Visually, the combustion ends just after the discharge is off.

The DL beam with diameter of 2 mm probed the zone inside the chamber located about 30 cm downstream from the electrodes and about 3 mm from the wall with the installed electrodes. At this position the combustion process is almost finished and most part of the fuel is consumed. The experimental results on the water vapor concentration in this zone are very important for estimation of the efficiency of the combustion process and fuel consumption. Preliminary results, obtained by Schlieren photography, has indicated more uniform distribution of the refractive index profile in this zone as compared to the zone of bright plasma emission.

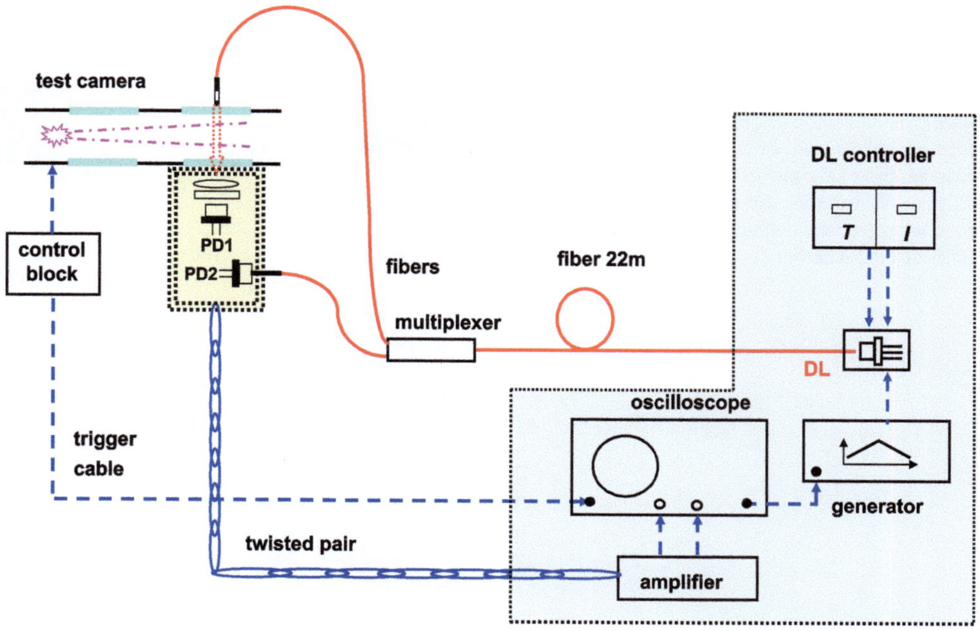

Fig. 5 Experimental set-up. DL—diode laser; PD1, 2 photodiodes; oscilloscope—Agilent 54621A; generator—saw-tooth voltage generator, DL controller (T—temperature, I—injection current)

During the experiments the mass of injected fuel and the discharge parameters were varied. The time scale of a single run is presented in the lower part of Fig. 4.

4.2 Optical scheme

The optical scheme was modified for the actual experimental conditions of the chamber. The results of the preliminary experiments have shown that DL beam transport to the test chamber of the aerodynamic tube using a free-space approach was absolutely unsuitable. Firstly, the strong acoustic vibrations of the facility caused serious problems in reproducibly coupling the DL beam to the focusing optics. Secondly, the absorption of the DL by the water vapor of laboratory atmosphere dramatically increased the absorption signal of the "room temperature" line 7189.344 cm^{-1}. In addition, strong electric noise of the pulsed plasma facility increased the instability of the DL drivers and, hence, the instability of DL operation. All these reasons dictated the location of the sensitive electronics (drivers, optical block with DL, recording system) in a separate room away from the facility and transporting the DL beam to the chamber by a fiber.

The final version of the experimental set-up is shown in Fig. 5. All parts of the set-up constrained by the dashed line in Fig. 5 were located in the separate room. The DL beam was routed to the chamber via a 22 m long single-mode fiber (core ∅ 9 μm). Besides this 22 m long fiber all other components were the same as in the laboratory experiments: fiber multiplexer, which divided the beam into two channels (signal and reference), end gradient collimators, collecting lens and photodiodes. The optical path inside the test chamber was 7 cm. The DL beam probed a cross section of about 2 mm within the combustion zone. The collimator of the signal beam was fixed in an optical head, which was solidly mounted at the input widow flange of the chamber. The head could be precisely translated in x-y directions and angularly aligned.

The collecting lens, photodiode and first preamplifier of the recording system were located in an electromagnetically isolated box. To minimize the free-path distance of the transmitted beam the box was mounted directly on the opposite window flange. The lens could be aligned in the same way as the collimator. An optical filter IKS-7 placed before the photodiode cut off the UV- and visible part of the spectra, transmitting the 0.9–2.6 μm spectral range. The use of the filter greatly reduced the broadband emission background of plasma and the combustion zone.

4.3 Recording system

The recording scheme described above was slightly modified. The output currents of the differential channel and second channel, proportional to the instant DL intensity, were recorded by a dual channel digital oscilloscope (Agilent 54621A). To decrease the strong electrical noise from the plasma generator both signals were transmitted to the recording system, located in the separate room, via an electrically isolated 22 m long twisted pair. The oscilloscope parameters are: 60 MHz bandwidth, 200 MHz sampling frequency, 2×10^6 point record length in each channel. The differential scheme greatly reduced the variations of the baseline, which enabled the use of the fast 8-bit ADC of the oscilloscope for signal digitalization.

The measurement cycle started with a trigger pulse, which switched on the system of gas flow, fuel injection and

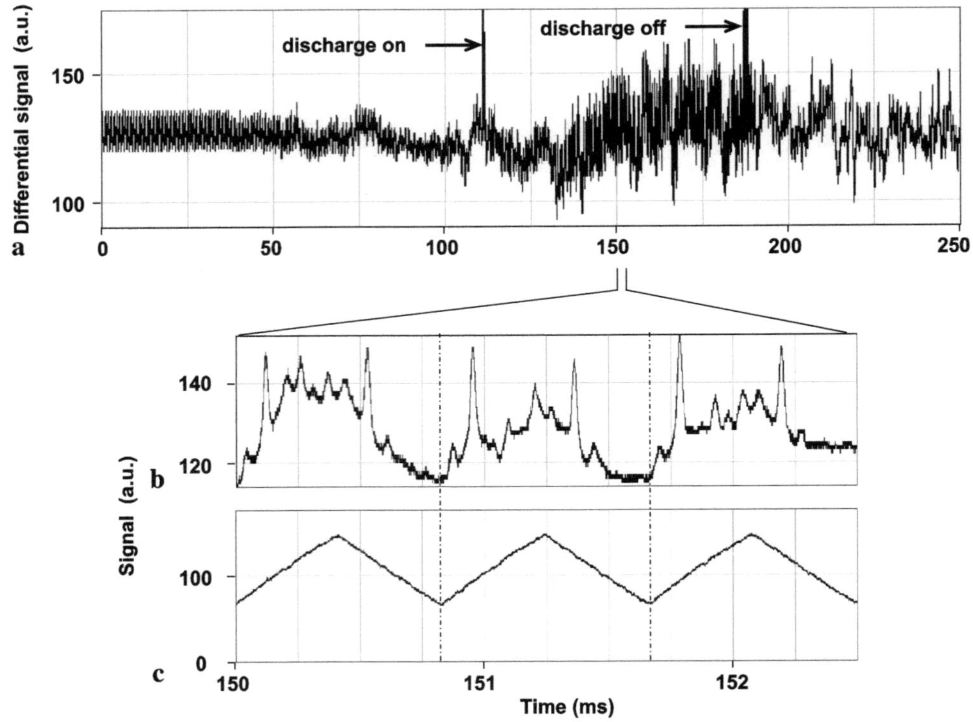

Fig. 6 (**a**) Oscilloscope trace of the first 250 ms of a run (differential output of the electronic scheme), (**b**) fragment of this run corresponding to three round-trip scans of DL wavelength, (**c**) the saw tooth DL intensity variation during these scans (second output of the electronic scheme)

plasma ignition (box 2 in Fig. 5). In this box a trigger pulse for the oscilloscope sweep was generated. The oscilloscope sweep was synchronized with the beginning of the DL injection current saw-tooth scanning.

4.4 Primary processing of the non-stationary absorption spectra

In these experiments hydrogen and ethylene were used as the fuel. The water vapor concentration was high enough to detect direct absorption with the appropriate SNR.

Two one-dimension files with 2×10^6 points each were detected in a single run. The full registration time was 500 ms. The first file contains about 600 scans of differential signals, the second one about 600 scans of the incident DL intensities, detected in the second channel. The raw data detected in the first 250 ms of a single run are presented in Fig. 6. The aerodynamic parameters of this run were: static pressure in pre-mixing zone, 130 Torr, fuel mass flow rate G_f 0.5 g/s.

The times of plasma on and off are indicated in Fig. 6. The run and the oscilloscope scan are started by a trigger pulse. 30 ms after the start, the air valve opens and 20 ms later the gas flow reaches the supersonic regime ($M = 2$). 120 ms after the start the discharge is ignited, the fuel (in this case hydrogen) is injected and the combustion process starts. At about 190 ms the discharge is off and the combustion ends.

In each scan the DL intensity varies during the round-trip cycle of wavelength tuning due to injection current modulation and the water vapor absorption. Short fragment of 3 scans (about 2.5 ms) is presented in Fig. 6b. This fragment corresponds to about middle of the run. Figure 6c shows 3 corresponding scans of DL intensity registered in the second channel. These temporal profiles of DL intensity were used to infer temperature and water concentration.

In the initial step of data processing the file in Fig. 6a was transformed into a 2D image. This procedure greatly simplified the general overview of the data and selection of the most important periods of the combustion process. In the following steps only the periods containing important information were processed. The algorithm used for the digital processing of the 2D images is based on the ImageJ software with open source code [20]. This algorithm enables comfortable visual presentation of the experimental data and significantly reduces the processing time. The result of such processing of the file from Fig. 6a is shown in Fig. 7. Figure 7a shows a 2D image of the differential signals detected during the first 250 ms of the run. The 2D image is constructed from the individual successive scans. In each scan the DL intensity varies according to the variation of the injection current (linear ramp) and H_2O absorption when the DL wavelength coincides with the wavelength of the absorption line. In this way the H_2O absorption spectra is registered in each scan. In the image presented in Fig. 7a the frequency ν of DL radiation changes from left to right (x axis). During the first half of the scan, ν decreases with the increase of DL injection current and vice-versa during the second half. All scans are arranged one below the previous one with the number of the scan n increasing from top to bottom (y axis).

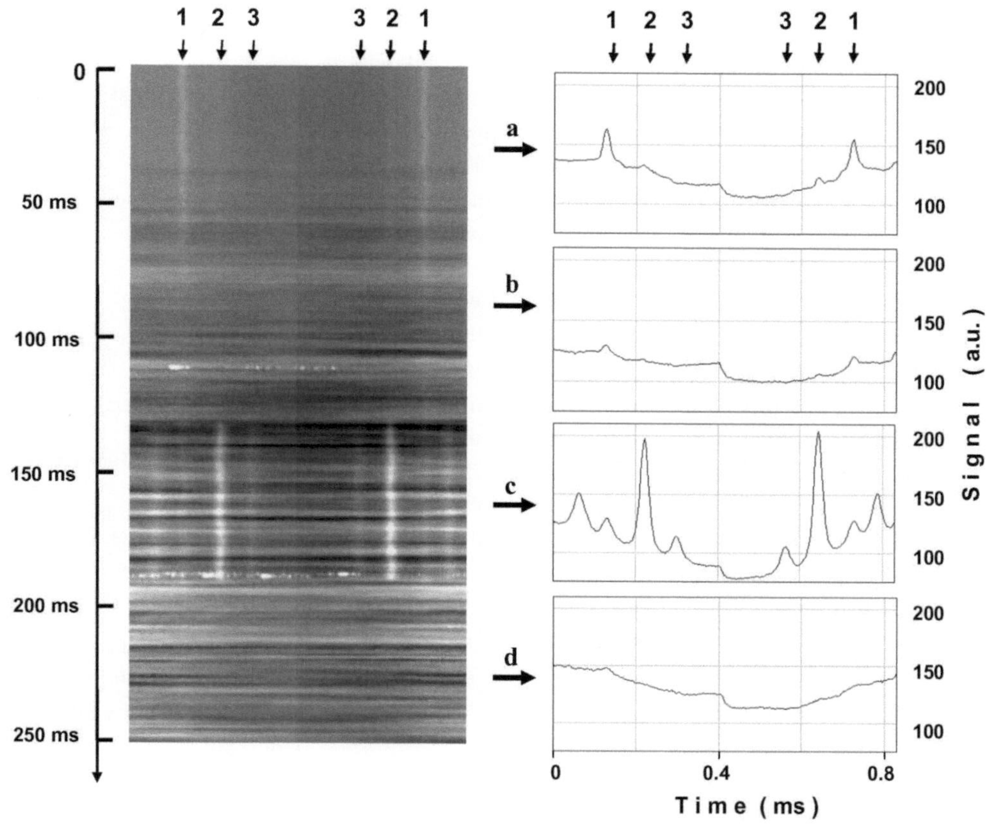

Fig. 7 2D image of the raw data (*left*) and H_2O absorption lines integrated over 30 scans (*right*). The time intervals (**a**–**d**) are explained in the text. The spectral lines used for processing are labeled by numbers. Line attribution: (1)—7189.344 cm^{-1}, (2)—7189.541 cm^{-1}, (3)—7189.715 cm^{-1}

Thus, the intensity F of the pixel of the image with coordinate (x, y) corresponds to the intensity of the transmitted DL radiation I at the moment t:

$$F(x, y) = I\big(n(t), \nu(t)\big).$$

The origin of this 2D image is the upper left corner. The y axis is directed in the vertical direction from origin down and corresponds to the number of a scan n, i.e. to the time from the beginning of the run ($t = 0$) to the end. The x axis is directed from the origin to the right. Thus, the picture constructed presents a 2D image of the time development of the absorption spectra of the probed zone. The interval of DL frequency tuning (axis x) is about 1 cm^{-1}, the full duration of one scan is about 830 μs, the duration of the presented part of the scans (axis y) is about 250 ms. The software allows one to select and process any part of the image (time interval) and to average any number of scans.

In the experiments with plasma-assisted combustion the baseline exhibited much stronger variations, as compared to the laboratory experiments, during one scan and from scan to scan because of strong turbulence in the probed volume. The dark horizontal lines in Fig. 7a are the result of strong deviation of the DL beam from the adjusted trajectory in a particular scan (or scans).

The absorption lines are well seen in the 2D image as the light vertical stripes. The different phases of the combustion evolution, denoted by the letters (a–d) can be distinguished. Figure 7 right presents the H_2O absorption spectra reconstructed from the 2D image in Fig. 7 left.

These spectra are the result of averaging over 30 scans in four different characteristic intervals of the run. During the time interval labeled by **a** in Fig. 7 the zone of observation is filled with air at room temperature. The "cold" line 7189.344 cm^{-1} is well seen both in the image and in the spectra (Fig. 7). At the end of the interval **a** air flow reaches the supersonic regime, which results in a decease of the gas temperature to ~200 K. The temperature drop causes freezing of water vapor. As a result, the intensity of the water absorption spectra starts to decrease and after approximately 20–30 ms disappears, which is clearly seen both in the image and spectra in Fig. 7 (interval **b**). About ~110 ms after the beginning the discharge is ignited and ~20 ms later the fuel is injected. The combustion process starts at this moment and the temperature rapidly increases. The temperature increase is well illustrated by the appearance of the H_2O "hot" lines: 7189.715, 7189.541 and 7189.199 cm^{-1}. In the same period the "cold" line 7189.344 cm^{-1} dramatically decreases. This structure of the spectra remains over a time interval of ~50 ms when the combustion is sustained (interval **c** in both figures). After the plasma is switched off the combustion rapidly ends and the water lines disappear (interval **d**). During this time interval the conditions of the supersonic flow are similar to those during interval **b**.

Note, that in the described experiments the deviations of the DL beam inside the chamber, variations of total and local concentration of water molecules, beam attenuation (scattering) due to aerosol formation inside the combustion zone and the tail of the flame were much stronger as compared to the laboratory experiments. Because of this the variations of the differential signal and variations of the baseline from run to run were rather strong. To keep the signal within the dynamic range of the ADC we had to decrease the amplification of the electronic system and, hence, the effective dynamic range of the absorption signal recording to about 5 bits (compare to 8 bits in the laboratory experiments). This loss of the linear dynamic range was partly compensated by the short time of digitizing of the used ADC (250 ns). The fast response of the ADC enabled to cut off high-frequency noises by a low-frequency digital filter applied at the first step of data processing. As a result, each point of the reduced file was an average of 16 raw points. In this way the level of effective digitization was improved to 7-bit.

4.5 Evaluation of the hot zone parameters

Based on the results of the laboratory experiments described in section 3.6, the SF model was used to fit the experimental spectra. For the run shown in Fig. 7 fitting gave a temperature of the probed zone of $T = 1050$ K, and a water vapor concentration (partial pressure) of 21 Torr. The developed model provided fitting with $\sqrt{\langle \varepsilon_i^2 \rangle} \sim 0.4\%$. In this example a linear polynomial was used for the baseline.

As was already pointed out in Introduction the developed technique and algorithm of data acquisition give the line-of-sight values for the temperature and H_2O concentration. In reality both parameters fluctuate along the optical path inside the camera. It means that the line shapes and widths of water molecules depend on the coordinate along the beam. The used technique can only provide "averaged" line shapes and infer "averaged" temperature and water concentration. It is well known that in any gas flow in a closed compartment there exists a wall layer of cold gas. The results of the preliminary investigations of the combustion zone structure by a shadowgraphic technique gave a rough estimation of this layer to about 2–3 mm. We did not account for this layer in our model.

To detect the variations of the parameters along the beam one needs to probe the combustion zone in the perpendicular direction and/or to use much more complex multi-beam scheme. Note that the Abel inversion technique cannot be used in our case because of irregular fluctuations of T, P, N distribution from scan to scan.

The total pressure in the zone was estimated from the width of the broadened lines. Broadening of the absorption lines by air and self-broadening by H_2O molecules were

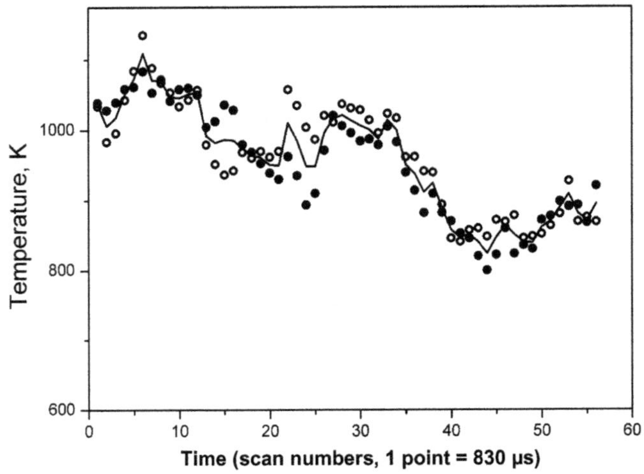

Fig. 8 Temporal profile of the probed zone temperature. The origin in the figure corresponds to the 130-th ms from the beginning of the run. The 60-th scan corresponds to the 180-th ms

taken into account using [11]. The coefficient of water self-broadening is 5–6 times greater than the coefficient of water line broadening by air. Accounting for self-broadening provided an accurate value of the total pressure in the probed zone $P = 200$ Torr, which is in quite reasonable agreement with the value ~210 Torr measured by the membrane pressure meter incorporated in the test chamber. The model overestimated the total pressure (about 270 Torr) when self-broadening was ignored.

The spectra in Fig. 7 were averaged over 30 scans. High signal-to-noise ratio enabled spectral fitting with fewer averaged scans. The temperature obtained as a result of such fitting is shown in Fig. 8. A total of 64 scans in the time interval from 135–185 ms were processed. Each point in the figure was obtained by averaging 9 scans with a "sliding" first number of the scan. The dark circles are the result of the processing of the first half of a scan (decreasing DL frequency), the open circles and the result from the second half of the scan (increasing frequency). The solid line is the average of the two. The values of the temperature inferred from both slopes coincide reasonably well.

The uncertainty of the temperature evaluation associated with the experimental errors and fitting procedure was estimated basing on the results presented in Fig. 8. We assume that during one scan (about 830 µs) the temperature is constant. Every value of the temperature $T_i^{+,-}$ is evaluated with an error $\varepsilon_i^{+,-}$ (signs + and − refer to two different slopes of a scan). The temperature difference $\Delta T = (T_i^+ - T_i^-)$ is equal to $(\varepsilon_i^+ - \varepsilon_i^-)$. Assuming that the values $\varepsilon_i^{+,-}$ are statistically independent with equal dispersions σ^2, one can estimate the standard deviation of T:

$$\sigma = \sqrt{\sum_1^m (T_i^+ - T_i^-)^2 / 2m}, \tag{5}$$

where m is the number of evaluated spectra. The precision thus estimated (statistical error) of the temperature measurements was $\sigma = 40$ K.

5 Conclusion

Diode laser absorption spectroscopy is applied for the remote measurement of temperature, total pressure and concentration of water vapor. The technique is based on the recording of absorption spectra of water molecules near 1.39 μm. The selection of the specific spectral lines was dictated by several reasons: the energies of low levels should be different, the lines should be reasonably resolved and the spectral interval should be free from lines of other gas components. The following H_2O absorption lines were selected: 7189.344 cm^{-1} ($E'' = 142$ cm^{-1}), 7189.541 cm^{-1} ($E'' = 1255$ cm^{-1}), 7189.715 cm^{-1} ($E'' = 2005$ cm^{-1}). All lines could be recorded in a single scan of the DL wavelength across a $\sim$1 cm^{-1} spectral interval.

An original differential registration scheme was developed to reduce baseline fluctuations and amplitude modulation of the DL intensity not connected with water vapor absorption. An optimal model of the experimental spectra fitting was found in laboratory experiments with stable temperature and pressure. The best coincidence of T and P values measured by TDLAS and by thermocouple and pressure sensor was obtained using the multi-line spectrum fitting model.

The developed and validated version of TDLAS sensor was applied for the diagnostics of the combustion process in a closed-type section of the supersonic aerodynamic tube located at the Joint Institute for High Temperature RAS. The combustion in the mixing supersonic flows of air and hydrogen or ethylene was ignited and sustained by a pulsed electric discharge. The air flow parameters were: Mach $M = 2$, static pressure 150–300 Torr, kinetic temperature 300 K.

In the initial step of data processing the transient absorption spectra were transformed into 2D images. This procedure greatly simplified the selection of the most important periods of the combustion process. The line-of-sight temperature and water vapor partial pressure of 1050 K and 21 Torr, respectively, were inferred from the dynamic absorption spectra. The estimated precision of the temperature measurement was $\pm$40 K. The temporal resolution of the developed technique was $\sim$1 ms.

Acknowledgements This work was supported by the grants of the Russian Academy of Sciences (OFN-10, Project 4.6 and Program N9, Project 9.2). The help of Dr. Ya. Ponurovskii during the initial wavelength calibration and valuable advices concerning the NEL DLs performance are greatly appreciated.

References

1. A. Rousseau, E. Teboul, N. Sadeghi, Plasma Sources Sci. Technol. **13**, 166 (2004)
2. M.P. Arroyo, R.K. Hanson, Appl. Opt. **32**, 6104 (1993)
3. M.G. Allen, Meas. Sci. Technol. **9**, 545 (1998)
4. S.T. Sanders, J. Wang, J.B. Jeffries, R.K. Hanson, Appl. Opt. **40**, 4404 (2001)
5. H. Teichert, T. Fernholtz, V. Ebert, Appl. Opt. **42**, 2043 (2003)
6. H. Li, A. Farooq, J.B. Jeffries, R.K. Hanson, Appl. Phys. B **89**, 407 (2007)
7. C.D. Lindstrom, K.R. Jackson, S. Williams, R. Givens, W.F. Bailey, C.J. Tam, W.F. Terry, AIAA J. **47**, 2368 (2009)
8. J.A. Silver, Appl. Opt. **31**, 707 (1992)
9. S.T. Sanders, J.A. Baldwin, T.P. Jenkins, D.S. Baer, R.K. Hanson, Proc. Combust. Inst. **28**, 587 (2000)
10. X. Zhou, X. Liu, J.B. Jeffries, R.K. Hanson, Meas. Sci. Technol. **14**, 1459 (2003)
11. L.S. Rothman, D. Jacquemart, A. Barbe, D.C. Benner, M. Birk, L.R. Brown, M.R. Carleer, C. Chackerian, K. Chance Jr., V. Dana, V.M. Devi, J.-M. Flaud, R.R. Gamache, A. Goldman, J.-M. Hartmann, K.W. Jucks, A.G. Maki, J.-Y. Mandin, S.T. Massie, J. Orphal, A. Perrin, C.P. Rinsland, M.A.H. Smith, J. Tennyson, R.N. Tolchenov, R.A. Toth, J. Vander Auwera, P. Varanasi, G. Wagner, J. Quant. Spectrosc. Radiat. Transfer **96**, 139 (2005). HITRAN website: http://www.cfa.harvard.edu/hitran/
12. N. Jacquinet-Husson, N.A. Scott, A. Chédin, L. Crepeau, R. Armante, V. Capelle, J. Orphal, A. Coustenis, C. Boonne, N. Poulet-Crovisier, A. Barbe, M. Birk, L.R. Brown, C. Camy-Peyret, C. Claveau, K. Chance, N. Christidis, C. Clerbaux, P.F. Coheur, V. Dana, L. Daumont, M.R. De Backer-Barilly, G. Di Lonardo, J.M. Flaud, A. Goldman, A. Hamdouni, M. Hess, M.D. Hurley, D. Jacquemart, I. Kleiner, P. Kopke, J.Y. Mandin, S. Massie, S. Mikhailenko, V. Nemtchinov, A. Nikitin, D. Newnham, A. Perrin, V.I. Perevalov, S. Pinnock, L. Regalia-Jarlot, C.P. Rinsland, A. Rublev, F. Schreier, L. Schult, K.M. Smith, S.A. Tashkun, J.L. Teffo, R.A. Toth, Vl.G. Tyuterev, J. Vander Auwera, P. Varanasi, G. Wagner, J. Quant. Spectrosc. Radiat. Transfer **109**, 1043 (2008). GEISA website: http://ara.lmd.polytechnique.fr/
13. SPECTRA website: http://spectra.iao.ru
14. X. Liu, J.B. Jeffries, R.K. Hanson, Meas. Sci. Technol. **18**, 1185 (2007)
15. D. Lisak, D.K. Havey, J.T. Hodges, Phys. Rev. A **79**, 0525071 (2009)
16. P.C.D. Hobbs, Appl. Opt. **36**, 903 (1997)
17. S.G. Rautian, I.I. Sobelman, Usp. Fiz. Nauk **90**, 209 (1966) [Sov. Phys. Usp. **9**, 701 (1967)]
18. J.A. Nelder, R. Mead, Comput. J. **7**, 308 (1965)
19. S.B. Leonov, D.A. Yarantsev, Plasma Sources Sci. Technol. **16**, 132 (2007)
20. W. Burger, M.J. Burge, *Digital Image Processing: An Algorithmic Approach Using Java* (Springer, New York, 2008). ImageJ website: http://rsb.info.nih.gov/ij/

Interest of absorption spectroscopy for the control of industrial processes. Application to H_2 massive production

L. Croizé · D. Doizi · B. Larousse · A. Pailloux · D. Reaux · C. Gallou · V. Dauvois · J.L. Roujou · Y. Zanella · P. Carles

Received: 15 September 2009 / Revised version: 4 November 2009 / Published online: 11 December 2009
© Springer-Verlag 2009

Abstract Absorption and TDLA spectroscopies find their applications in a lot of fields of research. The purpose of this article is to show how these methods can bring significant advances in chemical research projects. "H_2 massive production" using nuclear heat together with a thermochemical cycle is an important way to massively produce hydrogen, a potential energy vector. The sulfur–iodine cycle and the hybrid copper-chloride thermochemical cycles are some good candidates for water splitting.

In the case of the sulfur-iodine thermochemical cycle, the overall efficiency of the process essentially depends on the efficiency of HI section. Using optical techniques, such as a FTIR spectrometer for H_2O and HI concentrations determination, and a TDL spectrometer for I_2 measurements, it enables to get very significant results that will be useful to build a new thermodynamic model of the HI separation. This nonintrusive method has avoided any vapor change and prevented tedious experiments in harsh environments.

The same methodology is now applied for the study of the hydrolysis reaction of the thermochemical hybrid copper-chloride cycle. The study of this reaction is very important to assess the viability of this cycle because this reaction is not thermodynamically favored and it only occurs if a large excess of water is used. To better understand the influence of various parameters, such as water stoichiometry, temperature, reaction duration, an experimental setup has been designed and realized. The experimental setup uses two spectrometers to study the speciation of the gaseous phase and optimize the kinetics of the hydrolysis reaction. Concentrations of HCl and H_2O are obtained by fitting experimental FTIR spectra with calculated spectra. Parasitic reactions can appear leading to formation of Cl_2, measured by UV–Visible spectrophotometry.

The high temperature reaction at around 530°C is the only reaction of this copper-chloride cycle which is thermodynamically favored. A better understanding of its kinetics and the influence of the experimental parameters on this kinetics are needed. For this purpose, an absorption spectrometer able to measure oxygen is under study. This instrument is based on a high-finesse cavity and a DFB diode in order to access an oxygen absorption line spectra, near 1.3 μm. Preliminary results of this work will be presented.

1 Introduction

Absorption spectroscopy, in general, and tunable diode laser absorption spectroscopy (TDLAS) techniques, in particular, are used by a growing number of researchers and found their applications in several fields of research, e.g., atmospheric trace gas detection, isotopic composition measurement, industrial processes monitoring, combustion study, medical diagnostics, or study of an ecosystem, for example. The purpose of this communication is to present a full example where these methods bring significant advances in a particular chemical research project dedicated to H_2 *massive production.*

Hydrogen is a potential energetic vector, and water splitting ($H_2O \rightarrow H_2 + 1/2\,O_2$) is one way to massively produce it, using alkaline electrolysis, for example. However, this method has a poor efficiency, and production costs are significantly higher than when using fossil fuels to produce H_2. But in this latter case, greenhouse gases are produced. Developing a way to massively produce hydrogen

---

L. Croizé · D. Doizi (✉) · B. Larousse · A. Pailloux · D. Reaux · C. Gallou · V. Dauvois · J.L. Roujou · Y. Zanella · P. Carles
Department of Physical Chemistry, CEA, CE Saclay,
91191 Gif sur Yvette Cedex, France
e-mail: denis.doizi@cea.fr
Fax: +33-169-085411

without greenhouse gas emissions, with a low cost and a good efficiency is a challenge for the international community. Using nuclear heat together with a thermochemical cycle is an important way to massively produce hydrogen. The sulfur–iodine cycle and the hybrid copper-chloride thermochemical cycle are good candidates for water splitting and are under investigation in our laboratory [1–3].

For better forecast the industrial viability of these thermochemical cycles, absorption spectroscopies, such as Fourier transform infrared spectrometry (FTIRS) for HI, H_2O, HCl and ultraviolet–visible spectrophotometry for I_2 and Cl_2, and optical-feedback cavity-enhanced absorption spectroscopy (OF-CEAS) for O_2 and H_2 appears as powerful tools allowing online, real-time, contactless, nonintrusive, selective, and sensitive reaction kinetics, and speciation studies with compact and robust instrumentation. Both of those methods are based on the Beer–Lambert law.

The absorbance A at wave number ν is related to the transmitted light I_t through a gaseous medium, which is the measurable quantity:

$$A(\nu) = -\ln\bigl(I_t(\nu)/I_0\bigr)$$

with I_0 being the incident light.

The absorbance around a single isolated line can also be written as $A(\nu, P, T) = L.S^N(T).N.\phi(\nu - \nu_0, P, T)$ where L, P, and T are the absorption path length, pressure, and temperature, respectively, in the gas sample to be analyzed, $S^N(T)$ is the normalized line intensity (in $\text{cm}^{-1}/(\text{molecule cm}^{-2})$ at the sample temperature, ν_0 is the absorption line position, N is the number density of absorbing (molecules/cm^3), and $\phi(\nu - \nu_0, P, T)$ is the normalized profile of the line corresponding to the convolution of the line absorption profile and the instrument spectral response function. If the experimental parameters (total pressure, temperature, and path length) and the spectroscopic parameters (for example, extracted from the HITRAN08 database) are known, the absolute concentration of the active gas can be derived from the UV–Vis spectrometry, which monitors electronic transitions between electronic levels ($\pi \rightarrow \pi^*, \nu \rightarrow \pi^*$ transitions) and gives information on chromophoric compounds that absorb in the UV–Vis range (200–800 nm). FTIR spectrometry is a rovibrational spectrometry which gives information on vibrations and rotations in compounds. The solid–liquid or gaseous compound must have a dipole moment to be IR active. The spectral range varies from 0.78 to 1000 μm and delimits three major zones: near, medium, and far infrared (13000 to 10 cm^{-1}). Absorption spectroscopy instruments based on high-finesse cavity, like cavity ring down spectroscopy (CRDS) or optical-feedback cavity-enhanced absorption spectroscopy (OF-CEAS), exhibit very long optical path lengths (km) due to a resonant cavity and improved detection limits which allows detection of trace gas molecule (detection limits as low as 1 ppb can be achieved) or detection of a molecule that has weak absorption intensity as homonuclear diatomic molecules, like O_2 and even H_2.

2 The sulfur–iodine thermochemical cycle

The sulfur–iodine thermochemical cycle, which is schematically represented Fig. 1, is composed of three sections. The Bunsen section (Table 1 (1)) is an exothermic reaction in liquid phase which produces two aqueous nonmiscible acidic phases. The heavier liquid phase composed of HI, I_2, H_2O is named HI_x phase. In the HI section (Table 1 (2)), the HI_x phase is distillated and HI is dissociated into H_2. This incomplete and slow separation is the most important for the efficiency of the cycle due to the important calorific capacity of the HI_x mixture. The H_2SO_4 section (Table 1 (3)) decomposes sulfuric acid in order to extract O_2 and to recycle iodine. The overall efficiency of the process essentially depends on the efficiency of HI section and to improve this efficiency, CEA has chosen to use a reactive distillation column [4], a concept in which the ternary HI–I_2–H_2O mixture is distilled together with HI dissociation to produce hydrogen. The knowledge of the partial pressures of the species

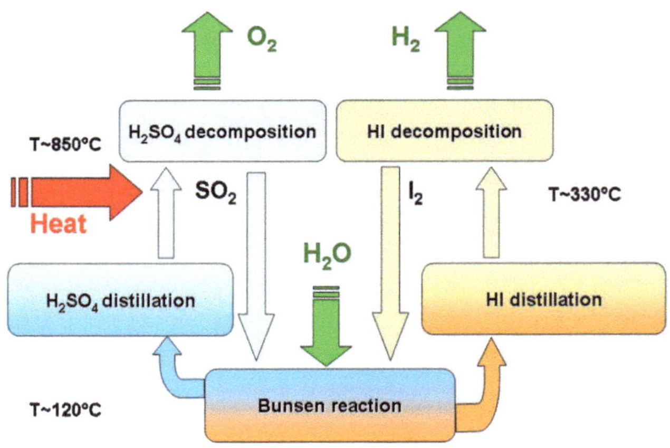

Fig. 1 General scheme of the sulfur–iodine cycle

Table 1 Reactions in the I–S cycle

Reaction		Temperature (°C)
(1) Bunsen section	$I_2 + SO_2 + 2\,H_2O \rightarrow 2\,HI + H_2SO_4$	120
(2) HI section	$2\,HI \rightarrow H_2 + I_2$	330
(3) H_2SO_4 section	$H_2SO_4 \rightarrow SO_2(g) + H_2O + 1/2\,O_2$	850

Fig. 2 The function of this experimental device is to confine the concentrated and corrosive environment at liquid (+solid)-vapor thermodynamic equilibrium

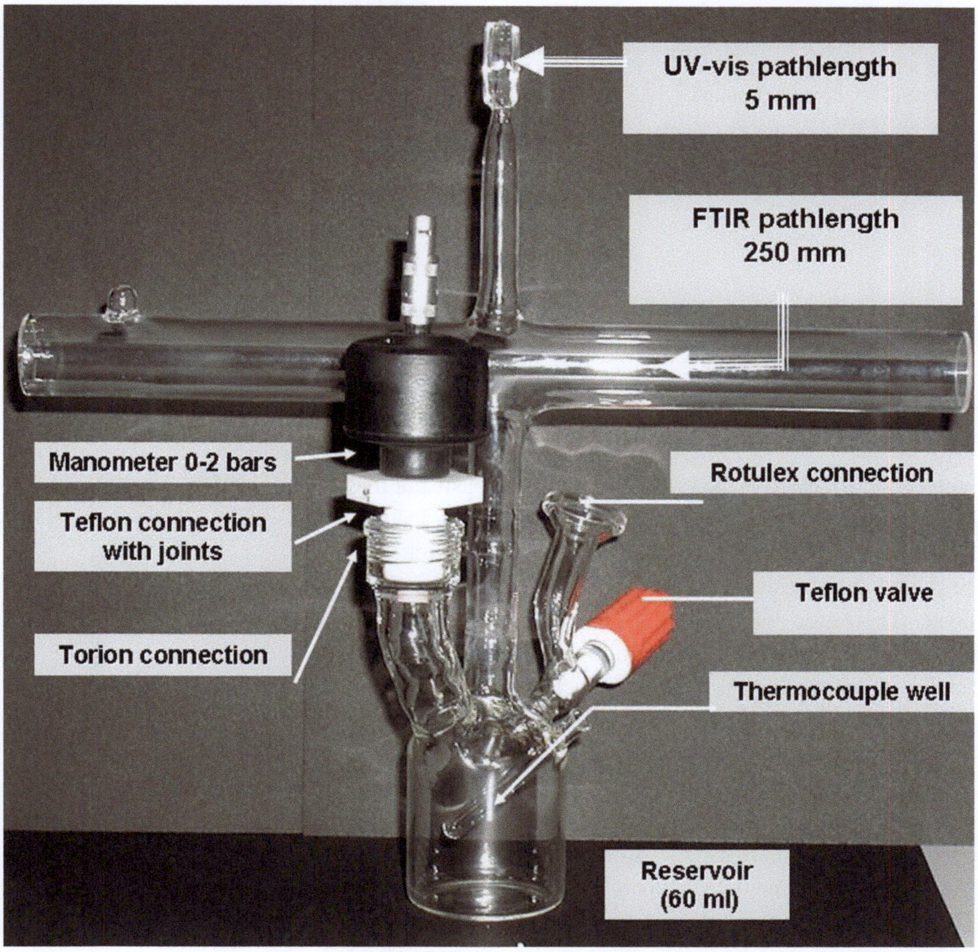

in the vapor phase for a wide concentration domain of the liquid phase is then necessary. One possible way is to take samples of the vapor phase to characterize each by titration after condensation but such methodology is tedious, time consuming, and implies the risk to alter the vapor composition. In our laboratory, optical "online" diagnostics have been used to measure HI, I_2, and H_2O in severe experimental conditions, involving concentrated and corrosive media and corresponding to two sets of experimental conditions: (1) in a temperature range between 20 and 140°C and at a total pressures up to 2 bar [5], (2) in a temperature range of 120–270°C and at a total pressure range up to 30 bar [6].

For measurements up to 2 bar, the instrumental setup (Fig. 2) is composed of a quartz and Pyrex reservoir containing the liquid mixture to be studied, equipped with a total pressure gauge (model EBRO VM 2000 ex), a Teflon valve and a connector to introduce products. K-type thermocouples and platinum resistance sensors are used to measure the temperature at different positions inside the oven. Two cells, with windows made of Suprasil quality quartz, are located in the upper part of the device and allow optical measurements of the gaseous phase concentrations of HI, I_2, and H_2O.

The first cell is 250-mm long for the measurement of HI and H_2O concentrations by infrared absorption spectrometry using a Bruker Vector 22 FTIR spectrometer with a spectral resolution of 0.4 cm^{-1}. The IR configuration includes a Globar source, the Vector 22 interferometer with a KBr beam splitter and a nitrogen-cooled InSb detector. The entire optical path is under an inert nitrogen atmosphere for both the part outside the spectrometer and Vector 22. The optical path length between the Globar source and the detector is longer than 1 m. In order to determine the HI and H_2O amounts in the probed vapors, a least-square procedure is used. Spectra are calculated as convolution of the absorption line profile (Lorentz model) and the FTIRS apparatus function (normalized to unity area, Gaussian with a half width at half maximum of 0.37 cm^{-1}). A least-square algorithm is used to adjust a theoretical spectrum on the experimentally measured spectrum and to minimize the sum of the squares of the residuals, experimental minus calculated spectrum. The starting point for the spectroscopic data used in the present work is the HITRAN molecular spectroscopic database [7], but significant changes and extensions have been made in order to treat the considered spectra. The uncertainty on the intensity and broadening para-

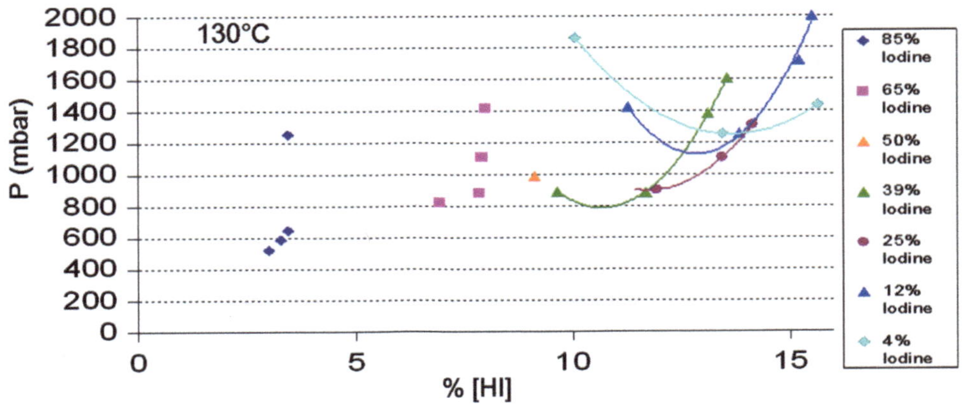

Fig. 3 Total pressures: Comparison with the results of the literature (Neumann [12], Knoche [13]). Pressure minima positions versus [HI] are in good agreement with the values of the literature

meters is difficult to evaluate on an individual basis but may be estimated, on average, to be around 5%. This leads to a minimum 5% uncertainty on HI and H_2O to which must be added the uncertainty resulting from the measurement noise. The latter, which all the more affects the retrieved concentrations when the absorption is small, is important for HI in mixtures containing relatively small amounts of this species.

The second cell is 5-mm long, it enables the determination of the I_2 concentration measurement by UV–Vis absorption spectrometry using an Ocean Optics HR 2000 spectrometer, spectrophotometer equipped with a halogen light source HL2000. The detector is a linear silicon CCD array. The spectral resolution measured with an Hg–Ar calibration lamp is better than 0.2 nm. The same methodology has been used to measure vapor liquid equilibrium up to the process domain [6]. An example of typical result is presented on Fig. 3 where the total and partial pressures have been measured at around 100°C up to 130°C for different HI concentrations and for various iodine concentrations (4, 12, 24, 39, 65, 85%). A minimum of four different HI concentrations are prepared for a fixed iodine concentration. This enables us to describe the total pressure curve and find the pressure minima.

Using those spectroscopic tools, very significant results (a complete set of experiments has been carried out for various iodine compositions describing the entire reactive distillation column, a first) that will be useful to build a new thermodynamic model of the HI separation.

3 The hybrid copper-chloride cycle: the hydrolysis reaction

The same methodology is now applied for the study of the hydrolysis reaction of the thermochemical hybrid copper-chloride cycle. The sulfur-based cycles are commonly considered to be the most important even trough they require heat at 850–900°C for the oxygen generation reaction.

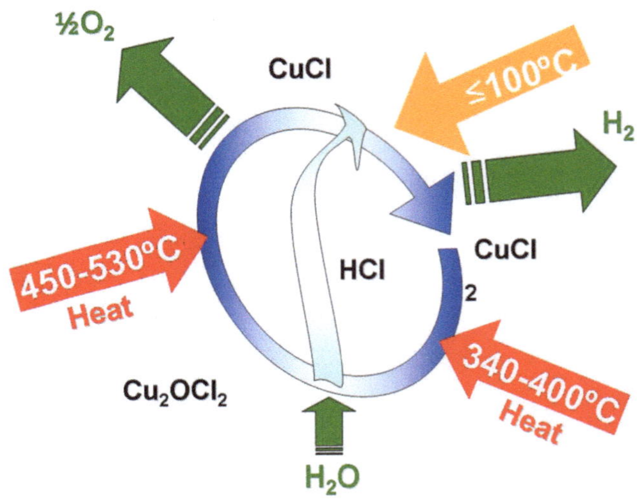

Fig. 4 General scheme of the copper-chloride cycle

Among the various nuclear reactors studied for generation IV, France has decided to promote the sodium fast reactor (SFR). The maximum temperature available with such a reactor is about 500°C. The production of H_2 using a thermochemical cycle coupled with a SFR needs to find a low-temperature thermochemical cycle. Thermochemical cycles for hydrogen production from water that would require heat input at temperatures less than 600°C would also allow reducing the thermal stress on construction materials and the family of hybrid Cu–Cl cycles was identified as the most viable for temperatures between 500 and 600°C [4]. The hybrid copper-chloride thermochemical cycle which is schematically represented in Fig. 4 is composed of three reactions. In the electrolysis reaction (Table 2 (1)) CuCl and HCl reacts electrochemically to produce cupric chloride ($CuCl_2$) at the anode and to release H_2 at the cathode. This reaction is being extensively studied at Atomic Energy of Canada Limited (AECL) and in Argonne National Laboratory (ANL), USA. The product $CuCl_2$ reacts with steam to produce copper oxychloride (Cu_2OCl_2) according to the hydrolysis reaction (Table 2 (2)). In the oxygen generation

Table 2 Reactions in the Cu–Cl cycle

Reaction		Temperature (°C)
(1) Electrolysis	$CuCl(s) + HCl(g) \to CuCl_2(a) + 1/2\,H_2(g)$	100
(2) Hydrolysis	$2\,CuCl_2(s) + H_2O(g) \to Cu_2OCl_2(s) + 2\,HCl(g)$	340–400
(3) Decomposition	$Cu_2OCl_2(s) \to 1/2\,O_2(g) + 2\,CuCl(s)$	450–530

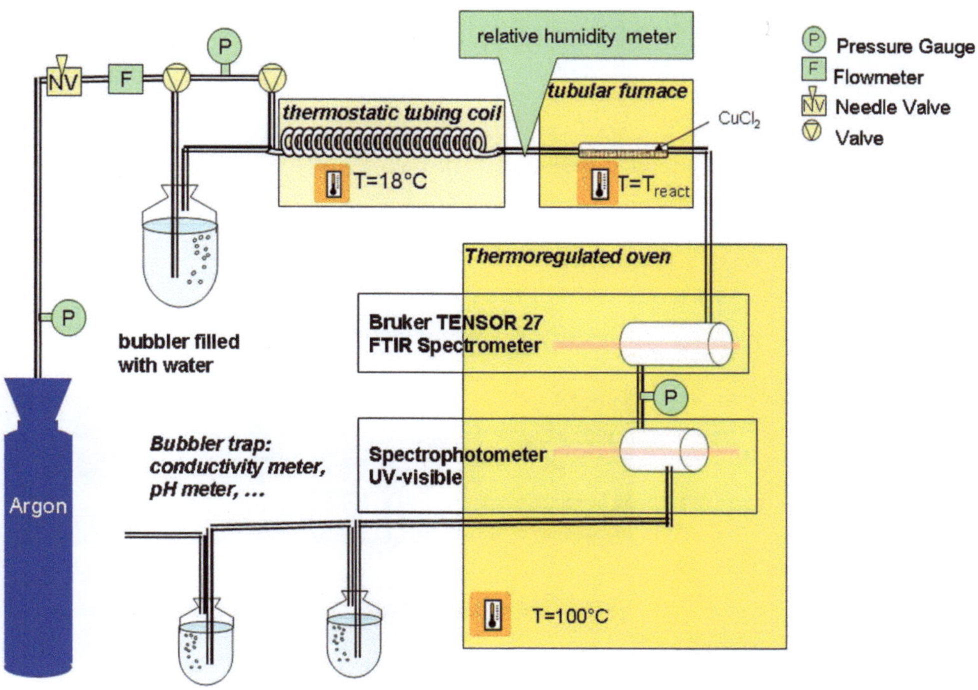

Fig. 5 Experimental setup for the cooper-chloride study

Reaction (Table 2 (3)), the Cu_2OCl_2 is decomposed to yield molten cuprous chloride (CuCl) and oxygen.

The study of the hydrolysis reaction is very important to assess the viability of this cycle because this reaction is not thermodynamically favored and high yields of the desired products are obtained only if a large excess of water is used, and yet this excess of water has to be separated in the final products requiring additional costs. To better understand the influence of various parameters, such as water stoichiometry, temperature, reaction duration, an experimental setup has been designed and realized. Unlike sulfur–iodine experimental setup, the study of the hydrolysis reaction is done with a dynamic experimental setup (Fig. 5) which means that reagents and reaction products are carried by a vector gas (99.999% argon). The reagent is placed in a basket in a horizontal glass reactor inside a tubular furnace. The carrier gas flow rate is controlled by a mass flow controller (Brooks Instrument) and is humidified by flowing it at 5 L/h through a bubbler filled with water at room temperature at 1 bar. The relative humidity is measured continuously at the inlet of the reactor to ensure a stable relative humidity. The temperature of the bed is measured with a thermocouple K type placed in the sample basket. The carrier gas passes through two absorption cells placed in a thermoregulated oven ($T = 100°C$). The first cell is a 250-mm long used for the measurement of HCl and H_2O concentrations by infrared absorption spectrometry performed using a Bruker Tensor 27 FTIR spectrometer with spectral resolution of 0.5 cm^{-1}. The IR configuration includes a Globar source, the tensor 27 interferometer with a KBr beam splitter and a nitrogen cooled MCT detector. The entire optical path is under an inert nitrogen atmosphere for both the part outside the spectrometer and tensor 27. H_2O and HCl concentration calculations follow the same principles that were used in the HI and H_2O calculations in the iodine–sulphur cycle study. Parasitic reactions can appear, in particular the formation of molecular chlorine according to the reaction $2\,HCl + 1/2\,O_2 \to Cl_2 + H_2O$ and $CuCl_2 \to CuCl + 1/2\,Cl_2$, measured by UV–Vis spectrophotometry (Ocean optics, Mikropack DH2000 deuterium lamp, HR 2000 spectrometer) between 300 and 400 nm (maximum absorbance around 330 nm) using a 10-cm path length optical cell placed in a thermally regulated oven. Figure 6 presents some typical results that have been obtained. At 14:41 the humidified carrier gas is passed through the tubular furnace. Immediately, hydrolysis takes place and HCl is produced. During the first part of the reaction, there is a strong correlation between produced HCl concentration and furnace tempera-

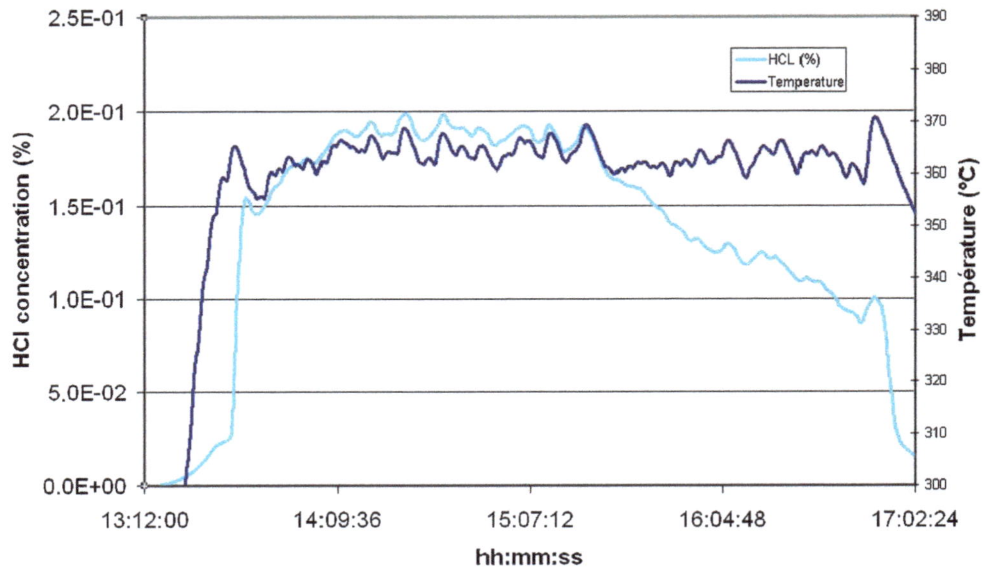

Fig. 6 Temporal evolution of HCl concentration and tubular furnace temperature: a strong correlation between HCl concentration and temperature is under investigation

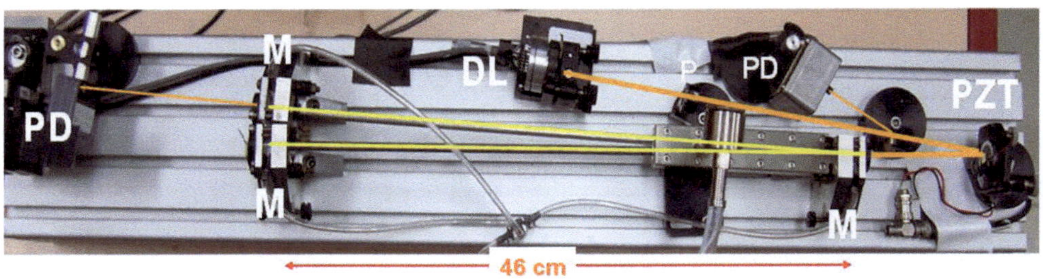

Fig. 7 Optical setup of our OF-CEAS

ture. Then, at 15:30, this concentration decreases because most of the reagent has reacted. This concentration becomes nearly zero when the humidified vector is replaced by dry argon, at 16:50. We have demonstrated that the hydrolysis reaction could be realized between 360 and 390°C without chlorine formation. For higher temperatures, optical density variations are shown around 330 nm indicating that Cl_2 has been formed.

4 The hybrid copper-chloride cycle: melanothallite decomposition

At last, the high-temperature reaction at around 530°C is the only reaction of this copper-chloride cycle which is thermodynamically favored. But a better understanding of its kinetics and the influence of the experimental parameters on its kinetics is needed. For that an online optical measurement of the oxygen due to Cu_2Cl_2O decomposition is required. During a preliminary experiment three flasks were sampled and analyzed with a mass spectrometer (Mat 271). The maximum oxygen concentration detected was about 60 ppm. Oxygen is a homonuclear diatomic molecule, so that electric dipolar transition is forbidden, and absorption line intensities are very weak. The strongest infrared absorption lines correspond to the so-called oxygen A-band with line intensities smaller than 9 cm^{-1}/molecule cm^{-2} [8]. Commercial spectrometers allow detection limit of about 10 ppm (for SNR = 1) and with an absorption cell volume smaller than 0.5 L can hardly be found and direct absorption spectroscopy is not sufficient. Detection limits can be improved using techniques frequency modulation or with instrument based on a high-finesse cavity, like cavity ring down spectroscopy (CRDS) [9] or cavity-enhanced absorption spectroscopy (CEAS). In collaboration with the University J. Fourier Grenoble a OF-CEAS (optical-feedback cavity-enhanced absorption spectroscopy—FR Patent 2 830 617 (2001)—J. Morville, D. Romanini, M. Chenevier, UJF) has been developed for the field detection of the industrial toxic hydrogen fluoride HF at 1.3 μm (Fig. 7). Because the laser source of this system can be tuned across an oxygen absorption line, it has been used to evaluate feasibility of an online optical measurement of O_2. But in this spectral range, absorption lines are more than two orders of magnitude smaller than in the oxygen A-band. The instrument is based on a V-shaped cavity that allows producing frequency selected optical feedback when the laser frequency enters in resonance with one of the cavity modes and the experimental setup have been demonstrated by Morville et al. [10]. The laser source is a distributed-feedback (DFB) diode laser with Peltier thermoelectric cooling. The beam is collimated via an aspheric lens with high numerical aperture, directly mounted on the housing and the collimated beam is then divided into two parts. The first part of the beam is directly focused onto an InGaAs photodiode and is used to monitor the power incident on the cavity. Feedback rate is controlled with variable attenuator, and a piezoelectric-mounted steering mirror is used to adjust the distance between the V-cavity and the laser cavity (a condition of resonance is that this length has to be a multiple of the length of the cavity arms length) and is controlled with a dedicated analog circuit. A second photodiode detects the signal at cavity output. The acquisition software written in Labview® enables efficient real-time data processing and a Voigt line-shape fit is then used to extract the sample concentration.

After a determination of spectroscopic parameters (and in particular of the argon-broadened half width of the absorption line profile), OF-CEAS quantitative measurements have been carefully characterized in laboratory conditions: sensitivity, detection limit, linearity, response time, and we have demonstrated that using the oxygen absorption line P19Q18 at 7822.9525 cm^{-1}, measurements at 0.64% of O_2 in argon were possible with this instrument. An intercomparaison has been done with a mass spectrometer (Thermo Mat 271) and OF-CEAS oxygen and mass spectrometry oxygen concentration measurement did not differ in more than 1.5% (in relative) at 0.65% of oxygen in argon and an article is under preparation to give more details about this work. With this preliminary work, we have shown feasibility of molecular oxygen optical measurement for a chemical reaction kinetic study. The next step will be to change the DFB diode and the mirror in order to access to the spectral range of around 760 nm where oxygen absorption lines are stronger in more than two orders of magnitude. Other prospect will be to use a DFB diode at around 4712.90 cm^{-1}, where the "strongest" (that means $S^N(T) > 10^{-26}$ $cm^{-1}/[\text{molecule}\,cm^{-2}]$) absorption lines of molecular hydrogen can be found [11], so that 0.25% of H_2 is expected.

5 Conclusion

In this paper, we have presented three examples where significant advances in a chemical study have been done thanks to a strong collaboration with optical spectroscopy laboratory: the potentialities of optical measurements have been presented through applications in the nuclear production of hydrogen by thermochemical cycles. When studying the sulfur–iodine thermochemical cycle, those nonintrusive methods have avoided any vapor change and prevented tedious experiments in harsh environments. In the case of the copper-chloride work, the kinetics study of the hydrolysis reaction can be followed using UV–Vis spectrometry which controls the occurrence of side reactions and FTIR spectrometry to determine the speciation of the gaseous phase. Use of these online optical techniques will allow a complete optimization of the hydrolysis reaction. In case of the melanothallite decomposition, the chemical species to measure is molecular O_2 which has very weak absorption. Such detection requires very low detection limits and OF-CEAS offers a great potential to measure it.

References

1. J.M. Hartmann, L.H. Coudert, D. Doizi, V. Dauvois, J.L. Roujou, V. Lorin, B. Larousse, P. Fauvet, P. Carles, Int. J. Hydrogen Energy **34**, 162 (2009)
2. D. Doizi, V. Dauvois, J.L. Roujou, V. Delanne, P. Fauvet, B. Larousse, O. Hercher, P. Carles, C. Moulin, J.M. Hartmann, Int. J. Hydrogen Energy **32**, 1183 (2007)
3. M.S. Ferrandon, M.A. Lewis, D.F. Tatterson, A. Gross, D. Doizi, L. Croizé, V. Dauvois, J.L. Roujou, Y. Zanella, P. Canes. Int. J. Hydrogen Energy, Ms. Ref. No. HE-D-09-01482R1 (in press)
4. M.A. Lewis, J.G. Masin, Int. J. Hydrogen Energy **34**, 4125 (2009)
5. D. Doizi, V. Dauvois, J.L. Roujou, V. Lorin, P. Fauvet, B. Larousse, P. Carles, J.M. Hartmann, Int. J. Hydrogen Energy **34**, 4275 (2009)
6. B. Larousse, P. Lovera, J.M. Borgard, G. Roehrich, N. Mokrani, C. Maillault, D. Doizi, V. Dauvois, J.L. Roujou, V. Lorin, P. Fauvet, P. Carles, J.M. Hartmann, Int. J. Hydrogen Energy **34**, 258 (2009)
7. L. Rothman, D. Jacquemart, A. Barbe, D.C. Benner, L.R. Brown, M.R. Carleer, C. Chackerian, Jr., K. Chance, V. Dana, V.M. Devi, J.-M. Flaud, R.R. Gamache, A. Goldman, J.-M. Hartmann, K.W. Jucks, A.G. Maki, J.-Y. Mandin, S. Massie, A. Perrin, C.P. Rinsland, M.A.H. Smith, R.A. Toth, J. Vander Auwera, P. Varanasi, JQSRT **96**, 139 (2005)
8. R.S. Mulliken, Phys. Rev. **32**, 880 (1928)
9. D.A.G. Deacon, A. O'Keefe, Rev. Sci. Instrum. **59**, 2544 (1988)
10. J. Morville, S. Kassi, M. Chenevier, D. Romanini, Appl. Phys. B **80**, 1027 (2005)
11. D.W. Ferguson, K.N. Rao, M.E. Mickelson, L.E. Larson, J. Mol. Spectrosc. **160**, 315 (1993)
12. C. Neumann, *Diplomaufgabe*. RWTH Aachen, Aachen (1987)
13. H. Engels, K.F. Knoche, Int. J. Hydrogen Energy **11**, 703 (1986)

Laser spectroscopic oxygen sensor using diffuse reflector based optical cell and advanced signal processing

J. Chen · A. Hangauer · R. Strzoda · M.-C. Amann

Received: 15 October 2009 / Revised version: 10 February 2010 / Published online: 25 March 2010
© Springer-Verlag 2010

Abstract In this paper, a VCSEL-based oxygen sensor applied for real-time combustion optimization in furnaces is presented. A diffuse reflector instead of a spherical mirror is used to avoid alignment problems during thermal cycles the sensor is exposed to in the in-situ measurement. In spite of the signal loss, this concept proves to be suitable for measurements in the exhaust of a gas furnace even if severe condensation at the optic surfaces occurs. The optic cell has a background level in the order of 10^{-5} in absorbance and is long-term stable. A real-time microcontroller-suited curve fitting algorithm is applied to further improve the long-term stable operation of the whole sensor, because a drift of the spectral background is compensated. Neither absolute wavelength scale, nor absolute wavelength difference between two measurement points is needed for concentration determination. Furthermore, the gas pressure can be simultaneously determined. Kalman filtering is also applied to reduce the concentration measurement noise without losing the reliability and fast response time of the sensor. The techniques applied in the sensor are not limited to oxygen sensing, but also can be applied for sensing of other gas molecules.

1 Introduction

For combustion applications in gas furnaces, optimization of the burning process is very important for energy savings and reduction of pollutants emission. Monitoring of the oxygen (O_2) content of the exhaust gas allows for an effective combustion control. For a large scale gas furnace with several MW power, the oxygen concentration is regulated to 1 vol%–2 vol% with utilization of an oxygen sensor and a control loop. Without using an oxygen sensor the burner has to be adjusted in a way that with varying gas quality and varying air density (summer/winter) the burning process is still complete, i.e. no high levels of carbon monoxide are generated. In this case the O_2 content in the exhaust can be 4 vol%–6 vol%. For wall-hanging gas boilers in households the oxygen content is adjusted to 4 vol%–6 vol%, compared to 7 vol%–8 vol% without regulation with an oxygen sensor [1]. According to Siegert's formula [2], by reducing the O_2 content in the exhaust by 4 vol%–5 vol% the efficiency of the furnace can be improved by approximately 1% [3] because the energy used to heat up this oxygen is no longer lost. An oxygen sensor thus enables optimum operation under different gas quality and ambient temperature conditions and eliminates the need for manual adjustment. This is especially important with a liberalized gas-market in the E.U. since the gas quality (i.e. calorific value) cannot be assumed to be constant.

The requirements for such oxygen sensors are fast time response (second range), high accuracy (0.1 vol% resolution), selectivity, in-situ measurement capability, and long-term stable operation even under harsh conditions in an exhaust gas pipe. The conventional technology—the ZrO_2-based lambda probes [4] are widely used in cars and also in medium or large scale heater appliances. However, these sensors suffer from several drawbacks.

- Despite the normal aging of the probe (up to three years are typically guaranteed in a furnace application), these sensors can be poisoned with heavy metals (e.g. Mn, Pb).

J. Chen (✉) · A. Hangauer · R. Strzoda
Corporate Research and Technology, Siemens AG,
Otto-Hahn-Ring 6, 80200 Munich, Germany
e-mail: jia.chen@wsi.tum.de

J. Chen · A. Hangauer · M.-C. Amann
Walter Schottky Institut, Technische Universität München,
Am Coulombwall 3, 85748 Garching, Germany

- Grids for flame protection are needed due to the high operating temperature of the sensor, to prevent ignition of unburned gas. This again increases the response time of the sensor.
- The high operating temperature of ZrO_2 sensor of above 600°C has to be well controlled, because any temperature variation will result in a change of the measured signal (Nernst potential).

In this paper, a tunable diode laser spectroscopy- (TDLS-) based sensor is presented as an alternative to the ZrO_2 sensor for combustion optimization. While complex and rather expensive TDLS sensor systems are already employed in process monitoring applications e.g. in power plants [5], they are not yet applied in medium or small scale burners e.g. for domestic heating appliances because there compact and low cost sensors are needed.

The challenges for the laser-optic oxygen sensor are as follows.

- The sensor must be long-term stable, which means the sensor optics must withstand the temperature cycles in the exhaust gas and must not get misaligned during the whole sensor lifetime (up to 10 years). And the sensor must also be robust against condensation or contamination on the optical surfaces. These may cause a certain decrease of the light power arriving on the photodiode.
- The sensor must be compact and based on computationally efficient algorithms.

Our solution to these problems is the utilization of a diffuse reflector and a microcontroller-suited on-line curve-fit algorithm, which is capable to evaluate gas concentrations in noisy signals. We employed wavelength modulation spectroscopy, with second harmonic detection [6] and least-squares curve fitting to improve long-term stability compared to fixed detection at the absorption line center [7]. These are explained in Sects. 2 and 3, respectively. Additionally, Kalman filtering is applied to further reduce the measurement noise of the concentration value without increasing the sensor response time. Furthermore, gas pressure is estimated simultaneously using the on-line curve fit algorithm (Sect. 4).

2 Design of the sensor optics: utilization of a diffuse reflector

The sensor probes an O_2 absorption line at a wavelength of 763 nm, where low cost vertical-cavity surface-emitting lasers (VCSELs) are available [8]. A structural diagram of the whole sensor is shown in Fig. 1. The sensor has a reflective geometry with laser and photodetector side by side and both directed at a reflector at 10 cm distance. A photograph of the sensor's optic cell is shown in Fig. 2. The laser

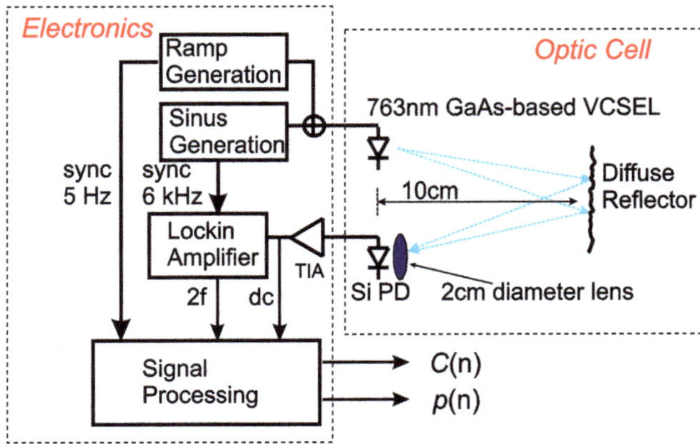

Fig. 1 Layout of the oxygen sensor consisting of an optic cell with diffuse reflective geometry and microcontroller-based electronics. $C(n)$ and $p(n)$ are the sensor outputs, which represent the n-th estimated gas concentration and pressure value

Fig. 2 Photograph of the sensor optic cell. The VCSEL emitted light is reflected by the diffuse reflector to a 2 cm diameter lens, which collects the light and images the speckle pattern to the 3 mm diameter silicon photodiode. The optical path length is 2 × 10 cm

beam passes through a window with Brewster angle and is scattered at the rough surface of the diffuse reflector, which is made of anodized aluminum with a surface roughness of 1–2 μm. The reflected light interferes at different positions on the photodetector, which results in a laser speckle pattern, i.e. a spatial distribution of interfering light modes on the surface of the detector. To collect as much light as possible, the 3 mm diameter silicon photodetector is equipped with a 2 cm diameter lens, which optically magnifies the detector area.

There are several sources of interferences in an optical cell which cause the cell background, i.e. a non-smooth transmission function of the nitrogen filled gas cell: reflections between laser and photodiode; laser speckle pattern and reflections between surfaces of optical components such as lenses or windows. In spherical mirror based cells, etalons are easily formed by the surfaces of laser and photodiode due to the high reflectivity of the mirror. These etalons have short spectral periodicity corresponding to a multiple of the physical length of the cell. For a diffuse reflector, the reflection is undirected, so that interferences with periodicities corresponding to a multiple of the cell length have a much lower amplitude than in a spherical mirror based cell. The second possible source for the cell background is the laser speckle pattern on the detector, which is wavelength dependent. A detailed analysis of the laser speckle pattern caused by utilization of a diffuse reflector can be found in [9, 10]. By applying the lens, many speckle points are integrated by the photodetector, with the effect that less intensity fluctuations appear on the transmission function when the laser wavelength is tuned [11]. Furthermore, only the light reflected at close positions on the diffuse reflector can interfere on the photodiode, because the detector is placed in the imaging plane of the lens. The interfering light beams have optical path length differences in $\leq$ mm range and result in fringes with long spectral periodicity. Reflections on the surface of optical components like lenses or windows of the laser and detector also cause fringes on the cell background. These components are needed for protection of the laser and detector from the combustion gas in the furnace. Their spectral period length depends strongly on the distance and thickness of these components.

The measured spectral background for the diffuse reflector-based cell is shown in Fig. 3. It has a peak to peak amplitude of 10^{-5}. This number refers to the intensity fluctuation on I_2/I_0, if no gas absorption is present. I_0 is the averaged intensity arriving at the photodetector and I_2 is the detected power of the second harmonic. The residual interference is a structure with a dominant periodicity corresponding to the distance between laser chip and the laser window (~1.5 cm). The detection limit is very likely to be improved if the optical cell is optimized with respect to the laser window. The baseline fringes due to diffuse reflector is probably below 10^{-5} and is also expected to stay stable during the time, which may not be the case for a spherical mirror based cell when a drift of adjustment occurs. The ultimate sensitivity limit is given by the photodetector noise, i.e. shot noise and dark noise, and is the order of 10^{-6} at 10 Hz bandwidth.

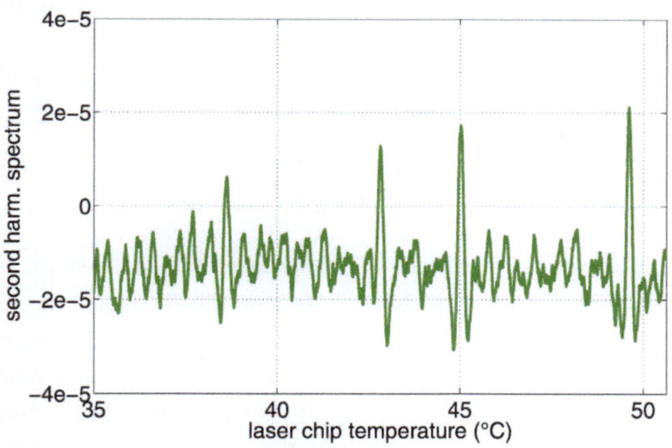

Fig. 3 The second harmonic background level of the optic cell. The high frequency etalon fringes have a peak to peak value of 10^{-5}. Residual O_2 absorption lines are present at 38°C, 43°C, 45°C, 50°C due to incomplete purging with N_2

The most significant advantage of a diffuse reflector lies in the simplicity of alignment: almost no alignment is required and optical setup changes due to thermal cycles can be tolerated. On the contrary, focusing the light on the photodetector by a spherical mirror is very alignment-sensitive, which is difficult to maintain with thermal cycles. The gas temperature in the exhaust gas pipe during operation varies from RT to 80°C (condensing gas furnaces) and from RT to 120°C–170°C (non-condensing gas furnace) and is monitored during sensor operation.

The hot and humid combustion gas causes condensation on cold surfaces like the laser window, lens and the reflector. We found that the diffuse reflector is robust against condensation, because the light is always partly reflected to the photodetector due to many possible reflection paths. The power decreases by a factor of 5 due to condensation. In comparison, applying spherical mirror the power decreases by at least a factor of 50 due to condensation. A brief comparison between diffuse reflector and spherical mirror is shown in Table 1. The disadvantage of the diffuse reflector is the limited reflectivity, which is approximately 1% resulting in several μW photodetector power, which is nevertheless sufficient for our application.

3 Design of the spectral evaluation algorithm—nonlinear curve fitting

Wavelength modulation spectroscopy with second harmonic detection is employed in this sensor. The VCSEL is tuned

Table 1 Comparison between diffuse reflector cell as described in this paper and spherical mirror based cell (as in [12]) applied in the exhaust gas measurement in gas/oil furnaces

	Diffuse reflector	Spherical mirror
Alignment	Almost none required	Difficult to maintain with thermal cycles, may result in total vanishing of the photo current
Power decrease due to condensation	Factor 5	Factor 50
Baseline fringes due to reflector	$\leq 10^{-5}$	$10^{-4}, \ldots, 10^{-5}$, can increase if alignment changes over time
Reflectivity (without condensation and dirt)	1% ... 2%	20% ... approx. 100%

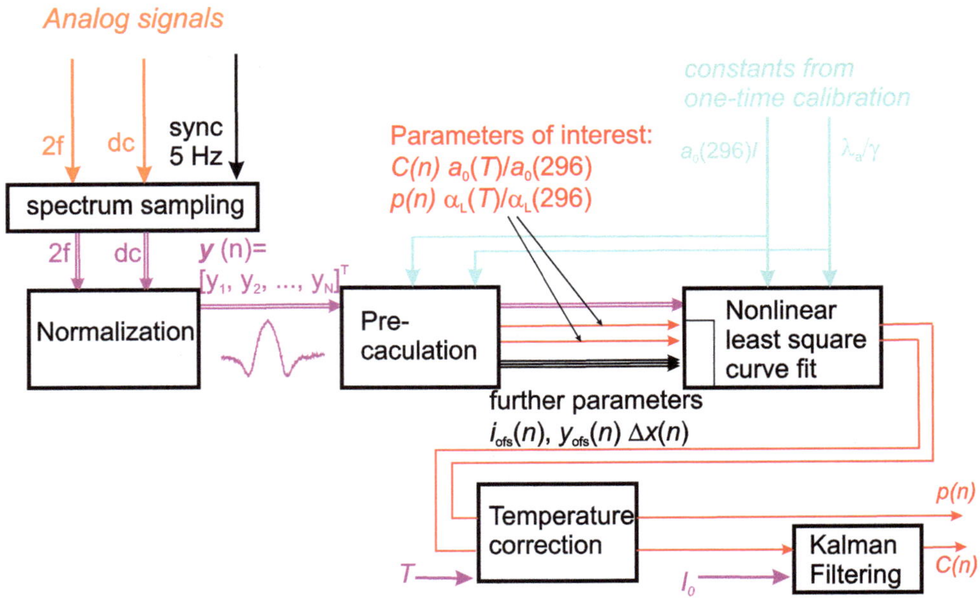

Fig. 4 Block diagram of signal processing chain for spectral evaluation

with current (Fig. 1), which is a ramp signal superimposed with a 6 kHz sinusoidal modulation. The measurement rate[1] of the one spectrum and gas concentration value is 5 Hz. The signal processing and control are done by a compact microcontroller-based electronics (board dimensions: 10 cm × 6 cm).

The task of signal processing is to extract the gas concentration and pressure from the recorded gas spectra, which is described with the following steps (Fig. 4): data recording, signal normalization, pre-calculation, nonlinear least-squares curve fit, temperature correction and Kalman filtering. After a brief introduction of signal normalization, the signal model used for curve fitting is described. Later on, the paper is focused on the pre-calculation to obtain good initial values for iterative nonlinear curve fit, whereas the curve fit algorithm itself and temperature correction to estimated concentration and pressure values are briefly discussed. Kalman filtering of the concentration values is briefly presented at last.

3.1 Signal normalization

The detected second harmonic power using a Lockin amplifier is dependent on the number density of absorbing molecules in the laser beam and also proportional to incident laser power which falls on the photodetector. Therefore, the second harmonic signal is normalized with the DC detector power after the transimpedance amplifier, so that the result is a pure measure of the gas absorption spectra.

If stray light falls additionally on the photodetector—even it is unlikely in the exhaust gas pipe—the first harmonic signal at the center wavelength of the gas absorption line or the modulated DC Light (difference signal between laser turned on and laser turned off) could be used to determine the incident laser power. The normalization with first harmonic is described in (3) in [7]. But in this case the ratio between the first harmonic and laser power has to be determined in a calibration at first. This ratio depends on the laser bias current and threshold current and the laser threshold current may change when the heatsink temperature of laser changes or when the laser ages.

3.2 Signal model for curve fitting

The O_2 A-band [13] at 763 nm consists of isolated lines, which do not interfere with species present in air or the exhaust gas, like H_2O, CO_2, CO, NO_x and so on [14]. The gas absorption line at ambient pressure has a Voigt profile with dominating Lorentzian halfwidth (pressure broadening) of ~89% of the Voigt halfwidth, in the range of room temperature to 170 degrees (temperature range in the exhaust of a gas furnace). There is no closed analytical form expression for the Voigt profile, neither for the absorption nor for the harmonic spectra. Therefore, computation of line shape is simplified by assuming a Lorentzian-based gas absorption line profile and the error caused by neglecting the Gaussian broadening is accepted. If the absorbance is low (usually less than 10%), the transmission for a single Lorentzian absorption line can be approximated by:

$$T(\lambda) = 1 - \frac{C\cdot l\cdot a_0}{1+(\frac{\lambda-\lambda_c}{\alpha_L})^2} + \left(\frac{C\cdot l\cdot a_0}{1+(\frac{\lambda-\lambda_c}{\alpha_L})^2}\right)^2 - \cdots \quad (1)$$

$$\approx 1 - \frac{C\cdot l\cdot a_0}{1+(\frac{\lambda-\lambda_c}{\alpha_L})^2} \quad (2)$$

[1] The measurement rate is 1/(time span between two measurements).

with C being the gas concentration in ppm, l the optical path length in m. a_0 is the absorption coefficient in $(\text{ppm}\cdot\text{m})^{-1}$ at center wavelength λ_c of the gas absorption line. α_L is the HWHM of the Lorentzian-shaped gas absorption line with the same unit as λ. The maximum absorbance in this system for gas furnace application is 0.5% (20% oxygen in 20 cm optical path length), the maximum relative error using this approximation is 0.25%.

The recorded second harmonic spectrum $\boldsymbol{y}$ consists of discretely sampled wavelengths:

$$\boldsymbol{y} = [y_1, y_2, \ldots, y_N]^T. \tag{3}$$

It is described with the following signal model:

$$y_i = C\frac{a_0(T)}{a_0(296)}a_0(296)lS_2\left((i - i_{\text{ofs}})\Delta x, \frac{\lambda_a}{\gamma p\frac{\alpha_L(T)}{\alpha_L(296)}}\right) + y_{\text{ofs}} + \epsilon_i, \tag{4}$$

with ϵ being white noise. a_0 and α_L are both temperature dependent. $a_0(296)$ and $\alpha_L(296)$ are the gas absorption coefficient and linewidth at ambient temperature $T = 296$ K. $S_2(x, m)$ is the second harmonic component of Lorentzian-shaped gas spectrum if the gas absorbance is one:

$$S_2(x, m) = \frac{1}{\pi}\int_{-\pi}^{\pi}\frac{1}{1 + (x + m\cos(z))^2}\cos(2z)\,dz, \tag{5}$$

with $z = 2\pi f_m t$. An analytical expression of $S_2(x, m)$ is given by Arndt [15], which is applied in the model (4). The modulation index $m = \lambda_a/\alpha_L$ is the wavelength modulation amplitude λ_a normalized with α_L. The variable of the S_2 function is the normalized wavelength scale $x = \frac{\lambda - \lambda_c}{\alpha_L}$.

The halfwidth $\alpha_L = p\gamma\frac{\alpha_L(T)}{\alpha_L(296)}$ itself is proportional to gas pressure p with proportionality factor γ and has a temperature correction term $\frac{\alpha_L(T)}{\alpha_L(296)}$. $p\gamma$ is equal to $\alpha_L(296)$. In this algorithm the temperature correction is applied after the curve fit, so that the parameter estimated in the curve fit is $p \cdot \frac{\alpha_L(T)}{\alpha_L(296)}$. By dividing the value by $\frac{\alpha_L(T)}{\alpha_L(296)}$ after curve fitting (see Sect. 3.6), the pressure p is obtained. The same holds for the concentration with the curve fit parameter $C\frac{a_0(T)}{a_0(296)}$.

The model based on the Arndt formula (4) has the following five parameters:

- x offset i_{ofs} is the index of the line center, which is not necessarily an integer
- the distance between two spectrum sample points in normalized wavelength: $\Delta x = x_{i+1} - x_i$
- gas pressure without temperature correction: $p\frac{\alpha_L(T)}{\alpha_L(296)}$
- spectral baseline offset: y_{ofs} (caused by laser residual amplitude modulation (RAM) [16])
- gas concentration without temperature correction: $C\frac{a_0(T)}{a_0(296)}$

The RAM present in the second harmonic signal is caused by the nonlinear current–intensity modulation behavior (PI characteristic) of the laser: the additive background on the second harmonic signal is caused by the second derivative of the PI curve. Since the cubic part of the PI curve for the utilized GaAs-based VCSEL is very small, the background is modeled with an offset only and higher order terms like a slope are omitted.

3.3 One-time calibration

One-time calibration with known gas concentration, pressure and temperature is done before the sensor is applied for field operation. In the calibration phase, the calibration factors for concentration $a_0(296)l$ and gas pressure λ_a/γ are determined. $a_0(296)$ can be also obtained from the HITRAN database [14] if the absorption line to be detected is known. Usually, ambient temperature and pressure ($T = 296$ K; $p = 1013$ hPa;) are applied in the calibration phase. The calibration can also be done during the sensor operation when the gas furnace is switched off, so the exhaust pipe is purged with fresh air with oxygen content 20.9% and ambient temperature and pressure.

3.4 Pre-calculation: determination of the initial values for parameters

Besides C and y_{ofs}, the other three parameters are nonlinear to the model, i.e. y_i changes nonlinearly with the parameters i_{ofs}, Δx and p. Therefore, an iterative curve fit is needed. To enable real-time curve fitting on a microcontroller, a pre-calculation method to determine good initial values for the model parameters has been developed using characteristic values of the measured spectrum. The procedure is shown in Fig. 5, and explained in the following steps.

- Step 1: the maximum and the adjacent two minima of the measured data are determined. The average of the minima positions as well as the position of the maximum give the index of the line center i_{ofs} (Fig. 5(a)).
- Step 2: the spectrum is averaged in a symmetrical manner around the maximum to determine y_{ofs}, because the second harmonic spectrum theoretically has a mean of zero (Fig. 5(b)) [17].
- Step 3: once the y offset is known, the maximum-to-minimum ratio R can be computed and m is obtained by applying the $m(R)$ relationship numerically determined from the Arndt formula (shown in Fig. 5(c)). The initial value of gas pressure without temperature correction is determined by

$$p\frac{\alpha_L(T)}{\alpha_L(296)} = \frac{\lambda_a/\gamma}{m}. \tag{6}$$

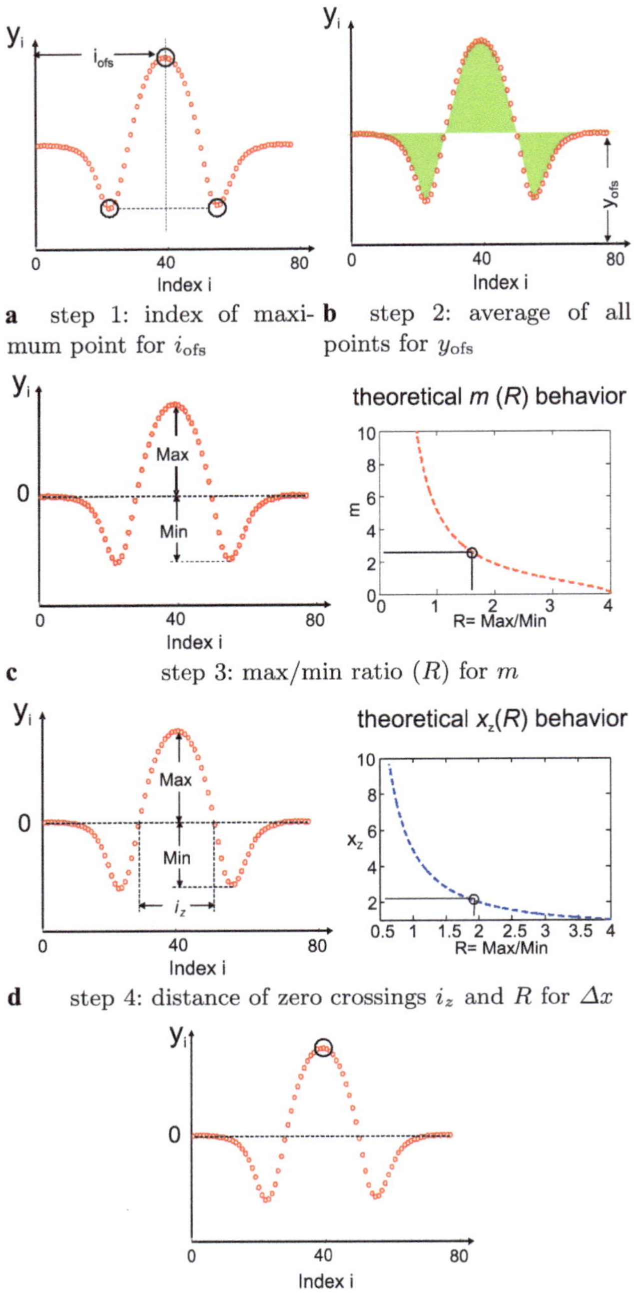

a step 1: index of maximum point for i_{ofs} **b** step 2: average of all points for y_{ofs}

c step 3: max/min ratio (R) for m

d step 4: distance of zero crossings i_z and R for Δx

e step 5: maximum point for C

Fig. 5 The procedure of pre-calculation: different steps to determine the initial values for different parameters

- Step 4: the initial value of Δx is determined by the distance between the two zero crossings. For this, the index difference i_z and its theoretical value x_z is divided, which gives $\Delta x = x_z/i_z$. x_z is given by applying the $x_z(R)$ relationship numerically determined from the Arndt formula. i_z is determined from the measured spectrum (see Fig. 5(d)).
- Step 5: the initial value of $C\frac{a_0(T)}{a_0(296)}$ is determined by using maximum point of the second harmonic spectrum divided by $a_0(296)lS_2(0, m)$, while $a_0(296)l$ is known from the one-time calibration.

An overview of the parameters and the determination method for their initial values are shown in Table 2.

3.5 Nonlinear least-squares curve fitting algorithms

With these proper initial values only one to three iterations of the Levenberg–Marquardt fitting routine [18] are required to obtain a good fit to the model given in (4). This is a nonlinear least-squares curve fit, which minimizes the square error between y and the signal model. The Levenberg–Marquardt algorithm requires the knowledge of the Jacobian matrix, which contains the derivatives of the model function with respect to each parameter [19]. To save computing time and capacity in the microcontroller, the Jacobian matrix is programmed analytically instead of estimating it numerically. The operating time is at least two times less compared to numerical estimation of the Jacobian matrix.

3.6 Temperature correction

After the nonlinear curve fit, a temperature correction according to theoretical calculations or calibration curves to the concentration and/or pressure is carried out. This is done by dividing the raw concentration and pressure values from the curve fit with the temperature correction functions $\frac{a_0(T)}{a_0(296)}$ (concentration) and $\frac{\alpha_L(T)}{\alpha_L(296)}$ (pressure).

According to HITRAN [14], we have:

$$\frac{\alpha_L(T)}{\alpha_L(296)} = \left(\frac{T}{296}\right)^{\eta}, \tag{7}$$

with absorption line dependent η in the range of 0.63–0.74, and

$$\frac{a_0(T)}{a_0(296)} = \frac{296}{T}\frac{Q(296)}{Q(T)}\frac{\exp(-\frac{E_L}{k_B T})}{\exp(-\frac{E_L}{k_B 296})}\left(\frac{T}{296}\right)^{\eta}. \tag{8}$$

$Q(T)$ is the total internal partition sum for the O_2 molecule [14], E_L is the lower state energy of the transition corresponding to the absorption line and k_B is the Boltzmann constant.

3.7 Post signal processing—Kalman filtering

After the curve fitting, residual noise from the sensor system is still superimposed on the concentration values. To separate the system noise from the real concentration variation, Kalman filtering is applied. Contrary to a conventional linear filter e.g. Wiener filter [20], the Kalman filter [21] is time-variant. It is a least-squares estimator in the context of time-variant models. Its "bandwidth" changes over time depending on the actual situation of the sensor and gas, e.g.

Table 2 The linear/nonlinear parameters of the model as well as the methods to determine their initial values. The two parameters $C\frac{a_0(T)}{a_0(296)}$ and $p\frac{\alpha_L(T)}{\alpha_L(296)}$ are the parameters of interest

Model Parameter	Definition	Spectrum characteristic values applied for initial value determination
i_{ofs}	x offset (unit: index)	The position of max/mins
y_{ofs}	y offset (unit: absorbance)	Mean value of the whole spectrum
$p\frac{\alpha_L(T)}{\alpha_L(296)}$	Gas pressure without temperature correction (unit: bar)	Max to min ratio R
Δx	Normalized wavelength scale	R and distance between zero crossings i_z
$C\frac{a_0(T)}{a_0(296)}$	Gas concentration without temperature correction (unit: ppm)	Maximum value

the variance of system noise introduced by the sensor itself and real gas concentration fluctuation. In this paper, the signal model given by Riris [22] is applied and it is assumed that real concentration fluctuations and the sensor system noise are two independent processes, i.e. their variations are independent. The variance of the sensor system noise is determined theoretically from the received DC detector power I_0. By subtracting it from the total variance of the estimated concentration, the variance of real concentration fluctuation is obtained. Details of this concept is planned to be outlined in a separate letter.

4 Sensor performance: concentration and pressure measurement

The sensor was mounted in the exhaust pipe of a wall-hanging condensing gas boiler with 10–20 kW heat power. The oxygen concentration was measured when the gas furnace was switched on and off for several times which is comparable to real operation. During on switching, the photodetector current decreases by a factor of ~3 due to condensation (Fig. 6). Contamination on optic surfaces turned out to be no issue in a gas furnace because of the high purity of the natural gas used for combustion. Reference measurements with electrochemical cell and lambda probe are carried out, compared to which, TDLS-based sensor shows a faster response time [23]. The real-time least-squares curve fit algorithm based on the advanced pre-calculation is implemented on a microcontroller with 24 MHz clock rate. The fitting result is shown in Fig. 7. The received detector power in this case is 500 nW and the detected power of second harmonic component is 500 pW, which is even two times lower than the lowest measured intensity with a wall-hanging gas boiler. The fit guarantees long-term stability of the sensor.

The standard deviation of the measurement i.e. the measurement uncertainty during boiler operation for one constant concentration value is 930 ppm (1 σ in 5 Hz) without Kalman filter. For 1 Hz measurement, this value is converted

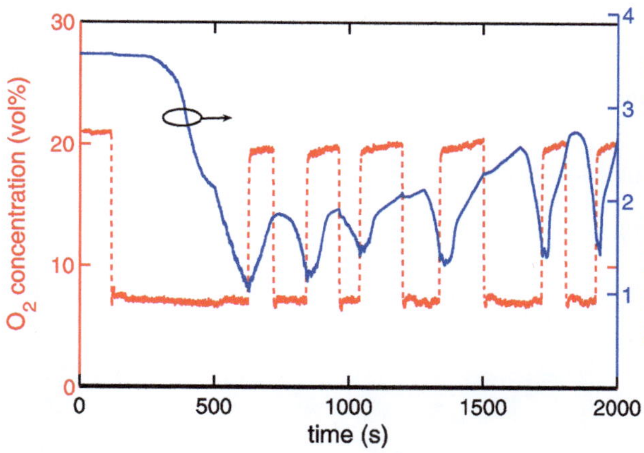

Fig. 6 The estimated concentration value during burner on/off switching. The photodetector current decreases by a factor of 3 due to condensation in this measurement. A decrease of factor 5 is observed on a longer time scale

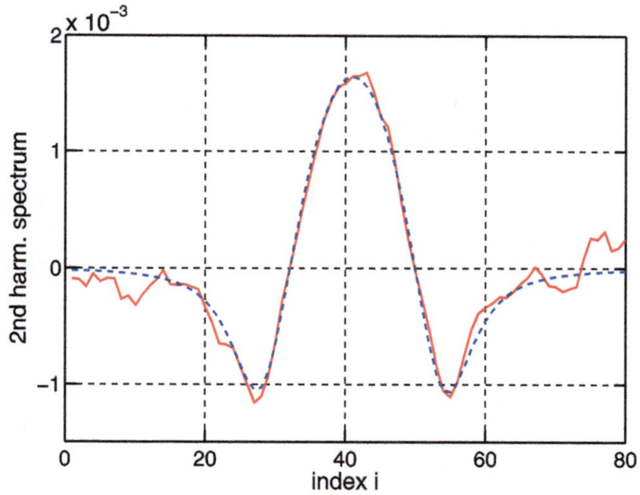

Fig. 7 Measured second harmonic spectrum (*solid curve*) and its minimal mean square curve fit with the model (4) (*dashed curve*). The detected photodetector power is 500 nW and the power of detected second harmonic is circa 500 pW

to 417 ppm by dividing by $\sqrt{5}$, which corresponds to an absorbance of 1.1×10^{-5}. With the utilization of Kalman filter, 250 ppm standard deviation with 5 Hz is achieved (Fig. 8).

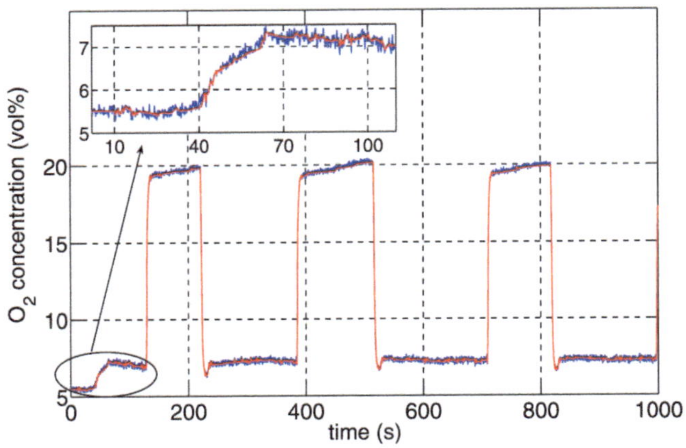

Fig. 8 The measured O_2 concentration (*blue*) and filtered result with Kalman filter (*red*). Unlike conventional lowpass filter, no detection delay is introduced additionally by the Kalman filter compared to the unfiltered sensor data

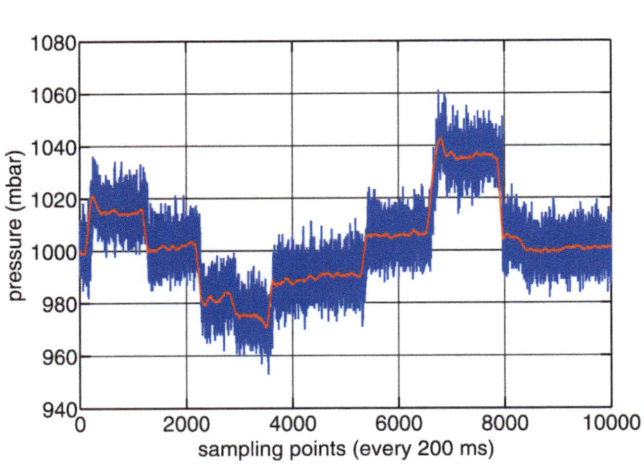

Fig. 9 Pressure measurement in every 200 ms (*blue*) and the 20 s average (*red*). The pressure values were set manually to 1000, 1030, 1000, 970, 962, 973, 1000, 1049 and 1000 mbar. The measurement is done with 100% oxygen and at ambient room temperature

No additional detection delay is introduced by the Kalman filter compared to the unfiltered sensor data, which is very important for measurement in furnace, where fast control loops are needed.

We also show the pressure measurement in Fig. 9, wherein a nonlinear response of the sensor to pressure values is observed. The deviation is lead back to the following reasons: the assumption of a Lorentzian-based absorption line profile is probably too approximate for a pressure determination with high precision; furthermore, the laser RAM influence [24] on the spectrum is only considered as an offset in the curve fit. Nevertheless, the pressure output values are reproducible, so a nonlinear correction can be done if gas pressure is to be estimated. The standard deviation of the estimated pressure measurement is 6 mbar at 5 Hz measurement rate. After nonlinear correction, the standard deviation would be approximately 9 mbar.

5 Conclusion and discussion

An oxygen sensor for combustion optimization in gas furnaces was developed, the measurement was realized in the exhaust gas pipe of gas furnace. Utilizing a diffuse reflector, the problem of possible misalignment of the optics due to thermal cycles, the condensation problem on optic surfaces and the problem of etalon fringes are solved. A real-time least-squares curve fit algorithm is presented, which improves the long-term stability of the sensor compared to fixed detection at the absorption line center, because the baseline drift is compensated. Real-time operation with a microcontroller is assured by determining good initial values using characteristic values of the second harmonic spectra (less iterations) and by programming the Jacobian matrix analytically (less computing complexity). There the maximum-to-minimum ratio R of the second harmonic curve is a very important characteristic value, it fully represents the curve shape, whereas the gas concentration is only a y-scale factor of the curve once the curve shape is fixed. Neither absolute wavelength scale, nor absolute wavelength difference between two measurement points is needed for concentration determination. Kalman filtering is applied to improve the detection sensitivity without losing the reliability and the response time of the sensor operation. Both concentration and pressure measurements are carried out in the gas furnace and the detection limit of concentration is 250 ppm in 5 Hz. Further application such as flame-on monitoring, which is an important safety issue in burners, is possible with TDLS-based O_2 sensors because of their fast response time and intrinsic self-checking capability in contrast to ZrO_2-based sensors [4]. The diffuse reflector based optical cell and described signal processing are not limited to oxygen sensing, but also can be applied for sensing of other gas molecules.

Acknowledgements The authors gratefully acknowledge the financial support by the Federal Ministry of Education and Research of Germany (Project 'NOSE', No. 13N877). The first author also would like to thank Dr. Ralf Meyer for careful proofreading and Armin Spitznas and Hans Link from Siemens Building Technology for technical consultation on combustion in furnaces.

References

1. A. Spitznas, H. Link, Siemens building technologies. Private communication (2008)
2. National Industrial Fuel Efficiency Service LTD, *The Boiler Operators Handbook* (Graham & Trotman Limited, 1989)
3. Buderus Heiztechnik GmbH (ed.), *Handbuch für Heizungstechnik* (Beuth Verlag GmbH, 2002)
4. E. Logothetis, in *Proceedings of the 8th Automotive Materials Conference: Ceramic Engineering and Science Proceedings*, vol. 1 (1980), p. 281
5. H. Teichert, T. Fernholz, V. Ebert, Appl. Opt. **42**, 2043 (2003)
6. J. Reid, D. Labrie, Appl. Phys. B: Lasers Opt. **26**, 203 (1981)

7. T. Iseki, H. Tai, K. Kimura, Meas. Sci. Technol. **11**, 594 (2000)
8. M. Grabherr, D. Wiedenmann, R. Jaeger, R. King, Proc. SPIE **5737**, 120 (2005)
9. D. Masiyano, J. Hodgkinson, R. Tatam, Appl. Phys. B: Lasers Opt. **90**, 279 (2008)
10. D. Masiyano, J. Hodgkinson, R.P. Tatam, Appl. Phys. B **100** (2010) This Issue. DOI:10.1007/s00340-010-4020-z. Special Issue of 7th Tunable Diode Laser Spectroscopy Conference
11. J.W. Goodman, *Laser Speckle and Related Phenomena* (Springer, Berlin, 1975), Chap. 2
12. J. Chen, A. Hangauer, R. Strzoda, M. Fleischer, M.C. Amann, Proc. Chem. **1**(1), 1383 (2009). Proceedings of the Eurosensors XXIII Conference
13. K.J. Ritter, T.D. Wilkerson, J. Mol. Spectrosc. **121**, 1 (1987)
14. L.S. Rothman, et al., J. Quant. Spectrosc. Radiat. Transf. **96**, 139 (2005)
15. R. Arndt, J. Appl. Phys. **36**, 2522 (1965)
16. P. Kluczynski, Å.M. Lindberg, O. Axner, Appl. Opt. **40**, 770 (2001)
17. A. Hangauer, J. Chen, M.-C. Amann, Appl. Phys. B **90**, 249 (2008)
18. D.W. Marquardt, J. Soc. Ind. Appl. Math. **11**, 431 (1963)
19. A. Bjorck, *Numerical Methods for Least Squares Problems* (SIAM, Philadelphia, 1996)
20. N. Wiener, *Extrapolation, Interpolation, and Smoothing of Stationary Time Series* (Wiley, New York, 1949)
21. R.E. Kalman, Trans. ASME—J. Basic Eng. 35 (1960)
22. H. Riris, C.B. Carlisle, R.E. Warren, Appl. Opt. **33**, 5506 (1994)
23. A. Hangauer, A. Spitznas, J. Chen, R. Strzoda, H. Link, M. Fleischer, in *Procedia Chemistry*, vol. 1 (2009), p. 955. Proceedings of the Eurosensors XXIII conference
24. P. Kluczynski, J. Gustafsson, Å.M. Lindberg, O. Axner, Spectrochim. Acta Part B: At. Spectrosc. **56**, 1277 (2001)

Laser applications in thin-film photovoltaics

R. Bartlome · B. Strahm · Y. Sinquin · A. Feltrin · C. Ballif

Received: 14 October 2009 / Revised version: 24 November 2009 / Published online: 14 January 2010
© Springer-Verlag 2010

Abstract We review laser applications in thin-film photovoltaics (thin-film Si, CdTe, and $Cu(In,Ga)Se_2$ solar cells). Lasers are applied in this growing field to manufacture modules, to monitor Si deposition processes, and to characterize opto-electrical properties of thin films. Unlike traditional panels based on crystalline silicon wafers, the individual cells of a thin-film photovoltaic module can be serially interconnected by laser scribing during fabrication. Laser scribing applications are described in detail, while other laser-based fabrication processes, such as laser-induced crystallization and pulsed laser deposition, are briefly reviewed. Lasers are also integrated into various diagnostic tools to analyze the composition of chemical vapors during deposition of Si thin films. Silane (SiH_4), silane radicals (SiH_3, SiH_2, SiH, Si), and Si nanoparticles have all been monitored inside chemical vapor deposition systems. Finally, we review various thin-film characterization methods, in which lasers are implemented.

1 Introduction

The terrestrial photovoltaic (PV) market has been dominated by the wafer-based crystalline silicon (c-Si) technology ever since the mid 1970s [1]. In the past decade, thin-film (TF) photovoltaics has attracted a growing interest both in academic and industrial environments. This has translated into a continuous gain of market share since 2004 (12.5% in 2008) and a constant increase of TF production volume (reaching a record-breaking 1.0 GW_p in 2008) [2]. The backbone of these TF technologies relies on hydrogenated amorphous silicon (a-Si:H), hydrogenated microcrystalline silicon[1] (μc-Si:H), cadmium telluride (CdTe), or $Cu(In,Ga)Se_2$ (CIGS) semiconducting materials. Despite lower efficiencies ($\leq$12% up to date[2]), commercial TF PV modules offer many advantages with respect to their wafer-based c-Si counterparts. Besides evident lower raw material costs, materials with different bandgaps can be engineered and stacked to exploit a larger part of the solar spectrum, thereby increasing the efficiency of the TF PV device. Moreover, films are deposited on inexpensive substrate materials, such as glass. Inexpensive and flexible modules may even be fabricated on metallic or plastic foils. Finally, large areas (>1 m^2) can be processed because (a) TF deposition facilities are usually scalable, and (b) the individual PV cells of a large module can be serially interconnected without the need of time-consuming wiring. To achieve this task, three patterns (p1, p2, p3) are micromachined during fabrication—principally by laser scribing. In this respect, lasers have become key-components in the production of TF PV modules.

Various applications are securing a bright future for lasers in TF photovoltaics. Laser scribing along with other laser processes for TF photovoltaics are reviewed under Sect. 2. Lasers are also integrated into various diagnostic tools to analyze deposition processes of Si thin films. Such applications are reviewed under Sect. 3. Finally, in the development stage of PV cells, lasers are applied in several TF characterization methods, reviewed under Sect. 4.

R. Bartlome (✉) · B. Strahm · Y. Sinquin · A. Feltrin · C. Ballif
Institute of Microengineering (IMT), Photovoltaics and Thin Film Electronics Laboratory, Ecole Polytechnique Fédérale de Lausanne (EPFL), Rue A.-L. Breguet 2, 2000 Neuchâtel, Switzerland
e-mail: richard.bartlome@epfl.ch

[1] This material is also known as hydrogenated nanocrystalline silicon (nc-Si:H).

[2] Total-area efficiency under standard test conditions.

2 Laser processes for TF photovoltaics

Among laser processes for TF photovoltaics, only laser scribing systems hardly suffer any competition from other techniques. Laser scribing systems have the ability to machine narrow patterns on a wide range of materials in a selective, precise, and cost-effective manner. It is a noncontact process, meaning that the mechanical stress applied on the micromachined layers is limited. Unlike the tip of a mechanical scriber that suffers mechanical wear [3], a laser scriber—when powered by a diode-pumped solid state laser (DPSSL)—can operate continuously for over 10 000 hours. Other laser processes, such as pulsed laser deposition and laser-induced crystallization, have not yet penetrated the TF PV market.

2.1 Laser scribing

As mentioned above, the serial monolithic interconnection of TF solar cell can be achieved by scribing three patterns during fabrication, a process developed by Nakano et al. [4]. The films are patterned after each layer deposition, as pictured in Fig. 1. The end-result is a PV module comprising serially interconnected strip-shaped solar cells. The number of cells and their width determine the ultimate open-circuit voltage and short-circuit current of the finished module, respectively. The scribed lines are narrow (20–100 μm) and closely aligned to each other in order to minimize the so-called dead area between the first and third patterns. The ideal cell width, of the order of 10^{-2} m, is a trade-off between the dead area and ohmic losses in the contacts. Furthermore, the ideal width of the p1 line is a trade-off between the dead area and the electric isolation between the first deposited contact of neighboring cells [5].

The laser setup depends on whether the films have been stacked in a *substrate* or *superstrate* configuration.

In the *superstrate* configuration, the substrate is transparent and facing the incoming sunlight. As pictured in Fig. 2, the first deposited layer is a transparent conductive oxide (TCO), followed by an absorber layer, and a back contact/reflector. Depending on the TF PV technology, the absorber can be a CdS/CdTe heterojunction, an a-Si:H p-i-n junction, or a micromorph tandem structure (a-Si:H/μc-Si:H) [6, 7]. The front TCO layer is scribed by infrared or ultraviolet laser radiation (p1 process). The industrial workhorses for this task have been Q-switched (ns pulses) DPSSLs emitting at 1064 nm or 355 nm (third harmonic), respectively. The active medium of these lasers is a neodymium-doped crystal (Nd:YAG or $Nd:YVO_4$). A frequency-doubled DPSSL emitting at 532 nm typically performs the p2 and p3 processes. During all three processes, the laser fluence on the workpiece is optimized in order to work within an ideal range situated above the ablation threshold. The threshold fluence of various relevant materials have been measured by Compaan et al. [8], whereas the pulse shape and the pulse overlap in TF Si technologies are discussed by Haas et al. [9, 10]. Molpeceres et al. compare the p1 process at 1064 nm and 355 nm using a tin-doped

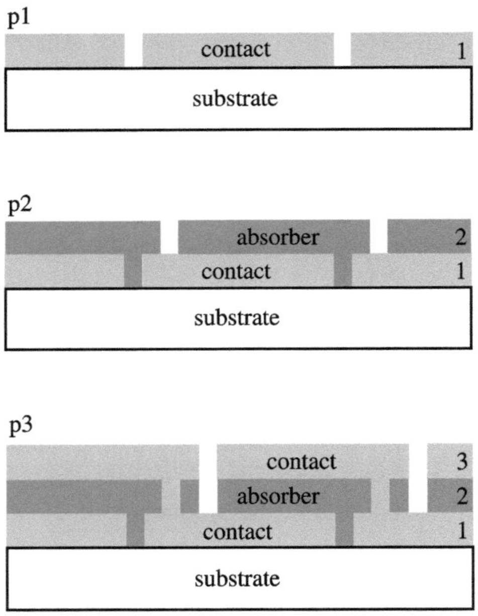

Fig. 1 Schematic view of the p1, p2, and p3 processes. The layers are numbered in the order in which they are deposited. They are micromachined by laser ablation, an established material removal process that has been extensively studied in literature [8, 16]. For this purpose, a short (ns) or ultrashort (ps or fs) pulsed laser beam is focused on the substrate. The repetition rate of the laser is synchronized to the feed rate of the workpiece. The first and second layers are micromachined during the p1 and p2 process, respectively. During the p3 process, the pulse energy is usually deposited in the second layer. The overlying volume is then removed by explosive ablation (caused by the pressure build-up in the second layer). The first layer must remain unaffected by the p2 and p3 processes

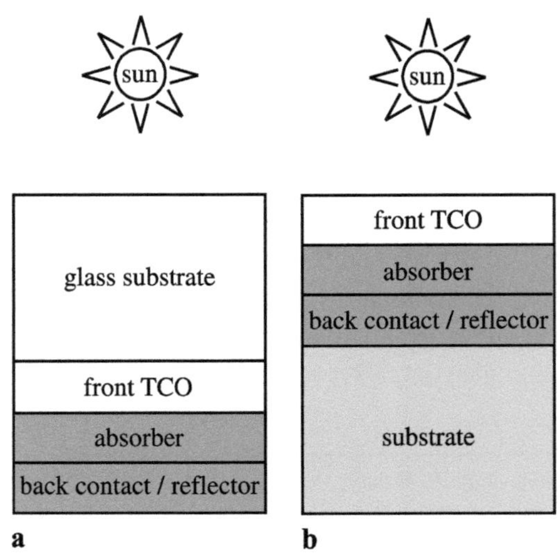

Fig. 2 Schematic view of a TF PV cell in the superstrate (**a**) and substrate (**b**) configuration

indium oxide as front TCO [11, 12]. In CdTe solar cells, all three processes can be in principle performed at 1064 nm, as the ablation threshold of CdTe at this wavelength—unlike a-Si:H or μc-Si:H—is lower than the ablation threshold of the TCO [8, 13]. Irrespective of the TF technology, the laser beam can be directed towards the PV module from the glass-side. In this manner, the ablated material is easily withdrawn from the opposite side. Furthermore, material ejection—which begins in a ps timescale—does not interact with the ns-long laser pulse.

In the *substrate* configuration, the PV module is fabricated in reverse order, i.e., from back to front with respect to sunlight (Fig. 2). This configuration is chosen to fabricate CIGS solar cells or TF PV cells on opaque substrates, such as steel and plastic foils. If the substrate is conductive, the individual PV cells may be monolithically interconnected by laser scribing, provided that an insulating layer is deposited first. Irrespective of the substrate material, the p2 and p3 processes must be conducted from the layer side. The p2 and p3 patterns of CIGS solar cells, which use molybdenum as a back contact, are often machined with mechanical scribers, but laser scribing systems are progressively taking over the task. To avoid molybdenum damage during laser processing of the overlaying layers, ultrashort laser pulses (ps or fs) are required [14, 15]. It is well known that the thermal dissipation beyond the volume that is ablated is then widely suppressed [16, 17].

In contrast to the smooth and hard surface of a glass substrate, a soft plastic foil is difficult to structure with a mechanical tool without causing damage. Furthermore, ns pulses have not been able to produce well-defined laser scribing grooves in flexible solar cells that incorporate metallic back reflectors such as silver or aluminum. As an alternative to removing the absorbing film entirely, analogous to mechanical or laser scribing, laser processes have been developed to locally transform the absorber into a conductive channel [18, 19]. These interconnection schemes, which replace the p2 process, are conducted at laser fluences below the ablation threshold.

Large industrial laser scribing systems, recently reviewed [20], process several lines simultaneously. The underlying concepts for beam delivery are further discussed in [21].

2.2 Laser-induced crystallization

Polycrystalline thin films on low-cost glass substrates are expected to combine the high efficiencies of wafer-based c-Si PV cells with the low manufacturing cost of TF Si PV cells. Different approaches have been investigated to overcome the conflicting requirement of a high temperature for crystallization and a low temperature for preventing damage on the low-cost glass substrate [22, 23]. Solid-phase crystallization of Si-coated glass has been the most successful approach so far in terms of device performance (10.4% efficiency on a 10 cm × 10 cm mini-module) [6, 24], but the process requires annealing at 600°C for several hours. Other crystallization methods have achieved higher efficiencies, however it must be emphasized that these processes require high-temperature-resistant substrates [22]. Laser-induced crystallization appeared once to be the only approach capable of forming polycrystalline Si films on commercially available glass substrates at low temperatures (<220°C). The technique simply consists in heating amorphous Si with ns or μs pulses above the melting point of Si (at fluences below the ablation threshold). During the rapid melting and resolidification of amorphous Si, the substrate does not have time to heat up. The process may be followed by epitaxial thickening. Substantially large lateral grain sizes (>10 μm) have been reported using a μs-pulsed Ar^+ laser system (cw beam modulated with an acousto-optic modulator) [25]. Most studies have been conducted with excimers laser and were reviewed by Adikaari et al. [26]. In spite of all these efforts, reports of device-grade material are still lacking. In the context of PV cells, early industrial investigations also seem to have been abandoned [27].

Recently, Nayak et al. reported nanocrystalline Si material following femtosecond-laser-induced crystallization of a-Si:H [28]. Nevertheless, the sub-bandgap absorption is so high that the material seems to have lost its semiconducting properties. Despite the amount of structural defects, which for the time being prohibits PV applications, the process produces remarkable light-trapping microstructures at the surface. Similar effects have been studied extensively by Her et al. on c-Si wafers [29].

2.3 Pulsed laser deposition

Deposition techniques constitute a vast technological field. In pulsed laser deposition (PLD), the output of a pulsed laser is focused onto a target material mounted inside a vacuum chamber. The ejected plasma plume expands away and impinges on the substrate of interest. The properties of the deposited film generally depend on the substrate temperature, the laser-target-substrate geometry, the presence of a background gas in the chamber, and several laser parameters (wavelength, pulse duration, fluence). PLD has been a competitive tool to grow thin films with well-defined complex stoichiometries. Photovoltaic-relevant materials (e.g., ZnO, CdTe, ZnS) have also been grown by PLD [30–32]. The chances of this technique entering the PV market, however, remain poor as upscaling is difficult. Furthermore, the ejection of macroparticles prevent many practical applications in photovoltaics [33].

Table 1 Laser-based techniques that have been implemented in PECVD systems to study silane-containing plasmas

Technique	Laser	Detected species	References
Infrared absorption spectroscopy	Pb-salt, QCL	SiH, SiH_3, SiH_4	[35–37, 43]
Cavity ring-down spectroscopy	Dye (ω, 2ω)	Si, SiH, SiH_3, Si nanoparticles	[34, 49–51, 93]
Laser-induced fluorescence	Dye (ω, 2ω)	Si, SiH, SiH_2	[52, 53, 55]
Laser light scattering	He–Ne, Ar^+, Nd:YAG (ω, 2ω, 3ω, 4ω)	Si nanoparticles	[56–64]
Laser-induced particle explosive evaporation	XeCl*, Nd:YAG	Si nanoparticles	[65, 66]

3 Laser diagnostics during deposition of Si thin films

In this section, we review how lasers are applied as diagnostic tools during deposition of Si thin films. Si deposition usually takes place in chemical vapor deposition (CVD) reactors. The most common precursor gases of Si thin films are silane (SiH_4) and hydrogen (H_2). Silane may either be dissociated by catalytic reaction or electron impact, as it is predominantly the case in hot-wire CVD (HWCVD) and plasma-enhanced CVD (PECVD), respectively. In Table 1, we summarize laser diagnostic tools that have been implemented in PECVD systems, which are by far the most common. Silane and its radicals have been monitored in such systems by fluorescence and absorption spectroscopy. These laser-based techniques are reviewed in Sect. 3.1. Under certain process conditions, the dissociation of silane in chemical reactors leads to the formation of nanoparticles. The latter can be detected by laser light scattering or laser-induced explosive evaporation. Both techniques are reviewed in Sect. 3.2.

3.1 Laser spectroscopy

Laser spectroscopy—as well as other spectroscopic techniques such as optical emission spectroscopy, mass spectroscopy, or Fourier transform infrared spectroscopy (FTIR)—have all been applied as plasma diagnostic tools. Their objective is to measure the density of molecules and radicals present in the plasma, which is of utmost importance to understand the relation between the plasma composition and the properties of deposited films.

The importance of a silane radical with respect to the growth of Si films is assessed on the basis of its density in the plasma and its surface reaction probability. For conditions under which device-grade material is grown, it is found that SiH_3 is the most abundant radical in PECVD reactors [34]. Radicals with a higher Si hydrogenation are, however, less reactive. Overall, SiH_3 is believed to be the most significant film growth precursor [34, 35].

3.1.1 Infrared absorption spectroscopy

By 1986 diode laser spectrometers had already detected SiH and SiH_3 in electric discharges [36, 37]. In these early studies, a cryogenically-cooled lead-salt laser was typically coupled into a multipass optical setup positioned along a discharge tube. Multipass arrangements, mostly of the White [38] or the Herriott type [39, 40], were used to increase the absorption path length by several meters, thereby enhancing the sensitivity of the detection scheme [41].

Further studies have been mainly dedicated to better understand the relevance of SiH_3 to the film growth mechanism. Sumiyoshi et al. [42] measured the SiH_3 density in a phenylsilane discharge. The latter was ignited between two cylindrical electrodes mounted in a glass tube. The parent gas was phenylsilane rather than silane to avoid possible interferences with silane absorption lines. The multimode diode laser was tunable around 4.5 μm, where Si–H stretching vibrations of SiH_3 take place (ν_3 band). Itabashi et al. [35] measured the SiH_3 density profile between the two electrodes of a parallel-plate reactor. From the infrared absorption at 720 cm^{-1} in SiH_4–H_2 discharges, they deduced a contribution of SiH_3 to the film growth of about 60%.

More recently, the authors have shown that silane (SiH_4) itself can be monitored in the pumping line of a CVD reactor [43]. Despite the low pressure in the pumping line (lower than the 1–6 mbars maintained in the reactor), the silane density was sufficient to be measured in a single-pass arrangement with a Peltier-cooled quantum cascade laser emitting around 4.6 μm. This wavelength range is particularly suitable, as it features intense roto-vibrational absorption lines of the ν_3 band of SiH_4. In the absence of powder particles, the authors could monitor in situ the TF Si deposition rate. Time-resolved silane density measurements inside the CVD reactor is a subject of current research.

3.1.2 Cavity ring-down (CRD) spectroscopy

Monitoring radicals in chemical reactors is a challenging task because (a) radical densities are low, (b) the optical path length available in laboratory-scale reactors rarely exceeds

1 m, and (c) reactor environments are harsh (high temperatures, aggressive chemicals). To enhance the sensitivity of a detection scheme, the effective path length may be increased by using a cavity ring-down (CRD) setup [41, 44, 45] or a multipass arrangement [38–41]—as mentioned above. Originally designed to measure precisely the high reflectivity of laser mirrors [46–48], the CRD technique is known today as a sensitive trace-gas detection scheme [41]. The technique consists in coupling a laser pulse into a high-finesse cavity formed by two highly reflective dielectric mirrors ($R > 99.99\%$) that are facing each other. Every time that the light is reflected back and forth inside the cavity, a small fraction of the light leaks out of the cavity. Instead of measuring the absolute value of the light intensity leaking out of the cavity, one determines the light intensity decay time. The higher the absorption inside the cavity, the smaller the measured decay time. The CRD technique is intrinsically insensitive to laser power fluctuations. A further advantage over other spectroscopic techniques is that the effective absorption path length of a CRD setup is extremely long (several kilometers).

Silane-containing discharges have been investigated by only a handful of CRD experiments. Exclusively designed in the visible range up to date, CRD cells have been coupled with the fundamental beam or the second harmonic of Nd:YAG-pumped dye lasers.

Kessels et al. [49] performed CRD measurements across an expanding thermal plasma (ETP). They were able to determine the absolute density of Si and SiH radicals in an Ar–H_2–SiH_4 plasma. The Si density, of the order of 10^{15} m^{-3}, was about 2 orders of magnitude lower than the SiH density. By means of a simple chemical and surface reaction model, they deduced that Si and SiH radicals contribute to the film growth mechanism by about 0.2% and 5% only, respectively. From time-resolved CRD measurements in the afterglow of an ETP, Hoefnagels et al. [34] showed that Si radicals have a higher gas-phase reactivity and, consequently, a shorter lifetime than SiH_3 radicals. Hoefnagels et al. also derived a surface reaction probability of >0.95 and 0.3 for Si and SiH_3, respectively. Finally, they demonstrate that SiH_3 is by far the most important film growth precursor in ETPs.

Under many plasma conditions used to deposit device-grade material, CRD measurements in the visible are influenced by the light scattering and absorption of Si nanoparticles. Nagai et al. [50] recently reported that the cavity loss in the 220–280 nm wavelength range is not only due to the expected $\tilde{A}^2A_1' \leftarrow \tilde{X}^2A_1$ broadband transition of SiH_3. In fact, a three-step mechanism was identified. In the first step directly following the discharge break-down, no extra cavity loss is measured. In the second step, the contribution of silicon particles increases rapidly as a function of time, and in the third step the cavity loss due to particles tends to reach a plateau. These three steps can be associated to silicon nanocluster formation, cluster agglomeration, and agglomerate-size growth by chemical sticking.

Grangeon et al. [51] had already applied CRD spectroscopy to detect Si particles in a large-area industrial parallel-plate reactor. In their work, they pointed out the importance of reactor conditioning prior to CRD measurements to avoid contamination of the mirror surfaces during the discharge. Mirror contamination can either come from direct film deposition or from chemical adsorption, e.g., of water. The contamination of CRD mirrors results in an unknown change of the mirror reflectivity and, consequently, in a change of the apparent extinction coefficient.

To summarize, the density and lifetime of Si, SiH, and SiH_3 radicals were determined by CRD spectroscopy. Gas-phase reaction rates and surface reaction probabilities could be derived from these measurements. Moreover, it was shown that CRD spectroscopy can detect the onset of Si particle formation.

3.1.3 Laser-induced fluorescence spectroscopy

The detection of Si, SiH, and SiH_2 has been reported in several laser-induced fluorescence (LIF) studies. The latter often require advanced acquisition schemes to discriminate against stray light emitted inside the CVD reactor. Some schemes average a large number of fluorescence pulses with gated integrators, and other perform afterglow measurements, where the LIF signal is acquired just after the extinction of a glow discharge.

Roth et al. [52] measured the Si density profile between the symmetric electrodes of a PECVD system. An Nd:YAG-pumped frequency-doubled dye laser tuned at 251.43 nm excited the Si radicals during a SiH_4–Ar discharge. The Si fluorescence was measured at 252.85 nm ($4s^3P_1^0 \rightarrow 3p^2\,{}^3P_2$ transition). These studies showed that the sheaths do not have any Si atoms, and that the SiH_4 dissociation is more efficient close to the radio-frequency-powered electrode.

Matsumi et al. [53] measured the density and spatial profile of SiH in pure silane discharges as a function of discharge parameters such as the pressure, the silane flow, and the radio-frequency (RF) power. The fluorescence of the SiH molecules at 413 nm was induced by a nitrogen-pumped dye laser tuned at 410 nm. SiH showed the same spatial profile as measured by Roth et al. for Si [52]. The SiH density in the ground state and in the excited SiH* state was estimated following simultaneous LIF and optical emission measurements.

Nozaki et al. [54] evaluated the absolute density of Si and SiH in an HWCVD system by comparing LIF intensities with the scattered laser light by Ar atoms. The Si and SiH radicals were excited by a tunable Nd:YAG-pumped frequency-doubled dye laser. The ns-long pulses delivered by the laser system were well adapted to detect SiH molecules, whose excited state lifetime is as high as 0.6 μs. The detection of Si atoms, however, was perturbed by the fact that the lifetime of excited Si atoms was comparable to the duration time of the laser pulse.

As demonstrated by Hert and Jolly [55], SiH_2 radicals could also be detected in silane-containing discharges using an Nd:YAG-pumped dye laser emitting at 580 nm. The induced SiH_2 fluorescence was detected at 618 nm. From time-resolved acquisitions at $T = 500$ K and $p = 9$ Pa, Hert and Jolly derived a reaction rate coefficient between SiH_2 and SiH_4 of $(2.3 \pm 0.5) \times 10^{-17}$ m^3 s^{-1} molecule^{-1}, and a surface reaction probability of SiH_2 of 0.6 ± 0.15, instead of 0.8 as previously assumed.

3.2 Nanoparticle detection

3.2.1 Laser light scattering

Particle detection by laser light scattering (LLS) in silane discharges was first reported in 1985 by Roth and coworkers [56]. They presented a detailed study of the particle spatial distribution in an RF-powered reactor [57]. Their setup consisted of ns-pulsed Nd:YAG lasers that were frequency-multiplied up to the fourth harmonic ($\lambda = 532$, 354, or 266 nm). An asymmetrical particle distribution across the inter-electrode gap was found following scattering measurements conducted at 532 nm. Shortly after this discovery, the interest in powder growth and dynamic grew steadily and led several research groups to conduct LLS experiments. Watanabe et al. [58, 59] studied the effect of a low-frequency modulation of the RF excitation in order to reduce the powder formation. The power of the He–Ne laser they used ($\lambda = 633$ nm) was only 1 mW. With 5-mW laser radiation at 633 nm, Bouchoule et al. [60] measured the scattered light intensity perpendicular to the incident beam (as in the previous work) and also the scattered light intensity within a solid angle of 6° with respect to the incident beam (forward scattering). This is the first extensive report on the effect of different discharge parameters such as the temperature, the silane flow, and the inter-electrode voltage on the silicon particle formation. It is also found that the temporal behavior of dusty discharges is unstable because of electron–particle interactions.

Ten years after Roth's discovery, first plasma physical/chemical models appeared. Howling et al. set up a quadrupole mass spectrometer jointly with an LLS device to demonstrate that negatively charged polysilicon particles are the precursors of larger powder particles [61], which are then detectable by LLS. In 1996, the same research group modeled experimental results obtained by LLS and derived different silicon powder formation mechanisms [62]. The experimental setup was based upon a polarized Ar^+ laser ($\lambda = 488$ nm) and a multiple-angle detection system (0°, 45°, 90°, 135°, 180°, and 270° with respect to the incident light). The laser light polarization could be adjusted at the laser output and at the entrance of the detector. With such an experimental arrangement, they reported a particle size detection limit of 2 nm. Despite this exceptional sensitivity—the best reported so far with LLS—the early stage of powder formation involving subnanometric particles could not be detected.

More recently, silicon nanoparticles have been detected by LLS in large-area industrial reactors, as opposed to all previous studies that were conducted in small laboratory-scale reactors (typical diameter of 20 cm). With the help of a low-cost high-power (500 mW) compact Nd:YAG laser, down-stream scattering experiments were directly conducted in the exhaust line of a large-area parallel-plate reactor [63]. Using an expanded Ar^+ laser beam (488 nm, 1.5 W, 1 mm × 25 mm) and a charge-coupled device camera to detect the scattered light by powder particles, Strahm et al. [64] mapped the powder distribution between the electrodes of a large-area parallel-plate reactor. It was found that the reduction of the inter-electrode distance shifts the appearance of powder in the discharge to higher working pressures.

3.2.2 Laser-induced particle explosive evaporation

Laser-induced particle explosive evaporation (LIPEE) is a technique developed by Boufendi et al. [65] in 1994 to study the early stages of cluster formation, when silicon particle are still too small to be detected by LLS. During LIPEE, silicon clusters are sublimated by an intense laser beam (typically 200 MW/cm^{-2}). Then, an optical emission spectrometer (OES) collects the light emitted by the excited states of the sublimated molecules. To prevent the plasma emission from saturating the OES detector, the laser pulse is usually shot a few milliseconds after the discharge is turned off. Silicon clusters have thus been sublimated with 25-mJ 20-ns-long pulses of a XeCl* excimer laser ($\lambda = 308$ nm) or with 15-mJ pulses of an Nd:YAG laser ($\lambda = 1064$ nm).

The kinetics of powder formation have been derived from time-resolved LIPEE measurements. For the first time, three distinct steps were evidenced: nucleation of crystallites, crystallites coalescence, and molecular sticking onto larger particles. For comparison, LLS is only able to detect the second and third steps in the particle growth mechanism. Transmission electron microscopy measurements confirmed the ability of LIPEE to detect the transition between these three particle growth steps [66]. LIPEE was further developed to study the radial distribution of crystallites in parallel-plate discharges and the influence of the reactor temperature on the formation of silicon crystallites in Ar–SiH_4 discharges. It has been demonstrated that lower reactor temperatures

Table 2 Laser-based characterization techniques applied in TF photovoltaics

Technique	Measured physical quantity	Deduced material properties	References
Micro-Raman spectroscopy	Inelastic scattering	Phase composition	[68–71, 73, 74]
Photoluminescence spectroscopy	Radiative recombination	Donor and acceptor binding energies	[67, 75–78]
Time-resolved photoluminescence	Photoluminescence decay time	Carrier dynamics	[67, 75, 79, 80]
Photothermal deflection spectroscopy	Laser beam deflections	Sub-bandgap absorption, defect densities	[81–83]
Light-beam-induced current	Photocurrent	Homogeneity	[67, 86, 87]
Internal photoemission	Photocurrent	Conduction band discontinuity	[88, 89]
Photoreflectance spectroscopy	Reflectance	Bandgap energies	[90]

promote crystallite formation, which in turn rapidly leads to the growth of coarser particles.

4 Laser-based characterization methods for TF photovoltaics

Lasers have played a significant role in the characterization of PV-relevant materials. An exhaustive review of all material properties that have been investigated by laser-based methods would go beyond the scope of the present work. The authors merely wish to provide a brief overview of characterization techniques that implement lasers. They are summarized in Table 2. For a detailed description of well-established techniques such as micro-Raman and photoluminescence spectroscopies, the authors refer to literature [67, 68].

In TF Si technologies, Raman spectra are routinely evaluated to determine the degree of crystallinity of μc-Si:H films [69–71]. The structure of device-grade μc-Si:H films consists of conical conglomerates of Si nanocrystals, where the micrometer-sized conglomerates and the nanocrystal grain boundaries are passivated by a-Si:H. The amorphous phase is associated with a broad Raman peak centered around 480 cm^{-1}, while the sharp peak centered around 520 cm^{-1} is characteristic of the c-Si phase. Micro-Raman spectroscopy has also contributed to better understand the complex nature of CIGS alloys. The bandgap of $Cu(In,Ga)Se_2$ increases as Ga is added to replace In. Record efficiencies are realized at $Ga/(Ga + In)$ ratios of 0.3 [72]. The Cu content of such high-efficiency solar cells may vary remarkably. In addition to the chalcopyrite mode centered around 177 cm^{-1}, Cu-rich CIGS exhibits a mode centered around 260 cm^{-1}, which is assigned to a detrimental highly-conductive $Cu_{2-x}Se$ phase [73, 74]. Irrespective of the TF PV technology, Raman spectra must be interpreted in light of the optical penetration depth of the laser within the sample. Ar^+, He–Ne, and frequency-doubled Nd:YAG lasers are common excitation sources found in micro-Raman spectrometers.

The photoluminescence spectra of TF PV materials present rather broad spectral features, and their interpretation is generally not straightforward. Furthermore, the high nonradiative recombination rates often require measurements to be performed at cryogenic temperatures. In spite of it all, various photoluminescence studies have contributed to unveil the nature of radiative recombinations in TF PV materials [75–78].

Depending on the TF technology, the photoluminescence decays in a ps to ns time scale [75, 79, 80]. These short decay times are typically resolved with a mode-locked laser and provide some insight into carrier dynamics. To a certain extent, the photoluminescence decay time of TF solar cells is correlated to device performances [79, 80].

At room temperature, nonradiative defect-mediated recombinations dominate in a-Si:H, CdTe, and CIGS devices. Structural defects are the source of the unwanted recombination centers in Si films. They are responsible for a low level of absorption below the bandgap. This sub-bandgap absorption cannot be measured with a traditional visible-near infrared spectrometer but instead requires a sensitive technique such as photothermal deflection spectroscopy [81]. In this technique, a modulated monochromatic beam is focused onto a sample that is immersed in a transparent liquid. Upon absorption of the monochromatic beam by the sample and diffusion of the generated heat into the liquid, a refractive index gradient builds up in the immediate vicinity of the sample and deflects a probe laser beam. The laser deflections are related to the absorption in the sample. The sub-bandgap absorption is related to the defect density [82, 83]. In photothermal deflection setups, a He–Ne laser is often chosen as probe beam due to its high beam quality and stability.

The commonly-reported current–voltage characteristic of a PV module gives a global view of its performance. A spatially resolved nondestructive assessment of the module performance can be undertaken with light-beam-induced current or thermographic systems [67, 84–86]. The light-beam-induced current technique (LBIC) maps the spatial distribution of the photocurrent in the module. Employed under short-circuit current conditions, it consists in recording the photocurrent while sequentially illuminating—typically with a laser beam—localized areas of the module under study. An LBIC analysis can, e.g., improve the yield of a production line through troubleshooting of the baseline process [87]. Multi-wavelength LBIC systems are even capable of selectively studying one junction within a multi-junction device [86].

Further specialized laser-based characterization techniques are reported in literature such as internal photoemission or photoreflectance spectroscopy. Internal photoemission consists in amplifying and measuring the external photocurrent of a heterojunction under optical pumping. After plotting the current versus the photon energy, one determines the threshold related to the conduction band discontinuity. Early experiments were limited to pumping from the top of the valence band in one material to the bottom of the conduction band in the other. With the development of high-power widely tunable infrared free-electron laser sources, it has become possible to directly pump a heterojunction interface across its conduction band discontinuity [88]. The free-electron laser-based internal photoemission technique was applied, e.g., to investigate the conduction band discontinuity of a CdS/CdTe interface [89].

Photoreflectance spectra are acquired during pump-probe experiments. While a modulated pump laser beam excites electron–hole pairs in a sample, the wavelength-dependent reflectance of the sample is measured with a widely tunable probe source. Instead of a laser beam, an AC voltage is often applied as modulating source. The derivative-like features of photoreflectance spectra correspond to interband transitions in the sample [90].

Finally, various lasers are applied to measure the angular distribution of scattered light on textured surfaces. The angular resolved scattering is then introduced into optical models to simulate the performances of TF solar cells [91, 92].

5 Conclusions

A wide range of laser-based methods are applied to characterize films in the development stage of PV modules. Furthermore, lasers have become key components in the manufacture of TF PV modules. They are applied as scribing tools to serially interconnect the individual cells of a TF PV module. While cells that are grown in a superstrate configuration are routinely processed by laser scribing systems, the monolithic serial interconnection of cells grown in a substrate configuration presents some difficulties. Other laser-based fabrication processes such as pulsed laser deposition and laser-induced crystallization are still marginal in the field of TF photovoltaics.

During deposition of Si thin films, laser systems have proven to be the only tools capable of detecting a wide range of chemical species in a nonintrusive manner. Silane (SiH_4), silane radicals (SiH_3, SiH_2, SiH, Si), and Si nanoparticles have all been monitored in situ in PECVD systems. While a considerable amount of effort has been undertaken to better understand the physics of chemical reactors, correlations between the plasma composition and TF properties are still lacking. In the authors' opinion, such correlations should be the target of future research.

References

1. R.M. Swanson, Prog. Photovolt. **14**(5), 443 (2006)
2. Photon Int. 170–206 (2009)
3. D. Bonnet, Thin Solid Films **361**, 547 (2000)
4. S. Nakano, T. Matsuoka, S. Kiyama, H. Kawata, N. Nakamura, Y. Nakashima, S. Tsuda, H. Nishiwak, M. Ohnishi, I. Nagaoka, Y. Kuwano, Jpn. J. Appl. Phys. **25**(12), 1936–1943 (1986)
5. K. Brecl, M. Topic, F. Smole, Prog. Photovolt. **13**(4), 297 (2005)
6. A.G. Aberle, Thin Solid Films **517**(17), 4706 (2009)
7. A.V. Shah, H. Schade, M. Vanecek, J. Meier, E. Vallat-Sauvain, N. Wyrsch, U. Kroll, C. Droz, J. Baillat, Prog. Photovolt., Res. Appl **12**, 113 (2004). doi:10.1002/pip.533
8. A.D. Compaan, I. Matulionis, S. Nakade, Opt. Lasers Eng. **34**(1), 15 (2000)
9. S. Haas, A. Gordijn, H. Stiebig, Prog. Photovolt. **16**(3), 195 (2008)
10. S. Haas, G. Schope, C. Zahren, H. Stiebig, Appl. Phys. A, Mater. Sci. Process. **92**(4), 755 (2008)
11. C. Molpeceres, S. Lauzurica, J.L. Ocana, J.J. Gandia, L. Urbina, J. Carabe, J. Micromech. Microeng. **15**(6), 1271 (2005)
12. C. Molpeceres, S. Lauzurica, J.J. Garcia-Ballesteros, M. Morales, S. Fernandez-Robledo, J.L. Ocana, J.J. Gandia, J. Carabe, F. Villar, J. Escarre, J. Bertomeu, J. Andreu, Appl. Surf. Sci. **254**(4), 1115 (2007)
13. J. Bohland, A. McMaster, S. Henson, J. Hanak, Specific PV-MaT R&D in CdTe product manufacturing—final subcontract report, March 2003. Subcontractor report NREL/SR-520-35177, National Renewable Energy Laboratory, 1617 Cole Boulevard, Golden, CO 80401-3393, January 2004
14. D. Ruthe, K. Zimmer, T. Hoche, Appl. Surf. Sci. **247**(1–4), 447 (2005)
15. J. Hermann, M. Benfarah, G. Coustillier, S. Bruneau, E. Axente, J.F. Guillemoles, M. Sentis, P. Alloncle, T. Itina, Appl. Surf. Sci. **252**(13), 4814 (2006)
16. D. Bäuerle, *Laser Processing and Chemistry*, 3rd edn. (Springer, Berlin, 2000)
17. P. Dittrich, R. Bartlome, G. Montemezzani, P. Gunter, Appl. Surf. Sci. **220**(1–4), 88 (2003)
18. F. Kessler, D. Herrmann, M. Powalla, Thin Solid Films **480**, 491 (2005)
19. A. Vijh, X. Yang, W. Du, X. Deng, Sol. Energy Mater. Sol. Cells **90**(16), 2657 (2006)
20. High-precision light—market survey on laser scribers for monolithic integration of thin-film modules. Photon Int. 258–272 (2008)
21. R. Bartlome, Y. Sinquin, S.A. Meyer, C. Ballif, A. Raible, Versatile thin-film photovoltaic laser scribing system, in *Proc. of the 24th European Photovoltaic Solar Energy Conference*, Hamburg, Germany (2009), pp. 2854–2855
22. R.B. Bergmann, Appl. Phys. A, Mater. Sci. Process. **69**(2), 187 (1999)

23. O. Nast, S.R. Wenham, J. Appl. Phys. **88**(1), 124 (2000)
24. M.A. Green, P.A. Basore, N. Chang, D. Clugston, R. Egan, R. Evans, D. Hogg, S. Jarnason, M. Keevers, P. Lasswell, J. O'Sullivan, U. Schubert, A. Turner, S.R. Wenham, T. Young, Sol. Energy **77**(6), 857 (2004)
25. G. Andra, J. Bergmann, F. Falk, E. Ose, H. Stafast, Phys. Status Solidi A, Appl. Res. **166**(2), 629 (1998)
26. A.A.D.T. Adikaari, S.R.P. Silva, Nano **3**(3), 117 (2008)
27. K. Yamamoto, A. Nakashima, T. Suzuki, M. Yoshimi, H. Nishio, M. Izumina, Jpn. J. Appl. Phys. **33**(12B), L1751 (1994)
28. B.K. Nayak, M.C. Gupta, Appl. Phys. A, Mater. Sci. Process. **89**, 663 (2007)
29. T.H. Her, R.J. Finlay, C. Wu, E. Mazur, Appl. Phys. A, Mater. Sci. Process. **70**(4), 383 (2000)
30. K. Matsubara, P. Fons, K. Iwata, A. Yamada, K. Sakurai, H. Tampo, S. Niki, Thin Solid Films **431**, 369 (2003)
31. S.K. Pandey, U. Tiwari, R. Raman, C. Prakash, V. Krishna, V. Dutta, K. Zimik, Thin Solid Films **473**(1), 54 (2005)
32. P.F. Luo, G.S. Jiang, C.F. Zhu, Chin. J. Chem. Phys. **22**(1), 97 (2009)
33. M. Hanabusa, Z.X. Liu, N. Nakamura, H. Hasegawa, Nucl. Instrum. Methods Phys. Res., Sect. B, Beam Interact. Mater. Atoms **121**(1–4), 367 (1997)
34. J.P.M. Hoefnagels, Y. Barrell, W.M.M. Kessels, M.C.M. van de Sanden, J. Appl. Phys. **96**(8), 4094 (2004)
35. N. Itabashi, N. Nihsiwaki, M. Magane, S. Naito, T. Goto, A. Matsuda, C. Yamada, E. Hirota, Jpn. J. Appl. Phys. **29**(3), L505 (1990)
36. P.B. Davies, N.A. Isaacs, S.A. Johnson, D.K. Russell, J. Chem. Phys. **83**(5), 2060 (1985)
37. C. Yamada, E. Hirota, Phys. Rev. Lett. **56**(9), 923 (1986)
38. J.U. White, J. Opt. Soc. Am. **32**(5), 285 (1942)
39. D. Herriott, H. Kogelnik, R. Kompfner, Appl. Opt. **3**(4), 523 (1964)
40. R. Bartlome, M. Baer, M.W. Sigrist, Rev. Sci. Instrum. **78**(1), 013110 (2007)
41. M.W. Sigrist, R. Bartlome, D. Marinov, J.M. Rey, D.E. Vogler, H. Wachter, Appl. Phys. B, Lasers Opt. **90**(2), 289 (2008)
42. Y. Sumiyoshi, K. Tanaka, T. Tanaky, Appl. Surf. Sci. **79/80**, 471 (1994)
43. R. Bartlome, A. Feltrin, C. Ballif, Appl. Phys. Lett. **94**, 201501 (2009)
44. M.D. Wheeler, S.M. Newman, A.J. Orr-Ewing, M.N.R. Ashfold, J. Chem. Soc. Faraday Trans. **94**(3), 337 (1998)
45. G. Berden, R. Peeters, G. Meijer, Int. Rev. Phys. Chem. **19**(4), 565 (2000)
46. V. Sanders, Appl. Opt. **16**(1), 19 (1977)
47. J.M. Herbelin, J.A. McKay, M.A. Kwok, R.H. Ueunten, D.S. Urevig, D.J. Spencer, D.J. Benard, Appl. Opt. **19**(1), 144 (1980)
48. B.A. Paldus, A.A. Kachanov, Can. J. Phys. **83**, 975 (2005)
49. W.M.M. Kessels, J.P.M. Hoefnagels, M.H.H. Boogaarts, D.C. Schram, M.C.M. van de Sanden, J. Appl. Phys. **89**(4), 2065 (2001)
50. T. Nagai, A.H.M. Smets, M. Kondo, J. Non-Cryst. Solids **354**, 2096 (2008)
51. F. Grangeon, C. Monard, J.-L. Dorier, A.A. Howling, Ch. Hollenstein, D. Romanini, N. Sadeghi, Plasma Sources Sci. Technol. **8**, 448 (1999)
52. R.M. Roth, K.G. Spears, G. Wong, Appl. Phys. Lett. **45**(1), 28 (1984)
53. Y. Matsumi, T. Hayashi, H. Yoshikawa, S. Komiya, J. Vac. Sci. Technol. A **4**(3), 1789 (1986)
54. Y. Nozaki, K. Kongo, T. Miyazaki, M. Kitazoe, K. Horii, H. Umemoto, A. Masuda, H. Matsumura, J. Appl. Phys. **88**(9), 5437 (2000)
55. M. Hertl, J. Jolly, J. Phys. D, Appl. Phys. **33**, 381 (2000)
56. R.M. Roth, K.G. Spears, G.D. Stein, G. Wong, Appl. Phys. Lett. **46**(3), 253 (1985)
57. K.G. Spears, T.J. Robinson, R.M. Roth, IEEE Trans. Plasma Sci. **14**(2), 179 (1986)
58. Y. Watanabe, M. Shiratani, Y. Kubo, I. Ogawa, S. Ogi, Appl. Phys. Lett. **53**(14), 1263 (1988)
59. Y. Watanabe, J. Phys. D, Appl. Phys. **39**, R329 (2006)
60. A. Bouchoule, A. Plain, L. Boufendi, J.Ph. Blondeau, C. Laure, J. Appl. Phys. **70**(4), 1991–2000 (1991)
61. A.A. Howling, C. Courteille, J.-L. Dorier, L. Sansonnens, Ch. Hollenstein, Pure Appl. Chem. **68**(5), 1017 (1996)
62. C. Courteille, Ch. Hollenstein, J.-L. Dorier, P. Gay, W. Schwarzenbach, A.A. Howling, E. Bertran, G. Viera, R. Martins, A. Macarico, J. Appl. Phys. **80**(4), 2069 (1996)
63. R. Bartlome, B. Strahm, A. Feltrin, C. Ballif, in *Proc. of the 34th IEEE Photovoltaic Specialist Conference*, Philadelphia, PA, June 7–12, 2009, paper 103.
64. B. Strahm, C. Hollenstein, Powder formation in SiH_4–H_2 discharge in large area capacitively-coupled reactor: a study of the combined effect of the inter-electrode distance and pressure. J. Appl. Phys. (2009 in print)
65. L. Boufendi, J. Hermann, A. Bouchoule, B. Dubreuil, E. Stoffels, W.W. Stoffels, M.L. de Giorgi, J. Appl. Phys. **76**(1), 148 (1994)
66. L. Boufendi, A. Bouchoule, Plasma Sources Sci. Technol. **3**, 262 (1994)
67. K. Durose, S.E. Asher, W. Jaegermann, D. Levi, B.E. McCandless, W. Metzger, H. Moutinho, P.D. Paulson, C.L. Perkins, J.R. Sites, G. Teeter, M. Terheggen, Prog. Photovolt. **12**(2–3), 177 (2004)
68. G. Gouadec, P. Colomban, Prog. Cryst. Growth Charact. Mater. **53**(1), 1 (2007)
69. E. Bustarret, M.A. Hachicha, M. Brunel, Appl. Phys. Lett. **52**(20), 1675 (1988)
70. M. Tzolov, F. Finger, R. Carius, P. Hapke, J. Appl. Phys. **81**(11), 7376 (1997)
71. E. Vallat-Sauvain, C. Droz, F. Meillaud, J. Bailat, A. Shah, C. Ballif, J. Non-Cryst. Solids **352**(9–20), 1200 (2006)
72. I. Repins, M.A. Contreras, B. Egaas, C. DeHart, J. Scharf, C.L. Perkins, B. To, R. Noufi, Prog. Photovolt. **16**(3), 235 (2008)
73. H. Miyazaki, R. Mikami, A. Yamada, M. Konagai, J. Phys. Chem. Solids **64**(9–10), 2055 (2003)
74. W. Witte, R. Kniese, M. Powalla, Thin Solid Films **517**(2), 867 (2008)
75. S. Komuro, Y. Aoyagi, Y. Segawa, S. Namba, A. Masuyama, D. Kruangam, H. Okamoto, Y. Hamakawa, J. Appl. Phys. **58**(2), 943 (1985)
76. G.Z. Yue, D.X. Han, D.L. Williamson, J. Yang, K. Lord, S. Guha, Appl. Phys. Lett. **77**(20), 3185 (2000)
77. M. Wagner, I. Dirnstorfer, D.M. Hofmann, M.D. Lampert, F. Karg, B.K. Meyer, Phys. Stat. Sol. A, Appl. Res. **167**(1), 131 (1998)
78. I. Dirnstorfer, M. Wagner, D.M. Hofmann, M.D. Lampert, F. Karg, B.K. Meyer, Phys. Stat. Sol. A, Appl. Res. **168**(1), 163 (1998)
79. W.K. Metzger, D. Albin, D. Levi, P. Sheldon, X. Li, B.M. Keyes, R.K. Ahrenkiel, J. Appl. Phys. **94**(5), 3549 (2003)
80. S. Shirakata, T. Nakada, Thin Solid Films **515**(15), 6151 (2007)
81. W.B. Jackson, N.M. Amer, A.C. Boccara, D. Fournier, Appl. Opt. **20**(8), 1333 (1981)
82. W.B. Jackson, N.M. Amer, Phys. Rev. B **25**(8), 5559 (1982)
83. N. Wyrsch, F. Finger, T.J. McMahon, M. Vanecek, J. Non-Cryst. Solids **137**, 347 (1991)
84. O. Breitenstein, M. Langenkamp, F. Altmann, D. Katzer, A. Lindner, H. Eggers, Rev. Sci. Instrum. **71**(11), 4155 (2000)
85. O. Breitenstein, J.P. Rakotoniaina, A.S.H. van der Heide, J. Carstensen, Prog. Photovolt. **13**(8), 645 (2005)
86. P. Vorasayan, T.R. Betts, A.N. Tiwari, R. Gottschag, Sol. Energy Mater. Sol. Cells **93**(6–7), 917 (2009)
87. K. Kushiya, Thin Solid Films **387**(1–2), 257 (2001)

88. C. Coluzza, E. Tuncel, J.L. Staehli, P.A. Baudat, G. Margaritondo, J.T. McKinley, A. Ueda, A.V. Barnes, R.G. Albridge, N.H. Tolk, D. Martin, F. Morier-Genoud, C. Dupuy, A. Rudra, M. Ilegems, Phys. Rev. B **46**(19), 12834 (1992)
89. K. Nishi, H. Ohyama, T. Suzuki, T. Mitsuyu, T. Tomimasu, Appl. Phys. Lett. **70**(26), 3585 (1997)
90. S. Theodoropoulou, D. Papadimitriou, A.G. Mamalis, D.E. Manolakos, R. Klenk, M.C. Lux-Steiner, Semicond. Sci. Technol. **22**, 933 (2007)
91. J. Krc, M. Zeman, O. Kluth, E. Smole, M. Topic, Thin Solid Films **426**(1–2), 296 (2003)
92. D. Domine, F.-J. Haug, C. Battaglia, C. Ballif, J. Appl. Phys. (2009 in print)
93. J.P.M. Hoefnagels, A.A.E. Stevens, M.G.H. Boogaarts, W.M.M. Kessels, M.C.M. van de Sanden, Chem. Phys. Lett. **360**, 189 (2002)

Laser-based sensor for a coolant leak detection in a nuclear reactor

T.-S. Kim · H. Park · K. Ko · G. Lim · Y.-H. Cha · J. Han · D.-Y. Jeong

Received: 15 September 2009 / Revised version: 18 February 2010 / Published online: 7 March 2010
© Springer-Verlag 2010

Abstract Currently, the nuclear industry needs strongly a reliable detection system to continuously monitor a coolant leak during a normal operation of reactors for the ensurance of nuclear safety. In this work, we propose a new device for the coolant leak detection based on tunable diode laser spectroscopy (TDLS) by using a compact diode laser. For the feasibility experiment, we established an experimental setup consisted of a near-IR diode laser with a wavelength of about 1392 nm, a home-made multi-pass cell and a sample injection system. The feasibility test was performed for the detection of the heavy water (D_2O) leaks which can happen in a pressurized heavy water reactor (PWHR). As a result, the device based on the TDLS is shown to be operated successfully in detecting a HDO molecule, which is generated from the leaked heavy water by an isotope exchange reaction between D_2O and H_2O. Additionally, it is suggested that the performance of the new device, such as sensitivity and stability, can be improved by adapting a cavity enhanced absorption spectroscopy and a compact DFB diode laser. We presume that this laser-based leak detector has several advantages over the conventional techniques currently employed in the nuclear power plant, such as radiation monitoring, humidity monitoring and FT-IR spectroscopy.

1 Introduction

In nuclear power plants, heavy water (D_2O) and light water (H_2O) are essentially utilized as a neutron moderator and a coolant in a pressurized heavy water reactor (PHWR) and a pressurized light water reactor (PWR), respectively. Although more improved safety technology has been implemented in those nuclear reactors, there are still possibilities that a coolant can be leaked into the environment or secondary coolant system through a steam generator as a result of the corrosion or the cracking in several reactor components such as pipelines and valves. In reality, a lot of leakage events which happened in nuclear power plants have been reported since 1970s [1]. Since these coolant leaks can induce sudden shutdowns of a nuclear reactor as well as a radioactive contamination of the environment, the nuclear industry strongly needs a reliable detection system to continuously monitor the coolant leaks during a normal operation of the reactors [1–6]. The regulatory guideline concerning the nuclear power plants recommends the standard nuclear plants to use two leak detection systems based on the different principles along with a radiation monitor, which can detect a low-rated leak at a very early stage, introducing the concept of leak-before-break (LBB).

Up to now, many methods have been developed and employed for the detection of coolant leaks in the nuclear power plants. Among them, radiation detection, acoustic emission detection, and humidity detection are widely used in both the PWR and the PHWR. However, those methods have been found to have some drawbacks through practical applications in the current power plants. For example, radiation detection enabled us to monitor the coolant leaks with a high sensitivity but it has some limitation to identify the exact location of the leaks [4]. The humidity monitor such as FLUS system, manufactured by Framatome-ANP, detects an increase in vapor content of air resulting from the leaks but suffers from both false alarm due to an unexpected change of environment humidity and a periodic replacement of the humidity sensors [6]. Similarly, since the sensitivity

T.-S. Kim (✉) · H. Park · K. Ko · G. Lim · Y.-H. Cha · J. Han · D.-Y. Jeong
Quantum Optics Division, Korea Atomic Energy Research Institute, 1045 Daedukdaero, Yuseong, Daejeon, Korea
e-mail: kimts@kaeri.re.kr

of acoustic emission detection varies strongly with a background acoustic noise, the method has some difficulties in being adapted under highly noisy circumstances. Alternatively, as specific methods for the detection of heavy water (D_2O) leaks in a PHWR, both tritium (^3T) detection using a liquid scintillation counter (LSC) and Fourier Transform infrared (FT-IR) spectroscopy are used. However, tritium detection needs high operation cost and it gives rise to a serious problem of nuclear waste disposal. Although the FT-IR spectroscopy is a good analytical method to measure heavy water without using any chemical reagents, the sensitivity is not enough to detect a low-rated leak.

In this paper, we report a new method as an effective tool for detecting a coolant leak by using a tunable diode laser spectroscopy (TDLS), which can be directly applicable to detect the D_2O leak in the PHWR. Also, we show some experimental results to test the feasibility of the method. Recently, laser absorption spectroscopy (LAS) has been used successfully to detect some constituents in vapor phase because the method permits accurate, non-intrusive and real-time monitoring of the vapors. Especially, TDLS provides several advantages over the system using the other laser sources since a diode laser is compact, less expensive and portable. Moreover, recent development of DFB (Distributed Feedback) diode lasers and fiber optics enables us to make a very compact detection system for practical application. In these backgrounds, we propose to use TDLS techniques for the application in the nuclear power plants since the method can be expected to provide a high sensitivity, which enables us to measure a very low-rated leak with non-intrusive and remote sensing capabilities. In order to evaluate the feasibility of this laser-based leak detection device, we established an experimental setup consisted of a near-IR diode laser, a home-made multi-pass cell and a sample injection part [7]. The sample injection part was made to mimic the real situation with some leaks existed in the coolant pipe lines of the nuclear power plants. The feasibility test was performed for the detection of the heavy water leaks, which can happen in a pressurized heavy water reactor.

2 Experimental setup

As mentioned previously, a newly devised leak detector for monitoring coolant leaks is based on the laser absorption spectroscopy. For the feasibility test of the device, we established an experimental setup which can mimic detection of the leaked coolant (D_2O) in the pressurized heavy water reactor (PHWR). Figure 1 shows a schematic diagram of experimental setup. The setup mainly consists of a laser source, multi-pass cell, a photo detector, and a sample injection part.

As a laser source for the absorption measurement, we used a single-longitudinal-mode tunable diode laser with an external cavity (Sacher, Model TEC500-1380). The oscillator of the laser was Littmann-type, which included diffraction grating at a grazing incidence angle with respect to a laser diode and a tuning module. As is known well, this structure enables the laser to be continuously tunable with minimal mode hops in the wavelength range of interest. This laser could be operated in the wavelength of near 1392 nm with a power of $\sim$3 mW. The spectral linewidth (FWHM) of the diode laser was less than 5 MHz.

The vacuum wavelength of the laser was measured by a wavemeter (Highfinesse, WS6). Additionally, the part of the diode laser beam was sent to a con-focal Fabry-Perot interferometer with a free spectral range of 750 MHz for the relative frequency calibration as the laser was scanned over the molecular transition lines. After the beam emitted from

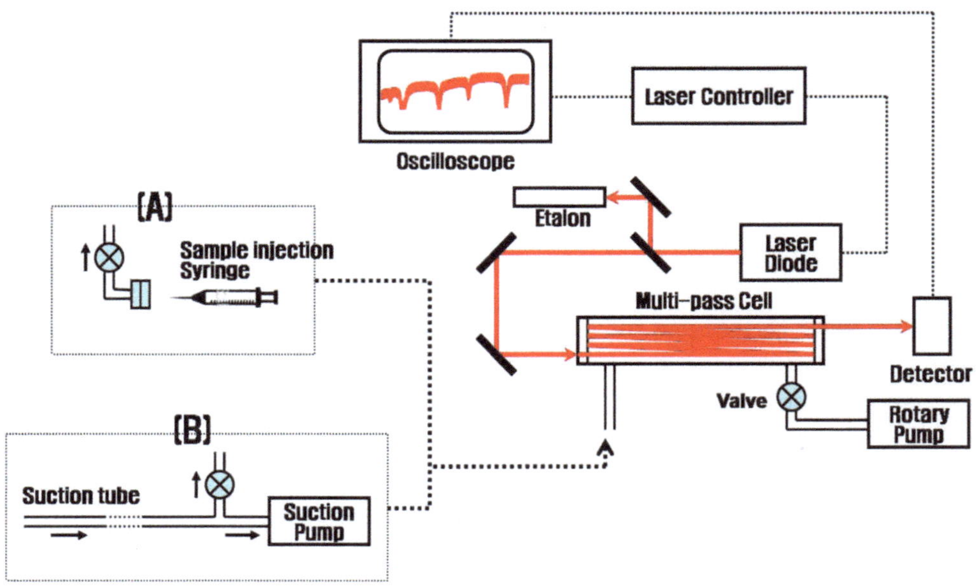

Fig. 1 Schematic diagram of experimental setup, which can mimic detection of the leaked coolant (D_2O) in the pressurized heavy water reactor (PHWR)

Fig. 2 The pictures of (**a**) the manufactured spherical-type Herriot multi-pass cell and (**b**) its reflection pattern by using a He-Ne laser

the diode laser passed through an optical isolator to protect the optical feedback from entering the ECDL, the laser beams was sent to the multi-pass cell through a series of mirrors with anti-reflection coatings. The laser wavelength was initially tuned close to a resonance transition of a target molecule. Then it was scanned back and forth through the resonance transition. The transmitted beam after absorption through the multi-pass cell was detected by using a sensitive photodiode and recorded by a digital oscilloscope.

In order to increase the sensitivity of the absorption, we used a home-made multi-pass absorption cell whose path length is 15.7 m with a total 26 passes. The manufactured multi-pass absorption cell is a spherical Herriot cell type and has two spherical mirrors with a focal length of 60 cm and a diameter of 5 cm was used. As is shown in Fig. 2(a), the mount block for the absorption cell was made by using aluminum and the cell body was made by a Pyrex glass with a diameter of 5 cm. Since the laser with the wavelength of 1392 nm is invisible, a He-Ne laser was used as shown in Fig. 2(b).

For the sample injection into the multi-pass cell, we used two methods, [A] and [B], consecutively as shown in Fig. 1. Firstly, in method [A], the heavy water (D_2O) was injected into the multi-pass absorption cell using a syringe for the preliminary detection of the D_2O sample. The sample injection system was similar to the structure of injection port in a typical gas chromatography. In order to prevent leaking of sample water when sample water was injected into a multi-pass absorption cell using a syringe, a silicone septum was used as a seal material. A silicone septum was designed to minimize low bleeds when a syringe was injected in the same spot every time. From the method, we could confirm that the injected D_2O sample could be detected very well by a laser absorption spectroscopy in multi-pass cell. Then, after we installed a compact mock-up system (method-B), mimicking the real situation with some leaks existed in the coolant pipe lines of the nuclear power plant, we performed the feasibility test for detecting leaks.

In the whole experiment, the heavy water sample of 99.9% was used and the pressure of the multi-pass cell was maintained to be about 7 torr by using a valve and a mechanical pump. In the case of a much higher pressure than 7 torr, the absorption spectrum could not be resolved because of the line broadening. On the contrary, in the case of lower pressure, the sensitivity was not enough to obtain a proper absorption signal. Also, the multi-pass cell was maintained to be room temperature (~23°C).

3 Results and discussions

For the feasibility test of the newly suggested leak sensor for the coolant (D_2O) in the PHWR, we tried to measure the HDO molecule instead of the D_2O molecule because the heavy water is changed to the HDO molecule in the atmosphere by isotope exchange reaction as follows:

$$D_2O + H_2O \underset{k_1}{\leftrightarrow} 2HDO \quad (k_1 = 1.1 \times 10^{-2}\ \mathrm{Pa}^{-1}\,\mathrm{s}^{-1}) \qquad (1)$$

If the coolant is leaked to the atmosphere, the D_2O molecule is converted to the HDO molecule with a very fast time, about 50 ms. Therefore, in order to detect the leaks with a very low rate, it is more effective to measure the HDO molecule than to measure the D_2O.

Table 1 shows several transition lines and their absorption cross sections for $H_2{}^{16}O$, $H_2{}^{18}O$ and HDO molecule in near 1392 nm. The data are all from the HITRAN database. In the meantime, there are three oxygen isotopes (^{16}O, ^{17}O and ^{18}O) in the H_2O and HDO molecule. Since the natural abundance of ^{17}O is very small, 0.037%, we did not take into account the molecular transition lines involving ^{17}O. On the other hand, we should take into account the absorption lines of $H_2{}^{18}O$. There are several absorption lines of the $H_2{}^{18}O$ in the spectral ranges I–IV (Table 1). We should select the absorption lines for detecting HDO with the consideration of the interference originating with $H_2{}^{18}O$ as well as $H_2{}^{16}O$.

Among the four region transition lines as described in Table 1, we chose the 7179.995 cm^{-1} in II region as a target transition line for the detection of the HDO molecule originated from the leaks of D_2O. For choosing the line, we considered the line intensity and minimum interference from nearby absorption lines of $H_2{}^{16}O$ and $H_2{}^{18}O$ molecule lines. Although the transition line 7191.039 cm^{-1} in IV region

Table 1 The transition lines and its absorption cross sections of water molecule in near 1392 nm

	Isotopomer	Frequency (cm^{-1})	Wavelength (nm)	Absorption cross section (cm^2, 0.02 atm)
I	$H_2{}^{16}O$	7176.197	1393.115	2.38×10^{-22}
	$HD^{16}O$	7176.419	1393.072	1.87×10^{-19}
II	$H_2{}^{18}O$	7179.180	1392.917	2.74×10^{-23}
	$H_2{}^{18}O$	7179.250	1392.903	9.32×10^{-23}
	$H_2{}^{16}O$	7179.752	1392.425	1.13×10^{-20}
	$HD^{16}O$	7179.995	1392.378	1.25×10^{-19}
III	$H_2{}^{18}O$	7183.360	1392.106	3.05×10^{-23}
	$H_2{}^{18}O$	7183.590	1392.062	2.64×10^{-22}
	$H_2{}^{16}O$	7183.879	1391.625	2.95×10^{-23}
	$HD^{16}O$	7183.973	1391.607	1.00×10^{-19}
	$H_2{}^{16}O$	7184.101	1391.582	7.79×10^{-23}
	$H_2{}^{18}O$	7184.460	1391.893	7.97×10^{-22}
IV	$H_2{}^{16}O$	7190.738	1390.298	1.75×10^{-20}
	$HD^{16}O$	7191.039	1390.240	2.06×10^{-20}
	$H_2{}^{18}O$	7191.360	1390.558	9.46×10^{-24}
	$H_2{}^{18}O$	7191.880	1390.457	1.71×10^{-22}

can be used to detect the HDO lines, we did not consider it because the transition line cannot be currently accessible with our diode laser system. On the other hand, considering the absorption strength of the HDO and H_2O, we hope that in the future, the transition lines in I and III regions are the best candidates for the leakage from the primary coolant system to a secondary coolant system through a steam generator, since the absorption cross section of H_2O is much lower than one of HDO.

Figure 3 shows a typical differential absorption spectrum for the H_2O and HDO in the sample inserted to the multi-pass cell while the diode laser is scanned through the II region in Table 1. The figure also includes the associated Fabry-Pérot signal to indicate the scanned frequency range. As can be clearly seen in the figure, the diode laser can be scanned through about 30 GHz continuously without mode hops. All data were recorded by a digital oscilloscope. To clarify the absorption spectrum, the signal was differentiated in the data processing. In Fig. 3(a) and (b), A is the position of the absorption spectrum peak of the HDO in the transition of 7179.995 cm^{-1} while B is the position of the spectrum peak of H_2O in 7179.752 cm^{-1}.

Figure 3(a) shows the absorption spectrum when the sample was injected into the multi-pass cell after the sample was collected from the location without D_2O leaks in the small mock-up system described in Fig. 1. In nature, hydrogen has two stable isotope ^1H and D (^2H) and their abundances are 99.985% and 0.015%, respectively. If the H_2O and HDO have similar absorption cross sections as described in region II of the Table 1, it is very hard to detect the absorption signal of the HDO molecule. As a result, whereas the absorption signal for the H_2O molecule is clearly shown in Fig. 3(a), the absorption signal corresponding to the HDO molecule is not shown within the sensitivity of our experiment. On the contrary, Fig. 3(b) shows the absorption spectrum when the injected sample is collected from the locations where the leaks of D_2O happened at the pipelines of the mock-up systems.

As mentioned earlier, in case that D_2O leaks into the atmosphere, the leaked D_2O is changed to HDO by the isotope exchange reaction with the ambient H_2O. The spectrum peak corresponding to HDO molecule appears clearly in Fig. 3(b) while the peak intensity of the H_2O absorption signal decrease due to the change of H_2O molecule into HDO molecule. From this result, if we measure the peak ratio between spectral peaks of the H_2O and HDO, we can assess whether the leaks of D_2O happens or not and how much the leaks of the D_2O occur.

Practically, when the D_2O leaks to the atmosphere from the main coolant pipe lines or valves in the PHWR, most of leaks happen in the form of heavy water aerosol of which the diameter is about from a few nm to a few *um*, not in the form of D_2O vapor. In that case, similar to the heavy water vapor, the heavy water aerosol diminishes the number of H_2O molecule with the isotope exchange reaction. Therefore, if we measure the decrease of the H_2O molecule as

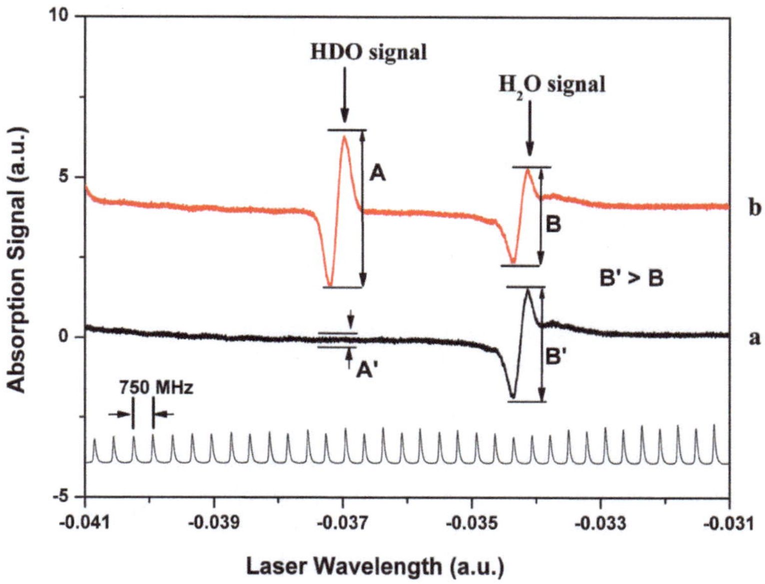

Fig. 3 A typical differential absorption spectrum for the H_2O and HDO in the sample inserted to the multi-pass cell while the diode laser is scanned through region II in Table 1. (**a**) Without D_2O leaks, (**b**) with D_2O leaks

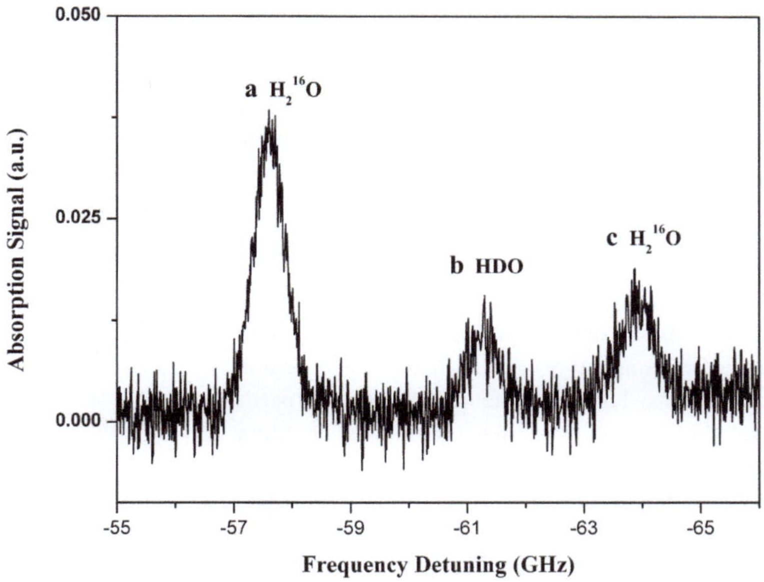

Fig. 4 A direct absorption spectrum for the $H_2{}^{16}O$ and HDO in the sample inserted to the multi-pass cell while the diode laser is scanned through region III in Table 1. (**a**) $H_2{}^{16}O$ in 7183.101 cm^{-1}, (**b**) HDO in 7183.973 cm^{-1}, (**c**) $H_2{}^{16}O$ in 7183.879 cm^{-1}, respectively

well as HDO molecule, we can expect to detect the leaks D_2O coolant effectively even in case that the leaks occur with the form of the aerosol.

Based on this experiment, we can conclude that this laser-based new leak detection method incorporating a tunable diode laser spectroscopy is very effective and convenient in the real detection for the leaks of the heavy water in PWHR. This detection method by using a near-infrared diode laser and a multi-pass cell can be applicable easily to the leak detection occurred at the pressure tubes and delayed neutron tubes as well as the main coolant pipe line of the PWHR. For the leak detection which happened at those parts, we can see that the sensitivity which can be obtained by using multi-pass absorption is sufficient for detecting a low-rated leak at a very early stage. Figure 4 shows the direct absorption spectrum when the sample was injected into the multi-pass cell after the sample was collected from the location without D_2O leaks in the small mock-up system described in Fig. 1(B). The measurement has been performed in the spectral range III, which has an absorption line for HDO at 7183.973 cm^{-1}. The absorption cross section of the line is similar to the one of the line of HDO at 7179.995 cm^{-1} in the spectral range II. Figure 4(a) and (c) are the absorption peaks of $H_2{}^{16}O$ in 7184.101 cm^{-1} and 7183.879 cm^{-1}. Figure 4(b) is the direct absorption peak of HDO in 7183.973 cm^{-1}. As is shown in Fig. 4, the HDO molecule with the concentration of about 300 ppm in natural water vapor can be easily observed by this tunable diode laser spectroscopy. If we consider the noise level, measurement accuracy and the influence of the nearby lines, we can presume that this method can detect the increase of the concentration of HDO molecule more than about 150 ppm. This estimated sensitivity is considered to be sufficient for the early detection of D_2O in primary coolant system in the PHWR since the coolant is composed of pure D_2O.

However, in order to detect the leaks which happened at a steam generator into the second coolant system, much higher selectivity is needed because the amount of the leaked heavy water is very small compared to the total quantity of the secondary light water (H_2O) coolants. For this reason, we have plans to develop an advanced leak monitoring device using a more sensitive technique based on the off-axis integrated cavity output spectroscopy (ICOS) [8–17]. The off-axis ICOS is known to be very promising and have several advantages over a multi-pass cell direct absorption or cavity ring-down spectroscopy (CRDS) because of its simplified alignment and low requirement for a mechanical stability [18, 19]. Then, if a new leak detector is developed based on the off-axis ICOS and a compact near IR diode laser, it will be very useful for routine monitoring of the leaks with the help of its compact size and stability. Moreover, after the device is loaded in a robot, it can detect some leaks which occurred at an inaccessible place due to the radioactivity or mechanical structure.

4 Conclusion

These days, there exist still possibilities that a coolant can be leaked into the environment or secondary coolant system in nuclear power plants. Therefore, the nuclear industry needs strongly a reliable detection system to continuously monitor a coolant leak during a normal operation of reactors. In these backgrounds, we proposed a new device for the leak detection of the coolant based on the tunable diode laser spectroscopy. Then, the feasibility test has been performed by using a diode laser with the wavelength of 1392 nm and a home-made multi-pass cell for the detection of heavy water in the PHWR. As an experimental result, the proposed device based on the TDLS is shown to be operated successfully in detecting a HDO molecule, which is generated from the leaked heavy water by an isotope exchange reaction between D_2O and H_2O even though the experiment has been performed qualitatively. Additionally, with the further work, we will improve the sensitivity and the stability by adapting the off-axis ICOS and a compact DFB laser. Finally we expect that this laser-based leak detector enable us to monitor the coolant leak successfully, having several advantages over the conventional techniques such as radiation monitoring, humidity monitoring and FT-IR spectroscopy.

Acknowledgements This work has been supported by the Nuclear Energy Research Program operated by the Ministry of Education, Science and Technology in the Republic of Korea.

References

1. N.C. Chokshi, M. Srinivasan, D.S. Kupperman, P. Krishnaswamy, in *18th International Conference on Structural Mechanics in Reactor Technology (SMiRT18)*, August 7–12, Beijing, China, 2005
2. U. Kunze, Nucl. Energy **34**, 213 (1999)
3. U. Kunze, B. Bechtold, Nucl. Energy **29**(3/4), 215 (1995)
4. S.Y. Choi, J. Choo, H. Chung, W. Sohn, K. Kim, Vib. Spectrosc. **31**, 251 (2003)
5. C. King, in *Materials Reliability Program: Survey of On-Line PWR Primary Coolant Leak Detection Technologies (MRP-187)*, EPRI, Palo Alto, CA, 2005, 1012947
6. N.Y. Lee, I.S. Hwang, H.-I. Yoo, Nucl. Eng. Des. **205**, 23 (2001)
7. J. Kim, S.E. Park, T.-S. Kim, D.-Y. Jeong, K.-H. Ko, Nukleonicka **49**(4), 137 (2004)
8. R. Vasudev, Appl. Spectrosc. **60**, 926 (2006)
9. R. Engeln, G. Berden, R. Peeters, G. Meijer, Rev. Sci. Instrum. **69**(11), 3763 (1998)
10. G. Berden, R. Peeters, G. Meijer, Int. Rev. Phys. Chem. **19**(4), 565 (2000)
11. J.M. Langridge, T. Laurila, R.S. Watt, R.L. Jones, C.F. Kaminski, J. Hult, Opt. Express **16**(14), 10178 (2008)
12. W.F. Drake, *Spectroscopic Techniques: Cavity-Enhanced Methods*. Handbook of Atomic, Molecular, and Optical Physics, vol. 43 (Springer, Berlin, 2005)
13. O.V. Naumenko, F. Mazzotti, O.M. Leshchishina, J. Tennyson, A. Campargue, J. Mol. Spectrosc. **242**, 1 (2007)
14. A. Campargue, F. Mazzotti, S. Beguier, O.L. Polyansky, I.A. Vasilenko, O.V. Naumenko, J. Mol. Spectrosc. **245**, 89 (2007)
15. S.-M. Hu, O.N. Ulenikov, E.S. Bekhtereva, G.A. Onopenko, S.-G. He, H. Lin, J.-X. Cheng, Q.-S. Zhu, J. Mol. Spectrosc. **212**, 89 (2002)
16. O.V. Naumenko, B.A. Voronin, F. Mazzotti, J. Tennyson, A. Campargue, J. Mol. Spectrosc. **248**, 122 (2008)
17. G.S. Engel, W.S. Drisdell, F.N. Keutsch, E.J. Moyer, J.G. Anderson, Appl. Opt. **45**(36), 9221 (2006)
18. J.B. Paul, L. Lapson, J.G. Anderson, Appl. Opt. **40**(27), 4904 (2001)
19. Y.A. Barhirkin, A.A. Kosterev, C. Roller, R.F. Curl, F.K. Tittel, Appl. Opt. **40**(11), 2257 (2004)

Index

MIX
Papier aus verantwortungsvollen Quellen
Paper from responsible sources
FSC® C105338

If you have any concerns about our products,
you can contact us on
ProductSafety@springernature.com

In case Publisher is established outside the EU,
the EU authorized representative is:
Springer Nature Customer Service Center GmbH
Europaplatz 3, 69115 Heidelberg, Germany

Printed by Libri Plureos GmbH
in Hamburg, Germany